World Geomorphological Landscapes

Series Editor:

Piotr Migoń

For further volumes:
http://www.springer.com/series/10852

Monique Fort • Marie-Françoise André
Editors

Landscapes and Landforms of France

Springer

Editors
Monique Fort
Geography Department, UFR GHSS
CNRS UMR 8586 PRODIG
University Paris Diderot-Sorbonne-Paris-Cité
Paris, France

Marie-Françoise André
Laboratory of Physical
and Environmental Geography (GEOLAB)
CNRS – Blaise Pascal University
Clermont-Ferrand, France

ISSN 2213-2090 ISSN 2213-2104 (electronic)
ISBN 978-94-007-7021-8 ISBN 978-94-007-7022-5 (eBook)
DOI 10.1007/978-94-007-7022-5
Springer Dordrecht Heidelberg New York London

Library of Congress Control Number: 2013944814

Printed on acid-free paper

Springer is part of Springer Science+Business Media (www.springer.com)

Series Editor Preface

Landforms and landscapes vary enormously across the Earth, from high mountains to endless plains. At a smaller scale, nature often surprises us, creating shapes which look improbable. Many physical landscapes are so immensely beautiful that they received the highest possible recognition – they hold the status of World Heritage properties. Apart from often being immensely scenic, landscapes tell stories which not uncommonly can be traced back in time for tens of million years and include unique events. In addition, many landscapes owe their appearance and harmony not solely to the natural forces. Since centuries, or even millennia, they have been shaped by humans who modified hillslopes, river courses, and coastlines, and erected structures which often blend with the natural landforms to form inseparable entities.

These landscapes are studied by geomorphology – 'the science of scenery' – a part of Earth sciences that focuses on landforms, their assemblages, and surface and subsurface processes that moulded them in the past and that change them today. Shapes of landforms and regularities of their spatial distribution, their origin, evolution, and ages are the subject of research. Geomorphology is also a science of considerable practical importance since many geomorphic processes occur so suddenly and unexpectedly, and with such a force, that they pose significant hazards to human populations and not uncommonly result in considerable damage or even casualties.

With this book focused on France, we launch a new book series *World Geomorphological Landscapes*. It aims to be a scientific library of monographs that present and explain physical landscapes across the globe, focusing on both representative and uniquely spectacular examples. Each book will contain details on geomorphology of a particular country or a geographically coherent region. The core of each book is a succinct presentation of key geomorphological localities (landscapes), representative for the geomorphic diversity of each country. Written in easy-to-read language, the landform evolution stories presented in each volume give together an overview of what each particular country has to offer. But they can also serve as a guidance for holidaymaking geoscientists as to where to go to enjoy the best of geomorphology.

The series is thus a unique reference source not only for geomorphologists, but all Earth scientists, geographers, and conservationists. It complements the existing reference books in geomorphology which focus on specific themes rather than regions or localities and fills a growing gap between poorly accessible regional studies, often in national languages, and papers in international journals which put major emphasis on understanding processes rather than particular landscapes.

The World Geomorphological Landscapes series is produced under the scientific patronage of the International Association of Geomorphologists – a society that brings together geomorphologists from all around the world. The IAG was established in 1989 and is an independent scientific association affiliated at the International Geographical Union and the International Union of Geological Sciences. Among its main aims are to promote geomorphology and to foster dissemination of geomorphological knowledge. I believe that this lavishly illustrated series, which however sticks to the scientific rigour, is a most appropriate means to fulfil these aims and to serve the geoscientific community.

Series Editor Piotr Migoń

Foreword

France presents a wide variety of landscapes, both natural and cultural, making this country the most visited in the world (Chap. 1). If we consider physical aspects of the landscape, this variety can be explained by a series of factors: (1) the topographical contrasts, from the highest summit of Europe down to the ocean; (2) a comprehensive collection of rock types (Fig. 1) and their related morphostructural units (sedimentary basins, basement uplands, active orogens, faulted systems with volcanism, coastal plains, deltas and marshes); (3) the long geological history and the succession of bioclimatic environments from tropical to glacial and periglacial environments, which left their imprint in the landscape; and (4) the large present-day bioclimatic diversity, at the junction of the Atlantic, Continental, and Mediterranean influences, which gives the geomorphological landscapes different atmospheres and lights (e.g. Brittany compared with French Riviera).

Prehistoric and historical (including Roman) legacies are also an integral part of French landscapes and landforms, often requiring a geoarcheological approach to better appreciate the role of early cultural groups in modelling their environment, exploiting their resources, and understanding the advantages of specific sites beyond their potential constraints or hazards. In that respect, two major characteristics should be highlighted. First, the abundance of limestone plateaus in the south of France explains both the number of beautifully ornamented caves and the wall paintings left by prehistoric humans as they sought sheltered sites during the last glaciation. Second, the combination of suitable soils and well-exposed slopes made France the country of viticultural 'terroirs', often developed on south- or southeast-facing scarps ensuring a good drainage and a sunny atmosphere whatever their structural origin: fault steps of Alsace, Burgundy, or Rhone Valley, cuesta of Champagne or thrust front of the western Jura.

This book is intended to provide the reader with an overview of French landscapes and landforms, reflecting as far as possible their diversity. To date, France is a country endowed with 34 sites inscribed as cultural heritage sites on the UNESCO World Heritage list, often backed by exceptional natural sites (e.g. Mont St-Michel (Chap. 5) or Paris and the banks of the Seine River (Chap. 2)). On the other hand, only three sites have been inscribed as natural heritage sites, including Corsican granitic landscapes (Chap. 23) and La Reunion volcanic landforms (Chap. 25), whereas the mixed 'Cultural and Natural Heritage' label is recognised in one site only (Gavarnie in the Pyrenees (Chap. 12)). Many chapters of this book deal with World Heritage Sites (WHS), either already listed by UNESCO or currently under consideration for UNESCO label recognition (e.g. Mont Blanc Massif, Chaîne des Puys volcanoes, Chauvet Cave – Pont d'Arc, and Camargue). Others have already received the label of 'Grands Sites de France': this is the case of Aven d'Orgnac (Chap. 19), Massif du Canigou (Chap. 13), and Cirque de Navacelles (Chap. 11). These UNESCO WHS and 'Grands Sites de France' are shown in Fig. 2.

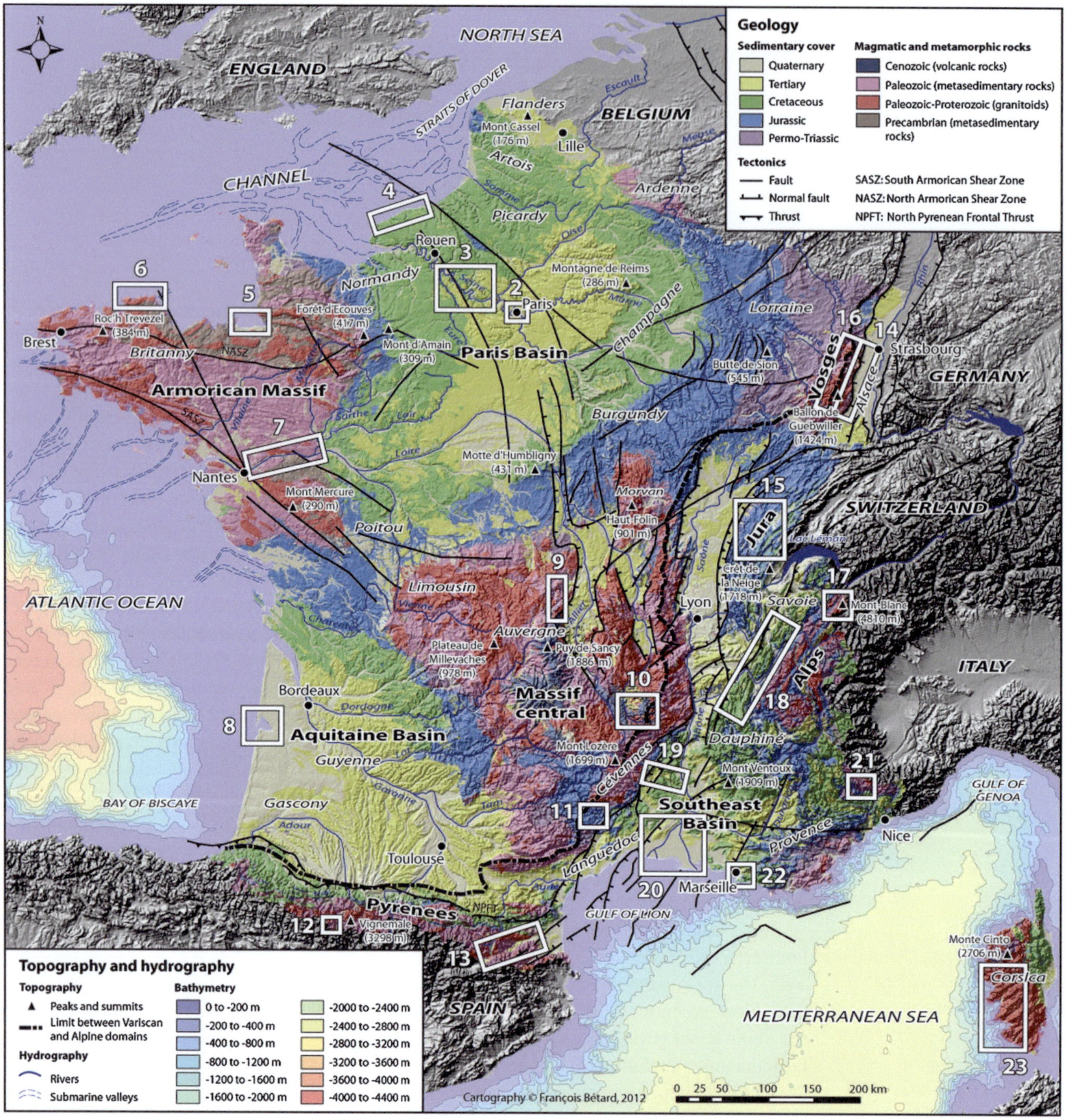

Fig. 1 Main geological frame of the different areas described in this book, with numbers refering to chapters (Drawing F. Bétard)

The book is organised in a sequence of chapters according to the following itinerary. Starting from Paris (Chap. 2), founded by the Romans, it is shown how in its early development the capital city benefited from links along the Seine River that favoured fluvial trade and from its substrate that offers abundant resource for building stones. Chapter 3 describes the famous entrenched meanders of the Seine River Valley downstream of Paris, developed under the alternation of glacial/interglacial periods: rich in mediaeval castles built on the chalk cliffs, the valley later attracted many Impressionist painters (Fig. 2).

The high Normandy coast and cliffs (Chap. 4), with their imposing verticality, their whiteness varying in tone with the ever-changing light and tide, and their ghostly shapes, also exerted a powerful attraction to painters and novelists. The instability of the cliffs has become a real concern for policymakers. Quite different is the Mont Saint-Michel bay (Chap. 5), characterised by its exceptional tidal range and the immensity of its tidal flat, overlooked by the Abbey perched on an island of which integrity is endangered by the progress of sedimentation. In contrast, the celebrated Pink Granite coast of northern Brittany (Chap. 6) illustrates both the influence of selective weathering under terrestrial conditions and the stripping of the saprolite by marine processes.

The Loire River Valley (Chap. 7), famous for its vineyards and royal châteaux, is much wider than the Seine Valley, and in its lower part it consists of a patchwork of geomorphic compartments (islands, separated by multiple branches of the river) that are remodelled under the influence of floods originating from the Massif Central highlands. Arcachon Bay (Chap. 8) is already part of southern France: bordered by the largest coastal dune belt in Europe, it is a typical semi-sheltered lagoon located on the Atlantic coast of France, facing the wave-dominated coast of the Bay of Biscay. It has become a famous oyster cultivation area and hosts various recreational activities (sport-fishing, sailing, and ecotourism) all year around.

The volcanic *Chaîne des Puys* range (Chap. 9) displays a wide spectrum of potentially active volcano types, very attractive for scientists, students, and visitors. Used for grazing as far back as the Neolithic period, and more intensively since the Roman Empire, volcanic slopes have been heavily deforested and eroded. With its status of French Regional Nature Park acquired in 1977, the *Chaîne des Puys* has been protected ever since. To the south, the Velay and its Vivarais rim (Chap. 10) combine many historical aspects of the Massif Central, from its Hercynian basement to the Tertiary tectonic and volcanic reactivations following the Pyrenean and Alpine orogenies, and from the Mesozoic and Tertiary planations to the Plio-Quaternary excavation of valleys. With the Cirque de Navacelles (Chap. 11), we have a perfect example of a meander incised by the Vis River in its gorge across the Causse du Larzac (limestone plateau), downstream from the metamorphic and granitic Cévennes Mountains. Yet it is much more complex than a simple bend in a watercourse, as karstic influence contributed to its evolution. Overlooked by panoramic viewpoints, it is listed on the 'Grand sites de France'.

The Cirque de Gavarnie (Chap. 12), a 1,500-m-high limestone theatre, is an exceptional landscape recognised in 1997 as a UNESCO World Heritage site for both its natural and cultural values. It results from a long evolution, including the formation of the Pyrenean range, the combination of both karstic and glacial influences, and a long history of human occupation and exchanges across the French-Spanish border. In the Eastern Pyrenees (Chap. 13), where the Mediterranean and Atlantic biomes meet, the Têt river catchment displays a spectacular suite of tectonic, glacial, periglacial, fluvial, and hillslope forms. Famous for its prominent fragments of smooth, pre-Quaternary Paleic relief remnants where glacial erosion has not serrated the interfluves, this Mediterranean part of the Pyrenees also preserves good records of post-orogenic landscape evolution and base-level changes since the early Neogene.

The Vosgian-Alsatian side of the Rhine Graben (Chap. 14) developed in response to Alpine orogeny: it is a unique, tectonically controlled landscape, famous for its vineyards and its traditional well-preserved architecture and culture. Nearby, the Jura Massif (Chap. 15) displays in its external part an original fold-and-thrust belt overriding the Bresse graben and bearing famous vineyards, in addition to scenic blind valleys, exsurgence springs and glacially formed lakes with Neolithic pile-dwelling settlements. The Vosges Mountains (Chap. 16) formed in uplifted Hercynian basement, affected by Quaternary glaciation. Upon its bald, rounded summit ridges, heathland and meadows with specific vegetal communities result from deforestation

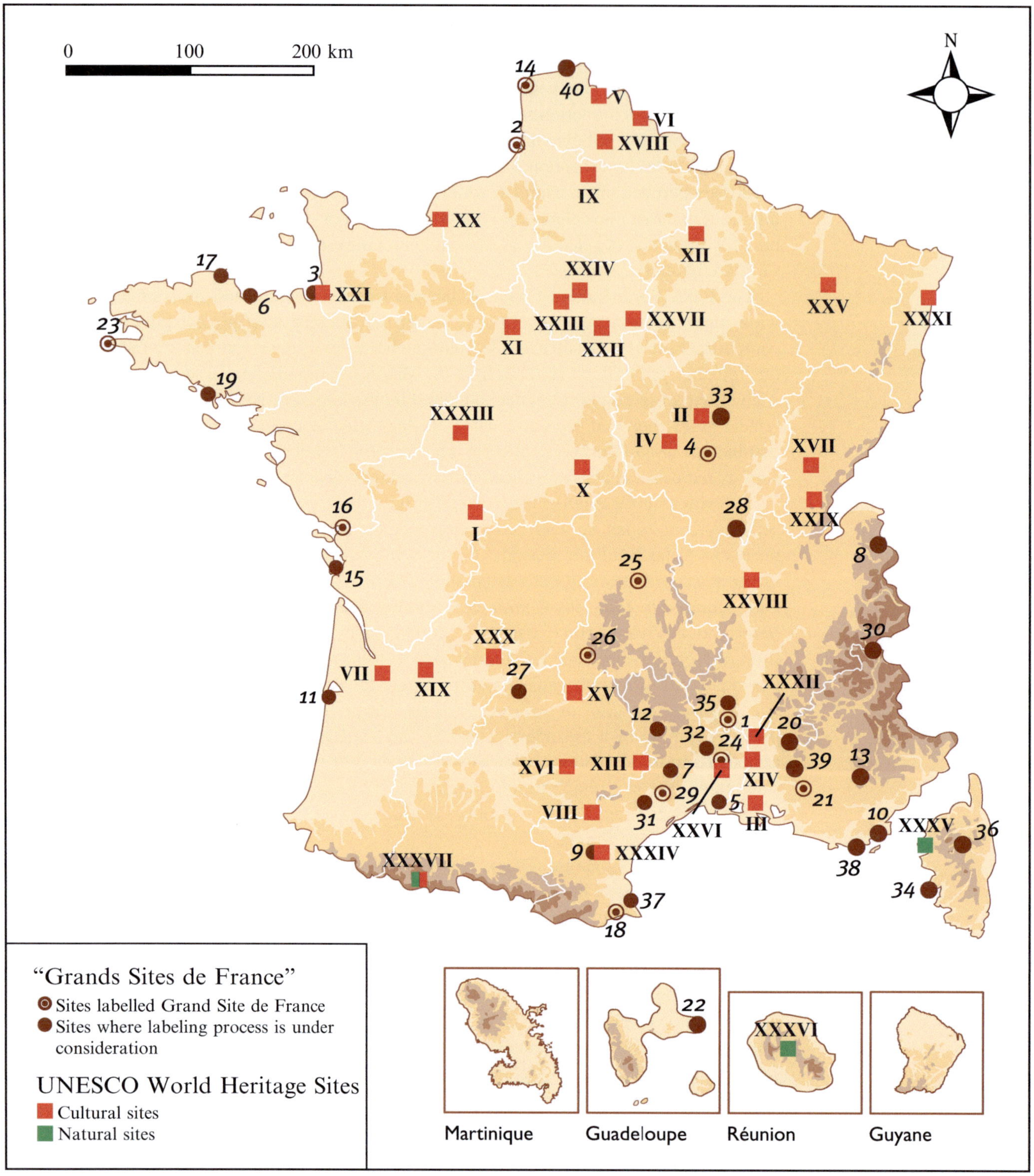

Fig. 2 Sites with Grands Sites de France Label (in arabic numbers). *1*: Aven d'Orgnac; *2*: Baie de Somme; *3*: Baie du Mont-Saint-Michel *; *4*: Bibracte-Mont Beuvray; *5*: Camargue gardoise; *6*: Caps d'Erquy-Fréhel; *7*: Cirque de Navacelles ***; *8*: Cirque de Sixt Fer à Cheval; *9*: Cité de Carcassonne *; *10*: Domaine du Rayol, Le Jardin des Méditerranées; *11*: Dune du Pilat; *12*: Gorges du Tarn et de la Jonte ***; *13*: Gorges du Verdon; *14*: Les Deux Caps Blanc-Nez, Gris-Nez; *15*: Marais et Place Forte de Brouage; *16*: Marais Poitevin; *17*: Abbaye de Beauport; *18*: Massif du Canigó; *19*: Massif dunaire de Gâvres-Quiberon; *20*: Mont Ventoux; *21*: Sainte-Victoire; *22*: Pointe des Châteaux; *23*: Pointe du Raz en Cap Sizun;

since the Bronze Age period. The protection and management of this natural heritage is ensured by a Nature Park and four Nature Reserves.

The Mont Blanc Massif (Chap. 17), the highest of the external crystalline massifs of the Western Alps, deserves a special place. Renowned for its extensive glacier cover, steep granite rockwalls and vertiginous peaks, the massif has attracted both tourists and scientists over several centuries. Accelerating glacier retreat and permafrost degradation both represent major threat to this emblematic landscape. Nearby, the French Prealpine geomorphological landscapes (Chap. 18) are excellent examples of sedimentary folded relief, and they host the four French members of the European and Global Geoparks Network including the Bauges Regional Nature Park, famous for its perched synclines and karstified relief. Great limestone cliffs overlying marly bedrock favour dramatic collapses like that of Mont Granier in 1248 AD.

Continuing southwards, Mediterranean influences become predominant. In the Lower Ardèche region (Chap. 19), the Messinian Salinity Crisis was the primary cause of the incision of the Ardèche Canyon and of the reorganisation of the Rhône River catchment. The end of this crisis contributed to the development of a specific karst system, including the Orgnac aven and the Chauvet cave, famous for its outstanding prehistoric wall paintings (the oldest in the world to date). The Lower Rhône River valley and its Delta (Chap. 20) have been a major axis of communication linking the Mediterranean to northern Europe since antiquity. Numerous archaeological excavations demonstrate human-environment interactions, emphasising the strong constraint of fluvial-deltaic environments, and how human societies, since the Greeks and the Romans, found some solutions to mitigate the fluvial risk. The landforms are now subject to different changes related to human activities, and the protection of the natural component has become a real challenge. The region of Mount Bego (Chap. 21) in the Southern French Alps, is an interesting place to observe scenic inherited landforms as well as cultural remains left by the ancient societies who lived there. The glacial imprint is widespread, and

Fig. 2 (continued) *24*: Pont du Gard *; *25*: Puy de Dôme; *26*: Puy Mary – Volcan du Cantal; *27*: Rocamadour **; *28*: Roches de Solutré-Pouilly-Vergisson; *29*: Saint-Guilhem le Désert et Gorges de l'Hérault **; *30*: Vallée de la Clarée et Vallée Étroite; *31*: Vallée du Salagou; *32*: Gorges du Gardon; *33*: Alésia; *34*: Iles Sanguinaires – Pointe de la Parata; *35*: Gorges de l'Ardèche; *36*: Vallée de la Restonica; 37: Anse de Paulilles; *38*: Presqu'île de Giens, Salins d'Hyères; *39*: Massif des Ocres; *40*: Dunes de Flandre. * *World Heritage UNESCO site;* ** *World Heritage UNESCO site on grounds of Saint-Jacques de Compostelle trails*; *** *World Heritage UNESCO site on grounds of Causses and Cévennes, Cultural Landscape of Mediterranean agropastoralism.* Note that this label 'Grand Site de France' is decerned by the French State to the Grand Site Manager for a period of 6 years. This label acknowledges the manager's action agrees well with sustainable development principles.

French World Heritage Sites (in Roman numbers). *I*: Abbey Church of Saint-Savin-sur-Gartempe; *II*: Cistercian Abbey of Fontenay; *III*: Arles, Roman and Romanesque Monuments; *IV*: Vézelay, Church and Hill; *V*: Belfries of Belgium and France; *VI*: Nord-Pas de Calais Mining Basin; *VII*: Bordeaux, Port of the Moon; *VIII*: Canal du Midi; *IX*: Amiens Cathedral; *X*: Bourges Cathedral; *XI*: Chartres Cathedral; *XII*: Cathedral of Notre-Dame, Abbey of Saint-Remi, & Palace of Tau, Reims; *XIII*: The Causses and Cévennes; *XIV*: Historical center of Avignon: Papal Palace & Avignon Bridge; *XV*: Routes of Santiago de Compostela in France; *XVI*: Episcopal City of Albi; *XVII*: Royal Saltworks of Arc-et-Senans; *XVIII*: Fortifications of Vauban; *XIX*: Jurisdiction of Saint-Émilion; *XX*: Le Havre, the City rebuilt by Auguste Perret; *XXI*: Mont Saint-Michel and its Bay; *XXII*: Palace and Park of Fontainebleau; *XXIII*: Palace and Park of Versailles; *XXIV*: Paris, Banks of the Seine; *XXV*: Place Stanislas, Place de la Carrière, and Place d'Alliance in Nancy; *XXVI*: Pont du Gard, Roman Aqueduc; *XXVII*: Provins, Town of Medieval Fairs; *XXVIII*: Historic Old Town center of Lyon; *XXIX*: Prehistoric pile dwellings around the Alps; *XXX*: Prehistoric Sites and Decorated Caves of the Vézère Valley; *XXXI*: Strasbourg – Grande Île; *XXXII*: Roman Theatre and its Surroundings and the "Triumphal Arch" of Orange; *XXXIII*: The Loire Valley between Sully-sur-Loire and Chalonnes-sur-Loire; *XXXIV*: Cité de Carcassonne – historic Fortified City of Carcassonne; *XXXV*: Gulf of Porto: Calanche of Piana, Gulf of Girolata, Scandola Reserve, Corsica; *XXXVI*: The Pitons, Cirques and Remparts of La Réunion; *XXXVII*: Pyrénées: Gavarnie-Mont Perdu (Drawing F. Bétard) the first human traces date back to the Neolithic, three millennia before Protohistoric societies left the outstanding petroglyphs of the so-called Vallée des Merveilles (Valley of Wonders). The Cosquer cave (Chap. 22) with its famous prehistoric paintings is located along a karstic coast submerged after the Last Glacial Maximum. Both lithology and tectonics explain the coexistence in the same area of terrestrial and submerged karst features that developed since the Neogene period.

The book concludes with three chapters about islands. First Corsica (Chap. 23) is like a granitic mountain in the sea, offering a large set of mesoscale forms and rock slopes carved by differential surface weathering, with contrasting landforms between the upper areas that were glaciated and the coastal landforms influenced by the sea. At local scales granite slabs and castellated tors, and at microscales, gnammas, and tafoni, are emblematic landforms that attracted early societies. The coral reefs and lagoons of French Polynesia islands (Chap. 24) illustrate well the pioneering theory of Darwin, revised by plate tectonics. These islands also offer an interesting example of combined volcanic geomorphology and karst development on an uplifted coral reef, affected both by high-energy events, and the generation of microforms by reef bioerosion. Finally, La Réunion (Chap. 25) is a volcanic island dating back at least 3 million years. It is celebrated as a Natural UNESCO Heritage site, thanks to its dormant volcano Piton des Neiges and its very active Piton de la Fournaise. Outstanding landforms include the steep *Remparts*, deep gorges, and waterfalls that favoured the development of endemic species that are now protected.

This book could not have been completed without the help and assistance of various individuals. We would like first to warmly thank Piotr Migoń, the director of the Landscapes and Landfoms Book series, who encouraged us to prepare this book in coordination with the Paris 8th International Conference of Geomorphology. Working with Elodie Tronche (Springer) was also a pleasure and we are most grateful to her for her involvement in the making of the book. Warm thanks also go to our French colleagues who agreed to prepare their contributions in a limited time period despite their many other commitments, and to additional partners who helped with the illustration, either by adapting existing figures or by kindly giving their permission for reproduction. Last, special thanks are due to Ian Evans, from Durham University, who greatly helped us in the final stage of English editing, and to François Bétard who provided the final general maps attached to this foreword and to Chap. 1. To all, we express our sincere gratitude.

Monique Fort
Marie-Françoise André

Contents

Short Biodata of Authors

Martine Ambert is a Geomorphologist in the Department of Geography of the Université Paul Valéry – Montpellier 3, and a member of the Geomorphological Heritage Commission of the French National Council of Geography (CNFG). Her research on natural sites known both for their scenic beauty and significant geomorphological features – geomorphosites – within the framework of her doctoral dissertation and teaching for the Masters/graduate program in Heritage Studies has given her the opportunity to take part in helping the public discover several remarkable sites in the Languedoc-Roussillon region.

Marie-Françoise André is Professor of Geomorphology at the Blaise Pascal University of Clermont-Ferrand where she belongs to the Laboratory of Physical and Environmental Geography (GEOLAB – UMR 6042-CNRS/UBP). She is a senior member of the *Institut Universitaire de France*. Her main scientific interest is in the field of rock weathering and erosion rates in Polar regions (Labrador, Svalbard, Lapland, Antarctic peninsula) and at World Heritage sites (Khmer and Maya temples of Cambodia and Mexico, Romanesque churches of the French Massif central).

Gilles Arnaud-Fassetta is Professor of Physical Geography at the University of Paris-Diderot (Paris 7), and a member of PRODIG Laboratory (UMR 8586 CNRS) in Paris, France. His research topics are applied hydrogeomorphology from mountains to coastal plains, geoarchaeology and palaeoenvironments reconstruction, and hydrological risk and environmental management. His countries of expertise are Algeria, France, Italy, Nepal, Pakistan, Portugal, Romania, Switzerland, and Syria.

Patrick Bachèlery is Professor of Volcanology and Petrology at the "Laboratoire Magmas et Volcans" at Blaise Pascal University in Clermont-Ferrand, France. His research interests focus on the petrology and evolution of intraplate volcanic islands. He was Director of the Piton de la Fournaise Volcanological Observatory, and participated in several marine expeditions.

Agnès Baltzer is Assistant Professor (HDR) in Marine Geosciences at the University of Caen, France. She has worked in deep waters on the gravity processes occurring on continental slopes of the north Atlantic, and now she is working on coastal areas, using side scan sonar imagery, very high resolution seismic profiles together with sediment cores. She is interested in the record by the sedimentary archives of the rapid climate changes (RCC) at different latitudes.

Frédéric Bertrand is Professor of Geography and Environmental Sciences, Natural Hazards and Risks, in the Department of Geography of Paris-Sorbonne University, and a member of PRODIG Laboratory (UMR 8586 CNRS). He has conducted several coastal research projects on both sides of the Atlantic Ocean (West Africa, South Portugal, South Brazil and France). His research interests include relations between coastal landforms and vegetation with respect to hydrological variability, coastal mapping and monitoring using remote sensing techniques and risk analysis. He is currently the co-conductor of the French Ministry of Environment, Ecology, Energy, Sustainable Development and the Sea (MEEDM) project on flood-risk assessment and management in the Arcachon Bay (BARCASUB project).

François Bétard is Assistant Professor in Physical Geography at Paris-Diderot University, and a member of the research team CNRS UMR 8586 PRODIG. His main research interests lie in the fields of geomorphology, pedology and ecology, with a special focus on relationships between geodiversity and biodiversity at various scales. He has worked in tropical and extra-tropical South America (Northeast Brazil, Southern Patagonia) but also in France, especially in the southern Armorican Massif and the Île-de-France region.

Pierre Boivin is an Emeritus CNRS researcher affiliated to the Laboratoire Magmas and Volcanoes at the University Blaise Pascal in Clermont-Ferrand, France. Experimental petrologist but also field geologist, he has been working on the origin and evolution of magmas, the oceanic crust and island arc of Archean age, the French cenozoïc volcanism (e.g., Chaîne des Puys and Velay), and on active volcanoes (the Réunion Island, Phlegrean Fields, Azores and Cameroon line).

Chantal Bonnot-Courtois, recently retired, was a CNRS Research Director in coastal sedimentology and geomorphology at the CNRS UMR 8586 PRODIG, attached to the Geomorphology and littoral environment Laboratory of Dinard (France). Her main research interest is in coastal sedimentary dynamics and landscape evolution in macrotidal environments. She is the author of many scientific papers on the Mount Saint-Michel bay, particularly on its present and Holocene sedimentological evolution.

Marc Calvet is Professor of Physical Geography and Geomorphology at the University of Perpignan (France) and head of the Laboratory on Geoenvironments "Médi-Terra", now integrated into the Prehistory and Quaternary team of the Muséum National d'Histoire Naturelle at Paris. He works on long term landscapes evolution on Mediterranean mountains, neotectonics, Quaternary geomorphology, using thermochronology and cosmogenic nuclides, and also on historic and Holocene fluvial landscapes.

Nathalie Carcaud is Professor of Physical Geography in the Landscape Department of Agrocampus Ouest, Angers, France. Her research focuses on fluvial geomorphology, specifically on the Loire Valley.

Jean-Michel Carozza is a Geomorphologist and Geoarchaeologist in Strasbourg University and a member of GEODE UMR 5602 CNRS in Toulouse (France). He first developed works on recent tectonic evolution of the Pyrenees, Himalaya and Taiwan. His present principal field of interest is Holocene landscape evolution and the way former civilizations interacted with their environment especially around the Mediterranean–Black Sea area.

Christine Chaussé is a Geoarchaeologist who works in the Institut National de Recherches Archéologiques Préventives (INRAP). She is also a member of the Laboratoire de Géographie Physique (LGP Meudon; UMR 8591 CNRS). She studies both aeolian and fluvial Quaternary deposits in the Seine Basin/the River Seine Basin in relation with ancient human occupations. Her researches document past environments (Pleistocene and Holocene) with the view to proposing both chronostratigraphic analyzes and landscape reconstitutions.

Jacques Collina-Girard is Professor in Aix-Marseille University. As a geologist and prehistorian, he is mainly interested in continental shelves, sea level rise and prehistoric archaeology. As a diver, he participated in Cosquer Cave Study in Marseilles and in geomorphological observations conducted, using scuba diving, in west Mediterranean Sea. He is the author of three books and 76 scientific papers.

Stéphane Costa is Professor of Physical Geography and Geomorphology at the University of Caen. He is a member of GEOPHEN, Physical and Environmental Geography Laboratory (UMR LETG Caen 6554 Géophen CNRS), France. His main scientific interest is in the field of coastal geomorphology and, especially, present-day rocky coast dynamics, sedimentary fluxes, and marine hydrology risks. His main countries of expertise are France and Great Britain.

Hervé Davodeau is Assistant Professor in Geography in the Landscape Department of Agrocampus Ouest, Angers, France. His research focuses on Landscape Policy, the evolution of the professional practices of landscape architects, and the consultation process in landscape architecture.

Emmanuelle Defive is a geographer specialised in geomorphology. Assistant Professor at the Blaise Pascal Clermont-Ferrand University (France), she belongs to the GEOLAB laboratory (UMR 6042 CNRS). She developed research in Velay and upper Loire catchment (French Central Massif), on long term valleys incision and Quaternary superficial deposits. She contributed to the 1/50,000 BRGM geological map of Le Monastier-sur-Gazeille (Velay, Mézenc). Its current research deals with Late Glacial and Holocene palaeoenvironmental studies and geoarcheology of this massif.

Philip Deline is a Geomorphologist at EDYTEM Lab (UMR 5204 Université de Savoie – CNRS, Le Bourget-du-Lac, France). He is Assistant Professor in Geography and Geosciences at Université de Savoie. His principal field of interest is rockfalls and rock avalanches in high mountain areas, permafrost degradation in rockwalls, Holocene glacier fluctuations, debris-covered glaciers, and glacial hazards. He works in the European Alps, in close collaboration with Alpine area scientists.

Magali Delmas is Assistant Professor of Physical Geography and Geomorphology at the University of Perpignan (France). She is a member of the laboratory Médi-Terra. Her Ph.D. (2009) was on timing and geomorphological impact of Quaternary glaciations in the Eastern Pyrenees. She now works on glacial, fluvial and karstic geomorphology, and Quaternary palaeoenvironments, using GIS, morphometry and terrestrial cosmogenic nuclides.

Samuel Etienne is a Studies Director at the Ecole Pratique des Hautes-Etudes, Paris, France. He leads the Laboratory of Coastal Geomorphology and Environment of Dinard, Brittany. He received his Ph.D. in Geography from the University of Paris-Panthéon Sorbonne and broadened his experience at Universities of Clermont-Ferrand and French Polynesia in Tahiti as an Assistant Professsor. He has participated in research projects in biogeomorphology, paraglacial geomorphology, coastal geomorphology, and stone monument conservation.

Monique Fort is Emeritus Professor of Geomorphology and Environmental Sciences, Natural Hazards and Risks, at Paris Diderot University, and a member of the research team CNRS UMR 8586 PRODIG. She worked extensively in various high mountains of the world (Alps, Central Asia, and Himalaya). Her research interests evolved from the relations of landforms with respect to geological structures, then to glacial and climatic fluctuations, and palaeoenvironmental reconstructions. Ongoing studies include current slope instabilities and flood hazards, their impacts and prevention (France, Marocco, Nepal, Pakistan, Mexico, Turkey). Former Vice President of the Intern. Assoc. of Geomorphologists (2005–2009), she is a member of the Commission on “Mountain Response to Global Change” of the Intern. Geographical Union (2008–2016). She is active President of the Groupe Français de Géomorphologie.

Christian Giusti is Assistant Professor in Physical Geography and Environmental Sciences at the University of Paris-Sorbonne (HDR). His main research interest is in history and epistemology of geomorphology. He is also an active member of the Mountain Geomorphosites Network and of the working group on geomorphosites at the IAG.

Yanni Gunnell is Professor of Physical Geography and Environmental Science at the University of Lyon, France. He studies the archives of environmental change and the timescales of continental denudation and material transfer across the landscape. His comparative approach ranges from cratonic environments to mountain belts and covers different climatic settings. He has worked in parts of India, Africa, Brazil, Peru, North America, Madagascar and Europe.

Fabien Hobléa is a Geomorphologist and is Assistant Professor in Environmental Geography at University of Savoie. His main present research interests are in karst and mountain georesources and geoheritage. He is working in the Alpine Range as well as in mountain areas all over the world (Patagonia, New Britain, Borneo, Chiapas, Saïan Mountains etc.). He is President or Member of Scientific Boards of several Nature Parks, Reserves and Geoparks (Bauges, Chablais, Chartreuse etc.).

Stéphane Jaillet is a Research Engineer at CNRS. Geomorphologist, speleologist and cartographer, he specializes in the study of karst and underground networks speleogenesis. At EDYTEM laboratory (University of Savoie), he develops the platform CTI3D (Cartography, Topography, Imaging and 3D). He is a member of the scientific team of the Chauvet cave, a scientific diving leader and also a member of the Ultima Patagonia expeditions.

Jacobus Kemp is Assistant Professor in GIS and Remote Sensing at the Department of Geography and Environmental Studies at Stellenbosch University, South Africa. His research interests include digital photogrammetry, synthetic aperture radar and other earth observation technologies applied to physical earth processes, geomorphology and geomorphometry.

Yannick Lageat is Emeritus Professor in Physical Geography at the University of Western Brittany (Brest). He is a member of the laboratory Geomer of the CNRS (UMR 6554-LETG). His principal fields of interest over the last 40 years are structural landforms and rocky coasts in various environments (France, Great Britain, Ireland, Poland, Scandinavia, Iceland, Groenland, Canada, United States, Philippines, India, Madagascar, South Africa, Namibia, Australia, and New Zealand).

Charles Le Coeur is Professor of Physical Geography at the University of Paris 1-Sorbonne and the former Director of the Laboratory of Physical Geography (CNRS-Meudon). He has studied long term and Quaternary geomorphology in intrusive ring complexes in Northern Ireland and Western Scotland. He has carried varied research on geomorphological responses to environmental change on Alpine slopes and rockglaciers, on Mediterranean rivers and coasts, and on numerous river restoration schemes.

Jerome Magail is an Anthropologist and an Administrator at the Museum of Prehistoric Anthropology of Monaco. In 2001 he submitted his thesis on the Bronze Age petroglyphs of Mont Bego situated in the south of France. Since 2006 he is the co-director of the Monaco-Mongolian joint archaeological expedition. He works in the center of Mongolia on an archaeological site of the Bronze Age. He is Associate Researcher in Traces Laboratory (UMR 5608) of the French National Center for Scientific Research.

Jean-Luc Mercier is Emeritus Professor from the University of Strasbourg. He has taught dynamic geomorphology. He is interested in weathering and soils, processes in geomorphology, watersheds, measures and modelization.

Ludovic Mocochain is a Geoscientist working at the Paris 6 University. Besides karstology, he studies the impacts of huge geological catastrophic events like the Messinian Salinity Crisis in the Mediterranean Sea.

Jean-Pierre Peulvast is Emeritus Professor of Geomorphology at the University of Paris-Sorbonne. He also teaches in the Master degree and Doctorate program at the Federal University of Ceará (UFC), Fortaleza, Brazil. His principal field of interest is structural geomorphology, but he is also involved in programs of geoarcheology (Egypt) and in studies on natural hazards and risks (Northeast Brazil). He mainly works on passive margins around the Atlantic ocean (Scandinavia, Canada, Brazil, Argentina), but also on various regions of France, Mediterranean Europe and Central Asia.

Alexandre Poiraud is a Geomorphologist, member of the Laboratory of Physical and Environmental Geography GEOLAB (CNRS-Blaise Pascal University). His main research topics deal with the relations of landslides and relief evolution in volcanic and post-orogenic areas and the integration of long-term geomorphology in actual risk mapping methods. His research questions the relation between quantitative statistical modelling and qualitative expert approach in scientific and technical practices.

Mireille Provansal is Emeritus Professor of Physical Geography in Aix-Marseille University, France, and is a member of CEREGE (CNRS UMR 7330). Her recent research topics are fluvial geomorphology and sedimentology of large rivers and deltas, impacts of engineering works on fluvial functioning and fluvial risks since 150 years, and palaeoenvironments in fluvial archeological sites. Her countries of expertise are south France, Romania and Vietnam.

Ludovic Ravanel is researcher in geomorphology working at the EDYTEM Laboratory (UMR 5204 Université de Savoie – CNRS, Le Bourget-du-Lac, France) on rockfalls, warming permafrost, paraglacial dynamics, and risks in high mountain areas. His main study area is the Mont Blanc region in the western Alps. He participated in various European projects including the recent PermaNET program (alpine-wide and long-term permafrost monitoring network). He is currently heavily involved in risk management in Chamonix.

Patrick Simon is Director of the Museum of prehistoric Anthropology of Monaco (Principality of Monaco). Being a geologist, his principal field of interest is Quaternary Geology and more particularly the relations between prehistoric lithic industry and raw materials (flint, quartzite, radiolarite, etc.). He is also editor of the Bulletin of Museum of Prehistoric Anthropology of Monaco.

Gilles Thomas is an underground historian and writer. He co-edited the work *Atlas du Paris souterrain* (Parigramme 2001) and is the author of *The Catacombs of Paris* (Parigramme 2011) and *Inscriptions des Catacombes de Paris* (Cherche-midi 2012). He produces an average of ten papers each year and also provides at the same time technical advices for journalistic, literary and academic projects, documentaries (BBC, Discovery Channel, National Geographic, etc.), and his name appears in film credits, including Pixar's *Ratatouille*.

Jean-Claude Thouret is Professor of Geomorphology and Geology in the Laboratory Magmas and Volcanoes at the University Blaise Pascal in Clermont-Ferrand, France. He has been working on high mountains and active volcanoes in the Andes, Indonesia and elsewhere. His principal field of interest is volcano growth and erosion, volcanic flows and associated hazards. He is author of 140 scientific papers and co-author of 30 books such as *Mountain Geomorphology* and *Geomorphological Hazards and Disaster Prevention*.

Nathalie Vanara is a Geographer, a Karstologist and a Speleologist. She is Assistant Professor (HDR) at the Department of Geography, Panthéon-Sorbonne University and a member of Meudon-Paris Physical Geography Laboratory. Known for her studies of caves in South-West China, the Georgian Caucasus and in the Pyrenean mountains, she has also written and co-written several books published in Karstologia Mémoires Series (numbers 8, 9 and 17).

Nicolas Villeneuve is Assistant Professor in the Department of Geography and member of the GeoSciences Laboratory at the University of La Réunion. His present research in dynamic geomorphology includes remote sensing and model methodologies applied to quantification of volcanic phenomenon, specifically: volcano–tectonics ground deformations, lava flow volume budgets, SO_2 degasing budgets, erosion and sediment transfers from the summit of volcano to the deep sea.

Anne-Véronique Walter-Simonnet is Assistant Professor of Geology at the Université de Franche-Comté, Besançon, France. Previously she was Assistant Professor at the Université de Basse-Normandie, Caen, France. Her principal field of interest is Quaternary, sedimentology and tephrostratigraphy. Since several years, she works on erosion processes and lacustrine sedimentation related to climate variations and anthropization.

Introduction: Landscapes and Landforms of France, A Large Diversity

1

Jean-Pierre Peulvast

The extraordinary diversity of landscapes observed in France is widely related to the geological and geodynamic characteristics of its territory, although it also represents the marks of past and present bioclimatic conditions and the results of the labor of hundreds of human generations. Among the physical conditions, geomorphic factors control, for example, the features of many urban sites and create strong constraints on the territorial organization or the rural activities, since they influence other characteristics of the environment (soils, local climates, hydrogeology, distribution of morphogenetic processes, etc.) and of the economic or everyday life (terrestrial or fluvial communications, hazards and risks, natural heritage, etc.). Among these controlling factors, one can find the stream patterns and the way they are inset into regional structures and landforms (e.g., sites of wind or water gap through folded ridges); the configuration of hill slopes, related or not with stepped systems of surfaces and structural asymmetries; or the contrasts in morpho-pedological and hydrological aptitudes between a karstic plateau and the surrounding depressions shaped into impermeable clay or marl.

At regional and continental scales, the distribution and characteristics of the large-scale geomorphic units are very influent. According to the classical works of E. Reclus, P. Vidal de la Blache, E. de Martonne, A. Demangeon, J. Brunhes, and R. Dion, the progressive construction of the French territory cannot be understood without taking into account the articulations between the large physiographic domains, directly controlled by the geology. For example, such articulations and contacts, including more or less practicable passageways between the main units, could be historically valorized as trade routes and commerce centers, as well as ways for invasions. Although the notion of determinism has been strongly criticized by human geographers in the last decades, this observation remains true nowadays. Indeed, it often refers to historical times in which technology could not provide to the societies the same degree of freedom as today with regard to the biophysical elements of their environment. But it appears that modern societies, even the most developed, do not escape from constraints and risks related to natural factors.

Therefore, among other environmental conditions, geomorphic factors are as influent as historical, political, or economic factors in the settlement of cities, in the development of an exchange route or another, or in the territorial management. What could be understood on the geography of France and of its regions without the classical reference to the "seuils" or thresholds between sedimentary basins and the "corridors" such as the rift system elongated from the Rhine graben to the Mediterranean Sea? What would be a regional or touristic geography which would neglect the characteristics of its mountains and systems of plains and low plateaus? The original landforms of some of these regions, folded, volcanic, or karstic, for example, are parts of the landscape heritage as well of the environment and are more and more frequently protected as such. In particular, this is the vocation of the programs of definition and inventory of geosites and geomorphosites.

The presence and distribution of these sites largely depend on the relief itself, more or less contrasted or vigorous, which is the result of combined and antagonist actions of internal geodynamics (emplacement of geological structures, uplift, stability or subsidence trend, neotectonics, volcanism, etc.) and of erosion *s.l.* (weathering, waste removal and transport, sedimentation). Understanding the landforms constructed this way, often over long geological periods, only appears through a morphostructural analysis, which considers the combination of these different factors. This knowledge is useful at continental scale as well as that of much smaller units such as the cuesta systems of Lorraine (Fig. 2, page x). This idea is well illustrated by the geomorphic maps of France at 1:2,500,000 and 1:1,000,000. At these scales, most of the master lines of the relief appear mainly controlled by the geological structure (Battiau-Queney 1993). This is the case of the mountain ranges, of tectonic

J.-P. Peulvast (✉)
Geomorphology, University of Paris-Sorbonne,
191 Rue Saint-Jacques, 75005 Paris, France
e-mail: jean-pierre.peulvast@wanadoo.fr

M. Fort and M.-F. André (eds.), *Landscapes and Landforms of France*, World Geomorphological Landscapes,
DOI 10.1007/978-94-007-7022-5_1, © Springer Science+Business Media Dordrecht 2014

origin, as well as of ridges and scarps shaped by selective erosion in the Armorican Massif and the Paris Basin (see Fig. between the Foreword and this Chapter). At medium scales, those of units of tens of kilometers represented on 1:50,000 and 1:100,000 maps, the structural control on space organization is still more evident, including in areas with moderate energy of relief.

Small-scale landforms and superficial regolith or soil mantles are influent as well. Mostly corresponding to the last stages of the morphological history and to the current dynamics, they give their mark to the landscapes and contribute to confer to them the potentials and constraints which weigh on land use and organization. The corresponding landscape diversity is exceptional. Owing to the situation of France at midlatitudes, the country underwent, during the last millions of years, tens of alternating periods of warm or cool temperate conditions and of cold climates, with repeated formation of continuous or discontinuous permafrost and of glaciers on the main mountains. Therefore, the territory displays an exceptional collection of active landforms and heritages of past morphoclimatic conditions. Good illustrations are given by features such as contrasted coastal landscapes, mountain glacial landscapes, or large fluvial valleys in regions of plains and low plateaus, the slopes and surroundings of which keep the strong imprint of the Quaternary periglacial dynamics.

The narrowing of the French "isthmus" (de Martonne 1942) increases the concentration in a relatively small territory of this large diversity of landscapes and geomorphic types, sometimes presented as a summary of the European landforms which only would be deprived of the Fennoscandian lakes and fjords (Mottet 1993). Two main types of landform systems may be distinguished, on the basis of the nature, number, and distribution of remarkable geomorphological landscapes and sites. To the north of the Pyrenees and northwest of a sinuous line which extends from the Montagne Noire to the crest of the Vosges massif through the crest lines of the Espinouse, Cévennes, Vivarais, Lyonnais, Beaujolais and Charolais mounts, and then the Côte d'Or, Langres and Upper Saône plateaus lies the France of plains, plateaus, and modest basement uplands. Except in the more elevated and partly mountainous massifs of Auvergne and Forez, it gently slopes toward the northwestern epicontinental seas (North Sea, English Channel) and the wide continental shelf, fringed by generally low coasts where abrupt cliffs (Pays de Caux, Bretagne) alternate with sand beaches. Only cut by a few passes or "seuils" (Langres, Charolais, Jarez), the master line coincides with the water divide between Atlantic and Mediterranean catchments and is also punctuated by most of the highest points of this part of the territory, between 500 and 1,700 m. The only exceptions are a few crystalline horsts (Forez: 1,640 m) and the highest volcanic mountains (Cantal, Puy de Sancy: 1,851 et 1,885 m), within the Massif Central.

To the east and southeast, the elevated rim of these highlands overlooks by a few hundreds of meters to 1,000 m a long corridor of plains and narrow plateaus which, from the Alsace plain to the Lower Rhône valley and the Bas-Languedoc through the Bresse plain and the Rhône corridor, belongs to the Alpine foreland. Similarly, the south of the Aquitaine Basin corresponds to the Pyrenean foreland. In the alpine domain (Alpes-Jura, Pyrénées-Provence, Corse, Pyrénées), the recent, mainly Cenozoic orogenic, processes formed very high mountains, with strong topographic contrasts, often over smaller areas. The distribution, altitudes (up to 4,810 m in the Alps, 3,300 m in the Pyrenees, 2,780 m in Corsica), and style of the landforms depend on these deformations and of the complex geodynamic history of the ranges which multiplicated structural orientations, divisions, and contrasts, between high alpine mountains, also represented in Corsica, narrow plateaus, highlands presenting jagged landforms in spite of modest altitudes (Corbières, Provence), small basins, wide coastal plains, and abrupt sea cliffs deeply scalloped by coves and "calanques." The drainage was progressively organized and reorganized toward the Mediterranean Sea (Rhône), the North Sea (Rhine), and also toward the Atlantic Ocean (rivers of the central and western Pyrenees), influenced by late deformations in the alpine foreland basins. Partly controlled by ample eustatic variations of the Plio-Quaternary times, they were more strongly marked, in the southeast, by the Messinian salinity crisis during which the Mediterranean Sea was temporarily reduced to a deep and dry enclosed basin, 5.4 million years ago. The depression of the base level down to −2,000 m triggered the fast incision of narrow canyons later fossilized during the Pliocene transgression (Rhône), as well as the formation of deep karstic systems, later drowned (Fontaine de Vaucluse).

References

Battiau-Queney Y (1993) Le relief de la France. Coupes et Croquis. Masson, Paris

de Martonne E (1942) France physique. In: Vidal de la Blache P, Gallois L (eds) Géographie Universelle, tome VI, La France. Armand Colin, Paris, 463 p

Mottet G (1993) Géographie physique de la France. Presses Universitaires de France, Paris, 655 p

2 Paris and the Seine River: Antic Sites, Underground Resources and Risks

Monique Fort, Christine Chaussé, Nathalie Vanara, and Gilles Thomas

Abstract

Paris City has developed on both sides of the Seine River. People began living there in the Mesolithic Age, as shown by the knowledge recently gained on early settlements and fluctuations of the river course thanks to excavations carried out with urban renewal projects. The very city (Lutetia) was founded by the Romans at the beginning of the Christian Era. The city benefited from its links with the river that favoured fluvial trade and from its substrate that offers abundant resource for building stones. This wealth explains an exceptional architectural heritage well preserved in the city. Yet, both the river and the old underground quarries are now a factor of potential risks that are carefully surveyed and managed. This chapter aims at reconstructing the different facets of unknown Paris, hidden resources and assets, as a complement to what is usually written on and praised for the City of Light.

Keywords

Palaeogeography • Underground Paris • Catacombs • Subsidence risk • Flood risk

2.1 Introduction

Paris is a magic place, attracting tourists from all over the world. Famous for the Notre Dame Cathedral, for its elegant, iron-made Eiffel Tower, Paris City has developed on both sides of the Seine River (Fig. 2.1). People began living there in the Mesolithic Age, but the very city – Lutetia – was founded by the Romans early in the Christian Era. This was a convenient site where the river could easily be crossed over by a ford and which favoured fluvial trade with the surrounding regions, as expressed by the Paris logo and maxim "Fluctuat nec mergitur". Paris is also surrounded by hills (*Montmartre, Passy, Montagne Sainte-Geneviève*), formed during the Quaternary by the Seine River incision through the Tertiary substrate (see also Chap. 3). The latter offers abundant resource for building stones that explain the exceptional architectural heritage still alive. Indeed, from the Mediaeval Age the population of Paris was the highest of any city of Europe, at the crossroads of trade routes between Spain and Flanders. This wealth favoured the development of significant religious edifices such as the abbeys of *Saint Germain-des-Prés*, *Saint Victor* and *Saint Martin-des-Champs* and the *Sorbonne*, one of the oldest universities in France.

During the following centuries, urban development involved spatial growth on the surrounding parishes and demolition of older buildings. Large renovation of the city beginning from 1853 was carried by Baron Haussmann who set a new grid of wide streets (like *Avenue de l'Opera*) to erase

M. Fort (✉)
Geography Department, UFR GHSS, CNRS UMR 8586 PRODIG, University Paris Diderot-Sorbonne-Paris-Cité, 5 rue Thomas Mann, 75205 Paris Cedex 13, France
e-mail: fort@univ-paris-diderot.fr

C. Chaussé
Inrap, 7 rue de Madrid, Paris, 75008, France

UMR 8591, Laboratoire de Géographie Physique, 1 place A. Briand, Meudon cedex 92195, France
e-mail: christine.chausse@inrap.fr; chausse@cnrs-bellevue.fr

N. Vanara
UMR LGP 8591, Paris 1 – Panthéon-Sorbonne University, Paris, France
e-mail: Nathalie.Vanara@univ-paris1.fr; nathalie.vanara@gmail.com

G. Thomas
SEADACC, Hôpital Cochin, 27 rue du Faubourg Saint-Jacques, 75014 Paris, France

M. Fort and M.-F. André (eds.), *Landscapes and Landforms of France*, World Geomorphological Landscapes, DOI 10.1007/978-94-007-7022-5_2, © Springer Science+Business Media Dordrecht 2014

Fig. 2.1 Ile de la Cité and Notre Dame. Around AD, the Seine channel was occupied by a series of mounds and small islands that formed natural bridge piles and determined what has become the very heart of Paris (Photo credit: Monique Fort)

insalubrious districts and make space for traffic and more residential buildings. Many old houses were pulled down.

Paris is now the centre of a very large agglomeration extending over more than 105 km^2, vulnerable to natural hazards like floods or indirect man-made hazards like cavity collapses. The city management has to adjust to these hazards; meanwhile, urban renovations provide exceptional opportunities to decipher the ancient traces of early settlements and environments. This chapter aims at reconstructing these earliest traces, highlighting both the Seine River and underground assets, resources and potential threats, often forgotten as Paris is so much considered as the City of Light.

2.2 An Ancient Site and History

Regaining the human history along the Seine in Paris is not an easy thing. The capital city, Paris, is also a city of art, and its heritage is protected by the *Conventions Culturelles* of 1954 and 1985, supplemented in 1991 by the inclusion of both riverbanks in World Heritage of UNESCO. If this regulatory framework ensures the protection of buildings, it also limits the renovation. Access to subsoil where archaeological remains are buried is therefore exceptional. However, this legal framework requires protective measures during the renewal of neighbourhoods. Excavations are then engaged when the remains discovered are of scientific interest.

Bounded on both sides of the *Île de la Cité*, the ancient and mediaeval city is now fairly well known thanks to its rediscovery during the renovation of the city in the nineteenth century. Outside this time and spatial slot, excavations have been few, and those illustrating the link between people and the river have been even less. However, over the past 20 years, the data were renewed thanks to new archaeological research illustrating both the more ancient past of the city and its river and the outskirts of the ancient Lutetia. These observations have been made on a few sites scattered along the river (Fig. 2.2a).

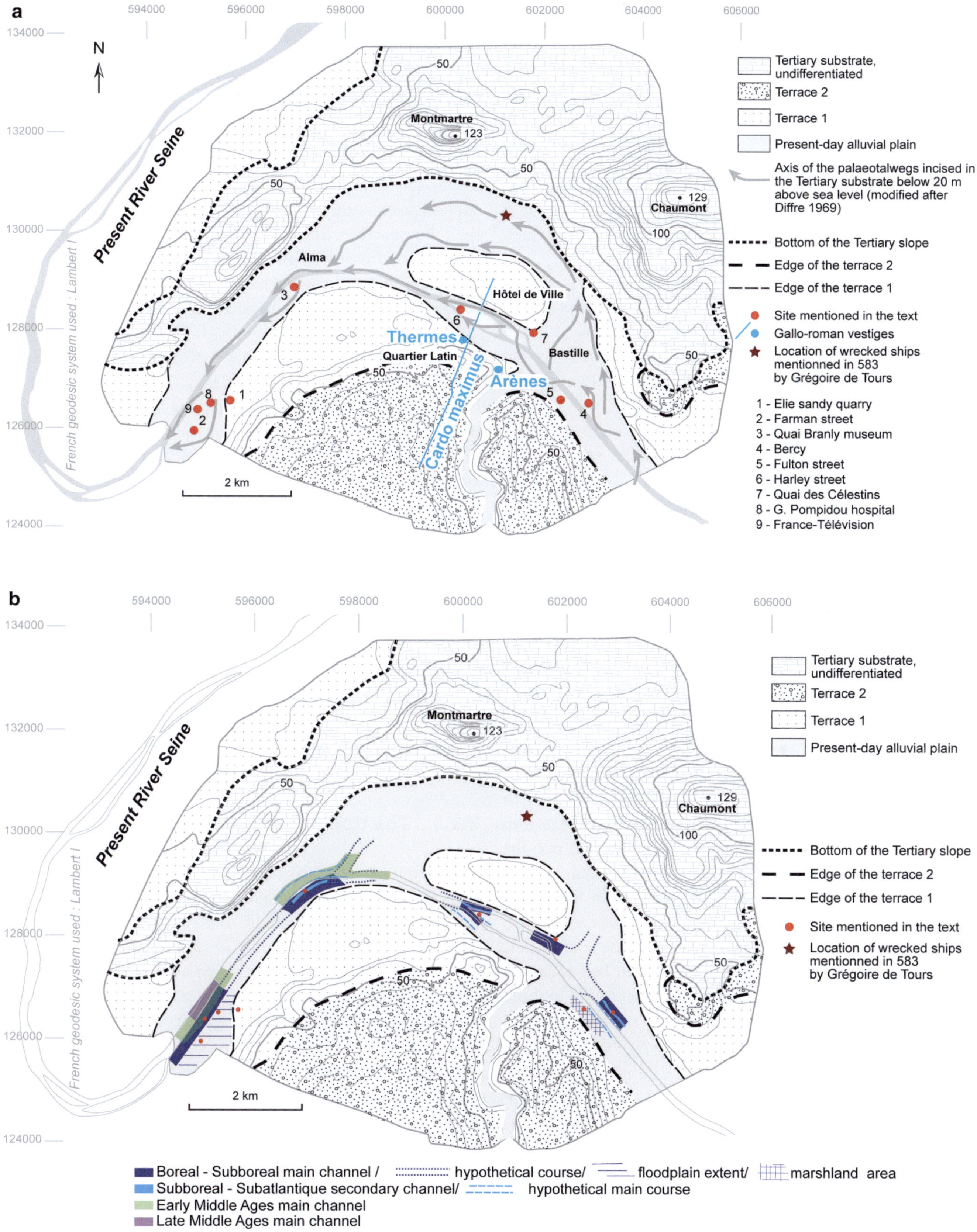

Fig. 2.2 (**a**) Simplified bedrock geology map of Paris and location of the excavations. (**b**) Schematic reconstitution of the course of the Seine River during the Holocene

2.2.1 The Parisian Space Prior to the Holocene

The Seine in Paris describes a wide meander, the first in a series that extends about 50 miles downstream. Its present bed is located in a floodplain (±30 m asl), 2 km wide at best that divides the city (Fig. 2.2a). On the right bank, it is connected to the marly limestone slope dominated by Chaumont and Montmartre buttes (>120 m asl). On the left bank, it is cut into a system of alluvial terraces (terraces 1 and 2), which outcrop between 35 and 65 m asl.

No data document these ancient fluvial formations, whose ages are currently ignored. However, compared with the more recognised system of terraces upstream of Paris (Chaussé et al. 2004), the formation of the terrace 1 could have occurred between the end of the Middle Pleniglacial (30 ky BP) and Weichselian Glacial Maximum (20–18 ky BP). In the nineteenth century, its sediments have yielded a set of lithic and faunal remains discovered in the sandpit Elie located southwest of the city (site 1, Fig. 2.2a). Artefacts would consist of handaxes and "*Mousterian flint*". The assemblage of large mammals combines taxa evolving both in cold and temperate environments (*Elephas primigenius, Cervus tarandus, Elephas antiquus, Rhinoceros mercki, Hippopotamus major*, etc.). Amongst these unstratigraphically located remains, human bones were discovered (Billy 1955). The site Elie could be the first evidence of prehistoric occupation in Paris dating back at best to Upper or Middle Pleistocene (>120 ky BP).

The setting up of the floodplain dates back to the Upper Pleistocene, i.e. last Weichselian cold stage (>30 ky BP) based on data collected from the upstream part of the Seine catchment (Chaussé et al. 2004). The plain is built upon coarse, 7–8 m thick fluvial deposits. These sediments can reach 12–13 m thick along palaeo-thalwegs incised into the Tertiary basement (Fig. 2.2a) (Diffre 1969). Their courses diverge at the Bastille area before joining again in Alma area. To the north, several palaeo-thalwegs flow in a wide loop along the foot of the Tertiary slopes, while in the south it cuts across the terrace 1 and prefigures the present bed course of the river. No archaeological remain has been found in these formations.

2.2.2 The Postglacial

The early Holocene history of the Seine River and that of its neighbours is a little better known. The first evidences date back to the Boreal (9,000–8,000 BP) and are located southwest of the city, in street Farman (site 2, Fig. 2.2a) where a Mesolithic site was recently found (Souffi et al. in press, 2012). The settlement located in the flood plain was bordered to the northwest by a channel (Fig. 2.2b). Although undated, this more or less deep channel could be linked to that identified at the Quai Branly museum (site 3, Fig. 2.2a), which was active between the Boreal and early Subboreal (±4,100 BP) (Chaussé et al. 2008). Posts associated with some ceramic shards represent the remains of a diffuse Neolithic settlement located on the lower bank of the channel. Such a channel was also found in Bercy (site 4, Fig. 2.2a, b) where its left bank supports a Neolithic village whose last occupants abandoned their dugouts (Fig. 2.3a) (Lanchon 1998). The Bercy channel was probably connected to a swamp as recognised at Fulton Street, developed in an old oxbow. The Neolithic groups then began to clear the area to allow growing crops (Leroyer 2006).

The next period begins just before the Subboreal and lasts until sub-Atlantic (2,700 BP). The river incised its bed, and the process was accompanied by a stronger hydrodynamic activity probably responsible for bank erosion as evidenced by the data collected in Harley Street, Quai Branly and Quai des Célestins (Fig. 2.2a). Along the Quai Branly and at Bercy, the channel shrunk gradually, hence was converted into a secondary channel being clogged (Lanchon 1998; Chaussé et al. 2008) (Fig. 2.2b). This process signs the migration of the main river bed to its current position. The banks of the palaeochannel were diffusely occupied by protohistoric populations. It was the same for the *Île de la Cité* which was until recently regarded as the cradle of the Gallic city. Recent excavations carried further downstream would instead locate in Nanterre the original Lutetia of the Parisii, a Gallic tribe mentioned by Julius Caesar (Viand 2008).

2.2.3 The Upper Holocene

The second part of sub-Atlantic beginning around 2,000 BP coincides with the appearance and development of Lutetia. It extended from the left bank to the right bank between the *Quartier Latin*, the *Île de la Cité* and *Hôtel de Ville neighbourhood*. The Gallo-Roman city was organised along a SSW-NNE axis represented by the *cardo maximus*, marked today by St. Jacques, Cité and St. Martin streets (Fig. 2.2a). Its creation dates back to the first century AD (Busson 2001; http://www.paris.culture.fr/en/). Later on, the mediaeval city grew from the first core. The configuration of the river at this time is not known due to lack of archaeological and geomorphological data. Only the work by Vacquer (Dupuy 1900) carried at the end of the nineteenth century proposes a reconstitution of the centre of Paris (Fig. 2.4), which since has hardly been improved.

It is rather on the outskirts of the ancient city that more recent data were collected. At G. Pompidou Hospital

Fig. 2.3 (**a**) Neolithic dugout discovered in the Bercy excavation (Photo credit: Carlos Valero, Inrap). (**b**) Mediaeval wattle discovered in the Quai Branly excavation (Photo credit: Patrick Pion, Université Paris X Nanterre)

and France Télévision building (sites 8, 9, Fig. 2.2a), southwest of Paris, investigations helped to identify the river course around the fourth to fifth centuries. It is a more than 7-m-deep main channel, extending over a width of about 70 m into which a second shallower (5 m) channel is cut and fill. It was fixed up in the Middle Ages and backfilled in the fourteenth century (Jugie 2011) (Fig. 2.2b). At Quai Branly (site 3, Fig. 2.2a, b), the main channel of the beginning of the Christian Era joined the present course of the river. But a secondary course was abandoned, transformed in fishery during the fourth and fifth centuries, as evidenced by post alignments and wattle discovery (Fig. 2.3b) (Pion et al. 2005).

The more recent history of the river and its riverside populations is not documented by archaeology. Written sources are taking over, and become abundant when the river expands over the city. The first written mention of the damages in Paris by a flood dates from February 583. *Gregory of Tours* tells when ship wrecks occurred in the north of Paris (Dupuy 1900). Obviously, the northern palaeo-thalweg along the Tertiary hill sides (Fig. 2.2a) remained active throughout the Holocene. Yet, no recent excavations have documented it to date.

2.3 The Dark Side of the City of Light: Quarries and Catacombs of Paris

Paris was founded and built upon its quarries and hundreds of kilometres of tunnels run under Paris, almost all of which are of artificial origin (Clément and Thomas 2001). The sedimentary, Tertiary substrate provided different types of rocks that were easily accessible and enabled the building of the city as we see it now in all its architectural glory. Yet, the abandonment of these underground galleries caused serious subsidence problems that had to be identified, consolidated and monitored. Now, the underground galleries of the capital have become an attraction.

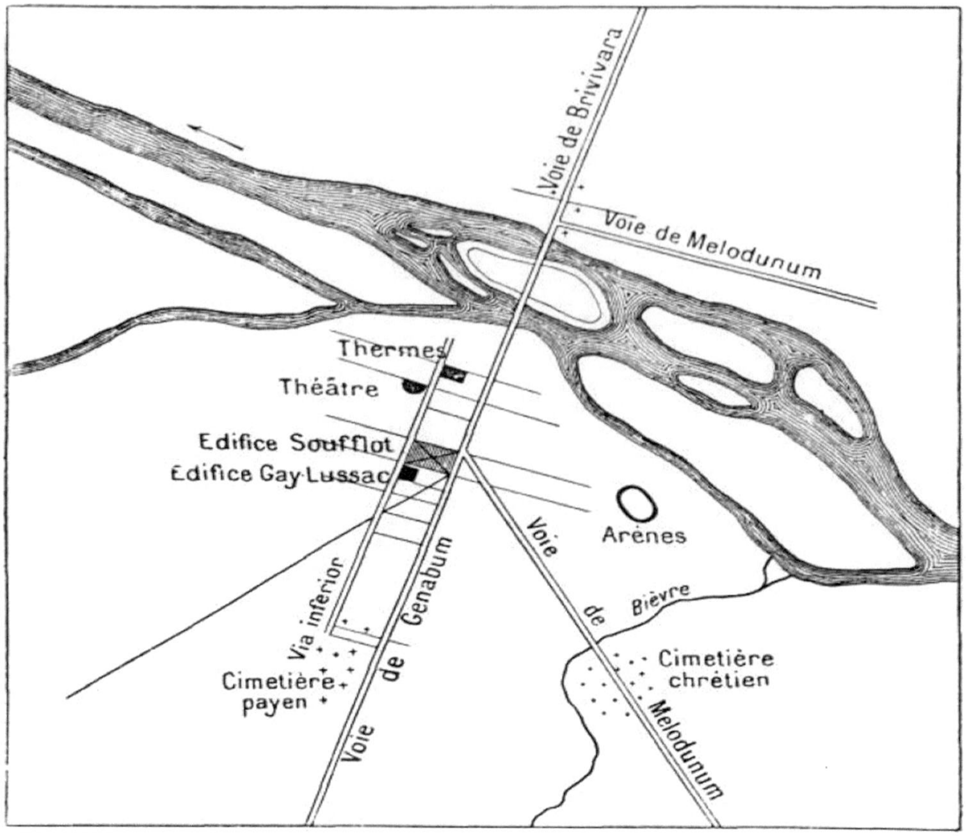

Fig. 2.4 Planche Vacquer: Palaeochannel network of the River Seine surrounding "Île de la Cité" during Gallo-Roman period, as distinguished by T. Vacquer, according to Dupuy (1900)

2.3.1 Paris Founded and Built Upon Its Quarries

Open sky quarries were exploited from Antiquity while underground quarries were exploited as from the end of the twelfth century only. In 1813, Imperial decrees forbade the opening of new quarries, and the last Parisian quarries ceased operating in 1910. This interdiction was extended to the entire Seine department on 9 May 1962. Today, it is estimated that one twelfth of Paris is undermined by ancient quarries and in the Île-de-France Region, 3,000 ha of land spread over 70 municipalities are affected by the presence of old quarries (Fig. 2.5).

Lutetian (=from Lutetia, Paris' Roman name) limestone or "*Banc Royal*" provided the building stone and was also used for delicate sculptures such as those of the porch of Notre Dame Cathedral due to its fine, regular grain (Fig. 2.6). Underground mining extended over 2,350 ha in Paris and the adjacent areas of Haut de Seine and Val de Marne departments. The digging of horizontal quarry entrances on hillsides gave access to galleries overlapping more or less at right angles. At intersections, pillars were left in place so that the rock overlying the mine remained stable. From the fifteenth century, shafts were dug to access valuable strata, which were hollowed out by longwall face advance. The voids were backfilled by mining waste or by imported surface earth. These were contained behind dry stone walls reinforced by stacked pillars (Fig. 2.7b).

Plaster of Paris made from Barthonien gypsum heated at 120 °C was exported worldwide through the multiplication of quarries from the seventeenth century: 800 ha were exploited, amongst which is the well-known quarry of Butte Chaumont. The rock cohesion required the use of explosives for its extraction. Underground galleries of ogival or trapezoidal shapes intersect at right angles, with abandoned pillars in place to ensure the strength of the whole structure.

Senonian chalk was used for the manufacture of lime and whitewash (*Blanc de Meudon*). It was extracted southwest of Paris (35 ha; Fig. 2.5). The quarries were first exploited by the method of irregularly abandoned pillars, with several tiered levels, which resulted in frequent collapses. Hence, a new operating technique was adopted in 1868 and imposed the use of regularly abandoned pillars. Once extraction was completed, the galleries were reshaped into regular arches for increased stability.

Ypresian plastic clay, used for bricks and tiles, was exploited in the 16th district of the city by shafts and radiating galleries.

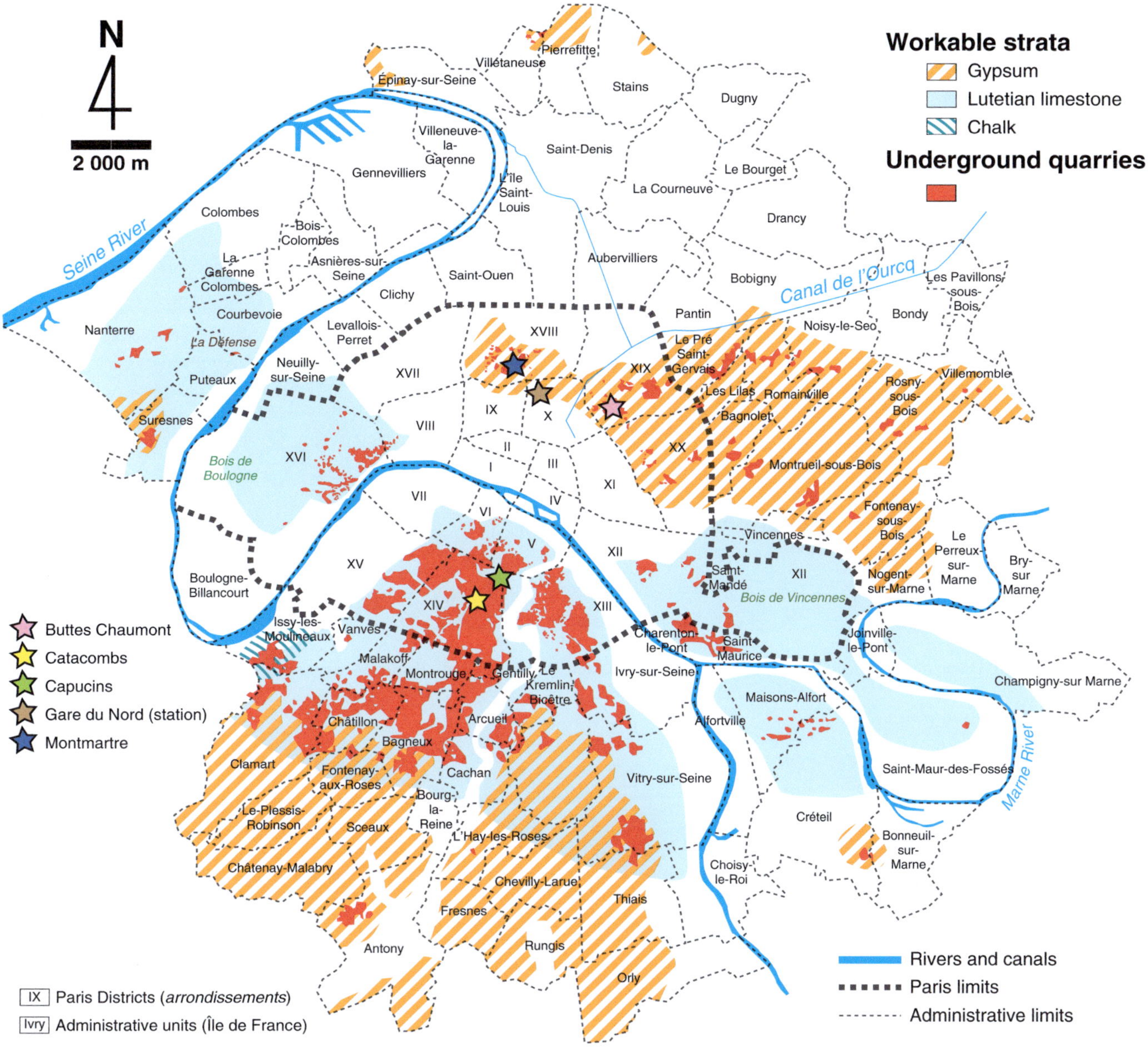

Fig. 2.5 Map of old quarries in the former Dept of Seine (Modified from the "Old Quarries Map", Ecole des Mines de Paris, and redrawn by N. Vanara)

Underground mining of supra-gypsum marls (upper Barthonien) and lignites (Ypresian false clays) are mentioned respectively in the 19th and 14th districts (Gerards 1908).

2.3.2 A City Under Surveillance: Subsidence, Collapses and Sinkholes

Paris lies on undermined bedrock (Caron et al. 1986). The fate of all abandoned cavities is to progressively subside with time. This slow and insidious work was revealed when sudden collapses of buildings occurred. Following several dramatic collapses, a Quarries Inspection Office was created in 1777; it carried out an efficient survey and reinforcement work of cavities. Surface hazards caused by underground disorders can be classified into three categories: (1) Subsidence was frequent in areas where quarries had been dynamited or backfilled (Buttes Montmartre and Chaumont). (2) Collapses are more dangerous. The most recent, largest and deadliest collapse occurred on 1 June 1961, between Clamart and Issy-les-Moulineaux where the roof fall of an old chalk mine wiped 23 buildings off the map. It was therefore decided in 1981 to carry out an early treatment consisting of almost systematic grout injection of residual cavities. (3) The origin of the subsidence is not always due to quarrying. In gypsum, groundwater can create pockets of dissolution evolving in sinkholes. The most important sinkhole was discovered in 1975 below the Gare du Nord train station, and its stabilisation has necessitated 7,300 m^3 of grouting.

Fig. 2.6 The Notre Dame porch delicately sculptured and made of Lutetian stones (Credit photo: Monique Fort)

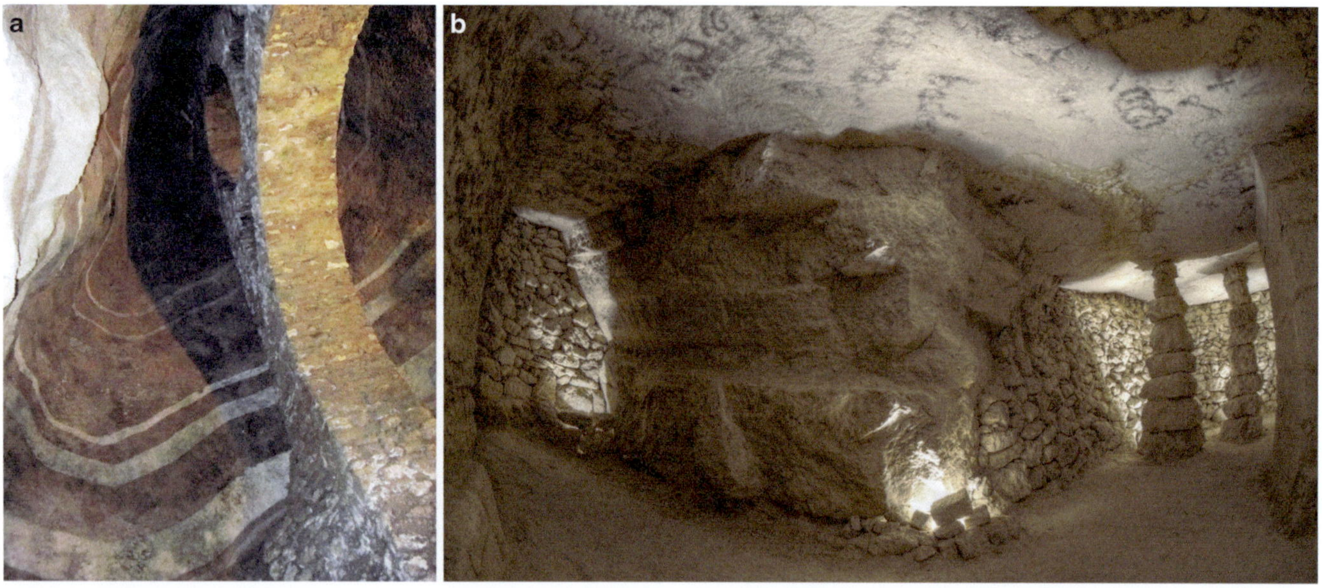

Fig. 2.7 (**a**) At the end of the Catacombs visit, the sinkhole has been cleared, cemented and painted with black and ochre lines representing the various geological strata of the site; it was then reinforced by a millstone arch which was set up in order to support the construction of a small building on surface (Credit photo: Gilles Thomas). (**b**) Crossroad of galleries below the Cochin hospital. Different types of supporting pillars exist to reinforce and strengthen the galleries' roofs: dry stone walls, abandoned pillar (in the centre of the photo), stacked pillars (along the middle of the right gallery) and reinforcement pillar (on the right of the photo) (Credit photo: Jean Martinez, SEADACC)

2.3.3 Mystery and Attractiveness of Underground Paris

Old underground quarries are amongst the many tunnels that constitute the body, arteries and intestines of underground Paris: sewer networks, canals and reservoirs, technical galleries (for electricity, communication, urban heating, etc.), crypts, wine cellars and caves (often transformed into concert halls or night clubs). These abandoned quarries were used by farmers to raise mushrooms; they were also a good place to protect civilian population from aerial attacks during World War II. Nowadays, underground Paris has its fans called "*cataphiles*". The Catacombs (14th district) of Paris is the most famous site (Fig. 2.5). These former limestone quarries were used from 1785 to 1933 to store insalubrious remains of all the cemeteries of Paris: the bones were arranged along the galleries and were set in romantico-macabre death scenes. In the access gallery to the ossuary, the west-east corridor has kept its abandoned pillars and dry stone walls which ensure the stability of old back filling. At the end of the visit, before the exit, two sinkholes were emptied and consolidated in 1874 and 1875: their walls are stabilised with cement and painted with black and ochre lines indicating the different geological strata of the cave section (Fig. 2.7a). The Capucins Quarries (where the 14th, 13th and 5th districts meet) are quite different (Fig. 2.5). All the different types of reinforcements used in Paris are displayed here (Fig. 2.7b). In addition, a well and a shaft are the most interesting of the site. The Capucins Fountain, built in 1810, is a well located in a gallery and its steps lead to the groundwater level; a water gauge allows the recording of water level variations. Nearby, the service shaft is a former sinkhole converted for staff access in 1841: its trapezoidal shape and square section with a height of 14.26 m is remarkable.

2.4 Paris under Water?

The threat of a major flood, such as it occurred in 1910, looms over Paris area. Awareness of this risk is relatively recent, and the vision of European cities drowned in the summer of 2002 (Vienna, Prague, Dresden), or American cities (New Orleans in 2005, New York in 2012), has only reinforced the fear that Paris could be paralysed by such a disaster.

2.4.1 The 1910 Flood

To measure its potential consequences, one can just compare the Paris City in 2010 with photos taken during the centennial flood of 1910, during which the water level rose to 8.62 m above the reference level (Fig. 2.8), forcing many Parisians to travel by boat (Radoux 1910). At that time, the

Fig. 2.8 The zouave of the Alma bridge is used as a benchmark to assess the magnitude of the Seine discharge across Paris. Alert state begins at 3.20 m (below the zouave's feet), and 4.30 m is the highest water level above which navigation along the Seine has to be stopped. Other benchmarks refer to floods of the last century, with the one in 1910 as the most disastrous. In 1658, the Seine rose up to 8.96 m (Fort 2003; Radoux 1910)

Seine waters flowed over the dikes, invaded its right bank and partially occupied the emplacement of its former prehistoric meander (Fig. 2.2a); hence, the *Bassin de l'Arsenal* passed through the *place de la République*, the St. Lazare railway station (Fig. 2.9) and the Provence and La Boetie streets before joining the current course at the place de l'Alma. On the left bank, the water had invaded the lower valley of the Bièvre (at the *Jardin des Plantes* and the *Gare d'Austerlitz*) (Fig. 2.10) and had spread into the old bed of the Seine River, at the ZAC Tolbiac, and between the river and the Boulevard St. Germain, and more widely in the 15th district.

If the damages caused by this flood were retrospectively evaluated to one billion euros, the cost of the same 100-year flood today would be much higher (20–40 billion euros, according to estimates) because the vulnerability of the city has increased considerably in the meantime. Since 1910 in fact, the urban landscape has changed a lot. Major rehabilitation operations have resulted in placing large equipment such as the Bibliothèque Nationale de France F. Mitterrand (13th) (Fig. 2.11), university campus of Paris-Rive Gauche (13th) and Jussieu (5th), Paris Omnisports Paris-Bercy (12th) and the Pompidou Hospital (15th) in areas that would be inundated by 1.5–2 m of water in case of a 100-year recurrence flood. Large areas (first, fourth and sixth in particular)

Fig. 2.9 The St. Lazare railway station during the 1910 flood. The station is situated in the Pleistocene Seine meander SW of *Butte Montmartre* (From Wikimedia Commons, the free media repository)

would probably have their basements and cellars flooded by a rise of the water table, while the streets should generally stay out of water due to dikes and protecting walls along the quays. Note also that since 1910 the number of networks installed underground (wiring and fluid flow, transport, sewers, parking lots), although theoretically isolated from groundwater could be swamped by the arrival of water from the surface, by possible upwelling, even by the collapse of fragile cavities (such as badly consolidated, old quarries). In total, it would be 20,000–50,000 people who should be evacuated from inundated terrains within Paris.

2.4.2 A Real and Foreseeable Risk of Inundation

In the suburbs, demographic pressure has everywhere made the urban pattern denser. Corridors that were natural floodplains of the Seine and its tributaries (Marne, Yerres, in particular) have been used by transport routes (railways, road and motorway networks) and industries and have been urbanised, so that these floodplains, which had the function of dispersing the flood wave and buffer the extreme flows, have been converted into impervious surfaces due to extension of asphalt and concrete coverings. If the flood of 1910 damaged 14,000 buildings and affected 200,000 people, a flood of similar magnitude today would affect the 230 communes settled in the Seine floodplain, with about 40,000 ha potentially inundated by water, and up to 800,000 victims (Fig. 2.10).

Aware of the problem, decision and policy makers in Paris (Prefecture of Police, City Hall, Water Unions) know they have little influence on natural causes whose conjunction is at the origin of the largest floods like the one in 1910. Amongst those causes, three of them are really significant (Fort 2003): (1) The configuration of the Paris Basin watershed leads the Seine and its tributaries (Yonne, Aube, Marne and the Grand Morin, Yerres) to converge to the Paris metropolitan area, which may result in the superposition of several flood waves and amplification of water rise in the centre of the "bowl", i.e. in Paris. (2) Significant rainfall across the entire basin may be another cause. (3) If rainfall occurs when soils are already saturated or made impermeable by frost, run-off may increase drastically and the response time to

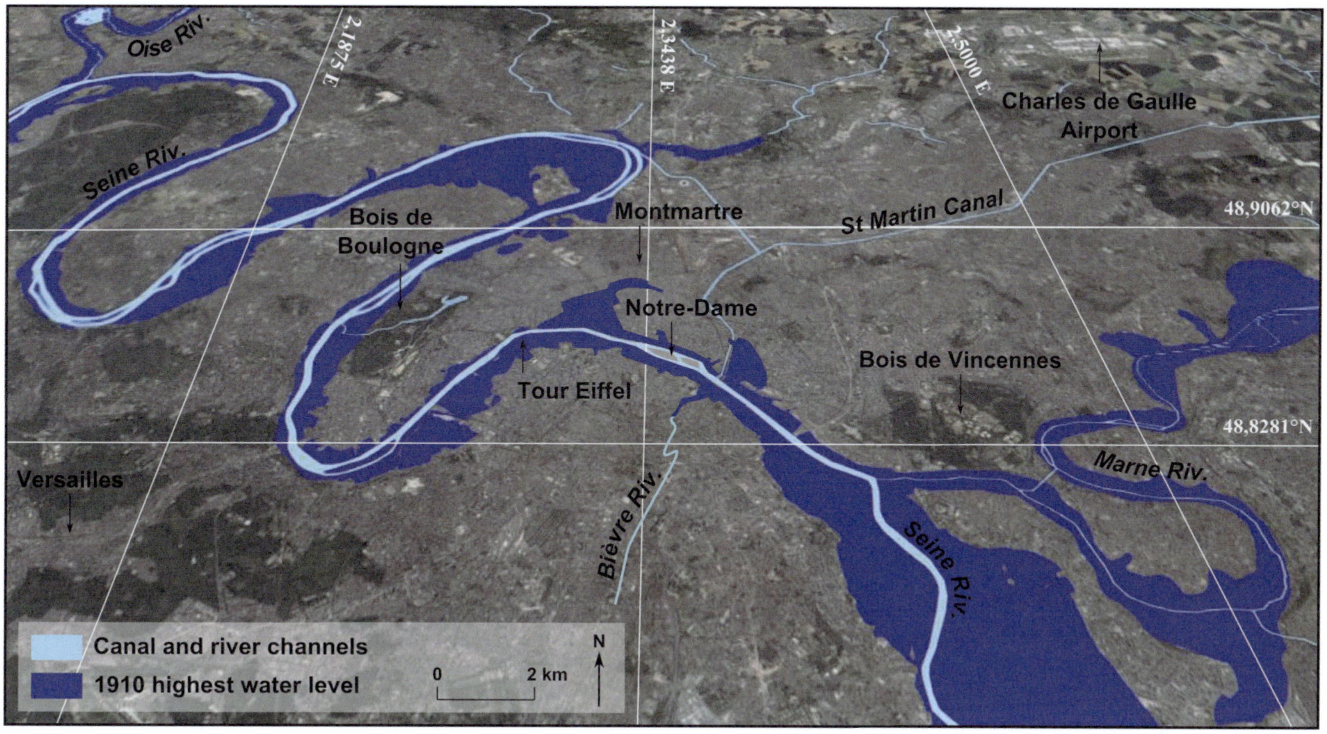

Fig. 2.10 Areas affected by the 1910 flood of the Seine River (Modified from PHEC of DRIEE, 2011, and completed by G. Melun, with permission)

discharge becomes very short. The uniqueness of the 1910 flood, caused by a particularly unfavourable combination of these three types of factors, helps explain why the discharge of the Seine rose up to 2,300 m^3s^{-1} (average Seine discharge in Paris is 328 m^3s^{-1}).

Protection measures do exist, either locally (containment of the bed of the Seine and its tributaries in highly urbanised areas) or regionally, as lake reservoirs built since 1949 upstream of Paris, on the Yonne River (Panessière), Seine River (Forêt d'Orient), Aube River and Marne River. Their function is twofold: sustain the flow of the Seine River during summer drought and retain excess water during winter. In reality, their total storage capacity (865 million m^3, barely one fifth of the volume concerned during a centennial flood) would allow a maximum 60 cm water level lowering, hence reducing partially the cost estimated to several billion euros of such an event. However, it would not prevent Paris from being flooded. Other solutions are being considered, such as the proposed flood expansion zone (2,000 ha) downstream of Montereau (south of Paris), which should allow, in case of simultaneous floods along the Seine and Yonne rivers, to absorb overflow waters of the Seine.

These achievements, valid on a long-term timescale, are completed by a protection scheme operating on the very short term. Floods in the Seine watershed are predictable, days in advance (Warning Flood Service), which thus allows enough time to get organised. Even if nobody knows when it will be realised, everyone is now convinced that the risk of a new centennial flood affecting again Paris and its suburbs is real, prompting officials from the region to invest right now in the protection and prevention, i.e. to anticipate without panicking concerned populations. Such a crisis scheme is now in existence, even if all the actors are not yet fully prepared. Priority will be given to people at risk (flooded neighbourhoods, patients in hospitals), and an evacuation scheme is already planned. Transportation will also be protected: RATP (Metro Agency) is the most advanced and has already identified along its network the 4,000 sites to seal with walls, which could be done in three days. SNCF (railways) must solve difficult problems in deeper stations like in St. Lazare and to a lesser extent at Gare de Lyon and Tolbiac-Massena train stations, as well as along the RER "C" line, which runs in the dike-contained channel of the Seine River. At EDF (Power), the emergency plan includes hundreds of generators, but potentially drowned networks could not operate again until the water has drained and the system secured, which could take weeks long. As for museums, they must arrange to put their reserves out of water reach and safeguard their national collections. The most difficult problem concerns the sewers, in order to avoid their flowing back and the resulting pollution. To sum up, it is likely that the life of millions of people and the economy of the Île-de-France would be paralysed for several months.

Fig. 2.11 Paris seen from the Sacré Coeur on the hill of Montmartre. The spread of urban development has nearly masked the Seine River valley, its adjacent terraces and hills. In the background, architectural landmarks of different periods: Notre Dame Cathedral (twelfth to fifteenth centuries) is hardly distinguishable from the Panthéon, erected (1764–1812) on the Montagne Sainte-Geneviève at the edge of the former Roman city of Lutetia. The Montparnasse Tower (1969–1972), 210 m high, and the Bibliothèque Nationale de France F. Mitterrand (1988–1996), 79 m high, are both emblematic of modern architecture (Credit photo: P. Migon, with permission)

2.5 Conclusion

The human occupation of Paris since the last thousand years took advantage of, and modified constantly, the natural site initially selected for its vicinity to the river. The *Île de la Cité*, together with a series of mounds and small islands, helped Lutetia to become a "bridge town". If some Roman remains are still visible (Arena, Cluny thermal spa) along the sides of the *Montagne Ste-Geneviève*, evidence of earlier occupation is hidden in the subsoil of the city, discovered by chance during large urban renewals. The left bank tributary of the Seine River, the Bièvre River (Fig. 2.10), famous for its tanneries, is now totally channelized underground, with its old course indirectly marked by low grounds and gardens (René Legall square, *Jardin des Plantes*) that would easily be submerged again in case of large floods. The garden of *Les Buttes Chaumont* was planned after the gypsum quarry exploitation was abandoned. Although mostly entirely hidden by buildings higher and higher, the very site of Paris is still well expressed by the contrasts between the banks of the Seine River, a Cultural UNESCO Heritage locally converted into recreational areas, and the top of hills like the *Butte Montmartre* or Trocadero, from where one can embrace the entire city and its very long history, with its architectural landmarks (Fig. 2.11). This landscape is no more natural, but attractive places like the Notre Dame crypt or the Catacombs caves tell us a lot on the strong links that have always attached the Seine River and its environment to the Parisian population.

Acknowledgements We would like to acknowledge all those who helped us in collecting and reproducing illustrations of this chapter – Richard Cottiaux (Inrap), Patrick Pion (Université Paris X), Carlos Valero (Inrap), Maryse Hubert and Gabriel Melun (University Paris-Diderot). We are grateful to Jean-Pierre Peulvast for his reading of an earlier draft of the manuscript.

References

Billy G (1955) Les restes humains de Grenelle et de Clichy. Bulletins et Mémoires de la Société d'anthropologie de Paris X-6(1–3):3–67

Busson D (2001) Paris ville antique. Editions du Patrimoine, Guides archéologiques de la France, Paris

Caron B, Dewolf Y, Diffre P, Gaudant J, Marvy J, Pomerol C (1986) Pourquoi Paris? Une métropole dans son environnement naturel. Association des géologues du bassin de Paris édit, Paris, p 32

Chaussé C, Voinchet P, Bahain JJ, Connet N, Lhomme V, Limondin-Lozouet N (2004) Middle and Upper Pleistocene evolution of the river Yonne Valley (France). First result. Quaternaire 15(1–2):53–64

Chaussé C, Leroyer C, Girardclos O, Allenet G, Pion P, Raymond P (2008) Middle late Holocene history of the river Seine in Paris (France); contribution of the "quai Branly" Museum data. The Holocene 18(6):967–980

Clément A, Thomas G (2001) Atlas du Paris souterrain. La doublure sombre de la Ville lumière. Parigramme édit, Paris, 193 p

Diffre P (1969) Hydrogéologie de Paris et de sa banlieue. Thèse de Doctorat de 3ème cycle de l'université de Paris 4, Paris

Dupuy P (1900, Réédition: 2013) Le sol et la croissance de Paris. Éditions des Régionalismes PyrèMonde. Annales de Géographie 9(46):340–358

Gerards É (1908, Réédition: 1991) Paris souterrain. DMI édit, Torcy, 667 p

Fort M (2003) Paris sous les eaux? Pouvoirs locaux 56(1):74–76

Jugie M (2011) Reconstitution des variations du tracé fluvial de la Seine dans la plaine alluviale de Grenelle (Paris, France), une approche géoarchéologique à partir d'enregistrements morphosédimentaires et les apports d'une cartographie diachronique. Mémoire de Master de l'Université Paris 1, Paris

Lanchon Y (ed) (1998) Le cadre naturel et les occupations néolithiques du site de Paris Bercy. Service Régional de l'Archéologie d'Île-de-France, Saint-Denis

Leroyer Ch (2006) L'impact des groupes néolithiques du Bassin parisien sur le milieu végétal: évolution et approche territoriale d'après les données polliniques. In: Duhamel P (ed) Impacts interculturels au Néolithique moyen. Du terroir au territoire : sociétés et espaces. Actes du 25e colloque interrégional sur le Néolithique, Revue Archéologique de l'Est, 25e supplément, Dijon, pp 131–149

Pion P, Chaussé Ch, Lecomte B, Raymond P, Cottiaux R, Rochard X, Maya LM, Bauchet O, Viré M, Allenet G, Yvinec JH, Dietrich A, Girarclos O, Oberlin Ch, Leroyer Ch (2005) L'histoire de Paris vue du fond. Une pêcherie mérovingienne peut-être associée à un moulin dans un chenal secondaire de la Seine à Paris – Quai Branly. In: Serna V, Galicé A (ed) Actes du colloque « La rivière aménagée: Entre héritages et modernité » Aestuaria 7, Orléans, pp 31–51

Radoux A (1910) Les crues de Paris (VIe-XXe siècles). Cause, mécanismes, histoire, dangers, la lutte contre le fléau. Berger-Levrault, Paris

Souffi B, Marti F, Griselin S, Bridault A, Chaussé Ch, David E, Drucker D, Granai S, Gosselin R, Hamon C, Leduc C (2012) Occupations mésolithiques en bord de Seine. Le site de Paris-15e "62 rue H. Farman". Premiers résultats. In: Actes des journée archéologiques d'Île-de-France, 2004, 2005, 2009, 2010, 2011, Service régional de l'Archéologie, Direction régionale des affaires culturelles, vol 2, pp 1–13

Souffi B, Griselin S, Bridault A, Chaussé Ch, David E, Drucker D, Granai S, Hamon C, Leduc, Vanhaeren M, Valentin F (in press) Mesolithic occupations in the Seine Valley: the site of Paris "62 rue H. Farman" (15th district of Paris, France). In: Arias P, Cueto M (eds) Meso 2010, the eighth international conference on the mesolithic in Europe, Santander, Spain. Instituto Internacional de Prehistoria de Cantabria (IIIPC), Cantabria University, 13–17 September 2010

Viand A (ed) (2008) Nanterre et les Parisii: une capitale au temps des Gaulois? Somogy Éditions d'Art, Paris

The Seine River from Ile-de-France to Normandy: Geomorphological and Cultural Landscapes of a Large Meandering Valley

3

Jean-Pierre Peulvast, François Bétard, and Christian Giusti

Abstract

Rich in historical and cultural heritages, the stretch of the Seine valley that links Île-de-France to Normandy, between Mantes-la-Jolie and the Eure and Andelle confluence zone, belongs to one of the major sets of entrenched meanders known in the world. The valley presents steep hillsides punctuated by white chalk pinnacles alternating with deep funnels, contributing to the picturesque landscape of the valley segment. The geomorphic history of the Seine valley is inseparable of the Quaternary bioclimatic history, with its alternating glacial-interglacial and stadial-interstadial periods. All along the Pleistocene, periglacial processes interacted with fluvial erosion, leading to the formation, deepening, enlargement, and migration of the large meanders. The resulting, present-day geomorphological landscapes are enriched by many cultural treasures. Its emblematic sites are the medieval castles of La Roche-Guyon and Les Andelys (Château-Gaillard) which were built on rocky promontories on the concave sides of two large meanders. The mid-Seine valley is also known as a high place of the impressionism, the founder and master of which, Claude Monet, settled here for the second half of his life and created the wonderful gardens of Giverny on the lower Epte River. He and many other impressionist and postimpressionist painters, sensitive to a certain harmony of the local landforms, magnified and immortalized the surrounding landscapes.

Keywords

Meanders • Chalk erosion • Periglacial features • Impressionism • Castles

3.1 Introduction

Among the regions of France where remarkable geomorphological landscapes also represent invaluable historical and cultural heritages, the stretch of the Seine valley that links Île-de-France to Normandy, between Mantes-la-Jolie and the Eure and Andelle confluence zone, is one of the most present in the collective memory. Due to moderate heights, it does not display the most spectacular landforms in any specific geomorphic type, although it belongs to one of the major sets of entrenched meanders known in the world. It rather owes its fame to events and constructions linked to its position on the most active way linking Paris to the sea and on the historical border of Normandy, as well as to the numerous literary and pictorial representations of its landscapes. The valley

J.-P. Peulvast (✉)
Geomorphology, University of Paris-Sorbonne, Paris, France
e-mail: jean-pierre.peulvast@wanadoo.fr

F. Bétard
Physical Geography, Paris-Diderot University, Paris, France
e-mail: francois.betard@univ-paris-diderot.fr

C. Giusti
Physical Geography and Environmental Sciences,
University of Paris-Sorbonne, Paris, France
e-mail: Christian.Giusti@paris-sorbonne.fr

M. Fort and M.-F. André (eds.), *Landscapes and Landforms of France*, World Geomorphological Landscapes,
DOI 10.1007/978-94-007-7022-5_3, © Springer Science+Business Media Dordrecht 2014

Fig. 3.1 Vétheuil (in the background, *right side*) and the concave side of the Moisson meander of the Seine valley, from Haute-Isle (Photo J.P. Peulvast, January 1982). Only the lower half of the slope is shaped into the Campanian chalk (pinnacles), whereas the upper half corresponds to Cenozoic layers. A winter flood underlines some lateral channels outside the main one (island on the *right side*)

presents steep sides punctuated by white chalk pinnacles and pierced by caves and troglodyte dwellings (Fig. 3.1). Its emblematic sites are the castles of La Roche-Guyon and Les Andelys (Château-Gaillard), ancient lookouts on the Seine valley, the latter having been harshly disputed in the thirteenth century between the Duke of Normandy and King of England, Richard the Lionheart, followed by John Lackland, and the King of France, Philippe August. But the mid-Seine valley is also known as a high place of the impressionism, the founder and master of which, Claude Monet, settled here for the second half of his life and created the wonderful gardens of Giverny on the lower Epte River. He and many other impressionist and postimpressionist painters magnified and immortalized the surrounding landscapes. Beyond the quality of the light, all these artists were sensitive to a certain harmony of the local landforms. A more scientific approach helps in understanding the layout of these landforms and discovering a rich geomorphological heritage which contributed to justify the preservation of part of them in the "Parc naturel régional du Vexin Français."

3.2 Geographical Setting

Sheltering two important cities, Mantes-la-Jolie and Vernon, the studied segment of the Seine valley crosses the border between Île-de-France and Normandy 50–100 km downstream from Paris. It includes two sets of large meanders separated by a 20 km long rectilinear segment, from Vernon to Gaillon. The confluence zone of the Epte River, on the right side, marks the entrance of the valley in Normandy (Fig. 3.2a). Upstream, the Seine River separates the Vexin Français (north), from the Mantois region (south); both units belong to the same plateau area.

Whereas the Seine River flows to NW, from 19 to 8 m a.s.l. downstream of Les Andelys, this plateau largely dedicated to farming gently slopes to southeast, from 150 to 160 m in the tabular Vexin Normand to 120–140 m around Mantes-la-Jolie. Therefore, increasing heights (up to 140 m) characterize the valley sides. More and more spectacular downstream, they form outstanding historic and touristic sites (Château-Gaillard, Côte-des-Deux-Amants; Fig. 3.3a, b). In the Vexin Français, north of Mantes-la-Jolie, narrow WNW-ESE alignments of wooded buttes overlook the plateau close to the valley, reaching 206 m a.s.l. near Vétheuil (Fig. 3.2a). The valley floor is generally wide (1–2 km), although it locally narrows to 600 m, between Giverny and Vernon. However, the presence of a few confluence zones and, above all, the meandering outlines of the main valley explain considerable variations in width as well as strong asymmetries in the valley sides.

Left-side tributaries remain scarce and short down to the wide confluence plain of the Eure River (Fig. 3.2a). Downstream of the narrow Vaucouleurs valley, which cuts through the low fluvial terrace forming the urban site of Mantes-la-Jolie, a few deep and mostly dry valleys only incise the rim of the otherwise weakly dissected plateau. On the right side, short tributaries also dissect the rims of the Vexin Français and Vexin Normand. Only some of them are drained by permanent creeks, in their downstream reaches.

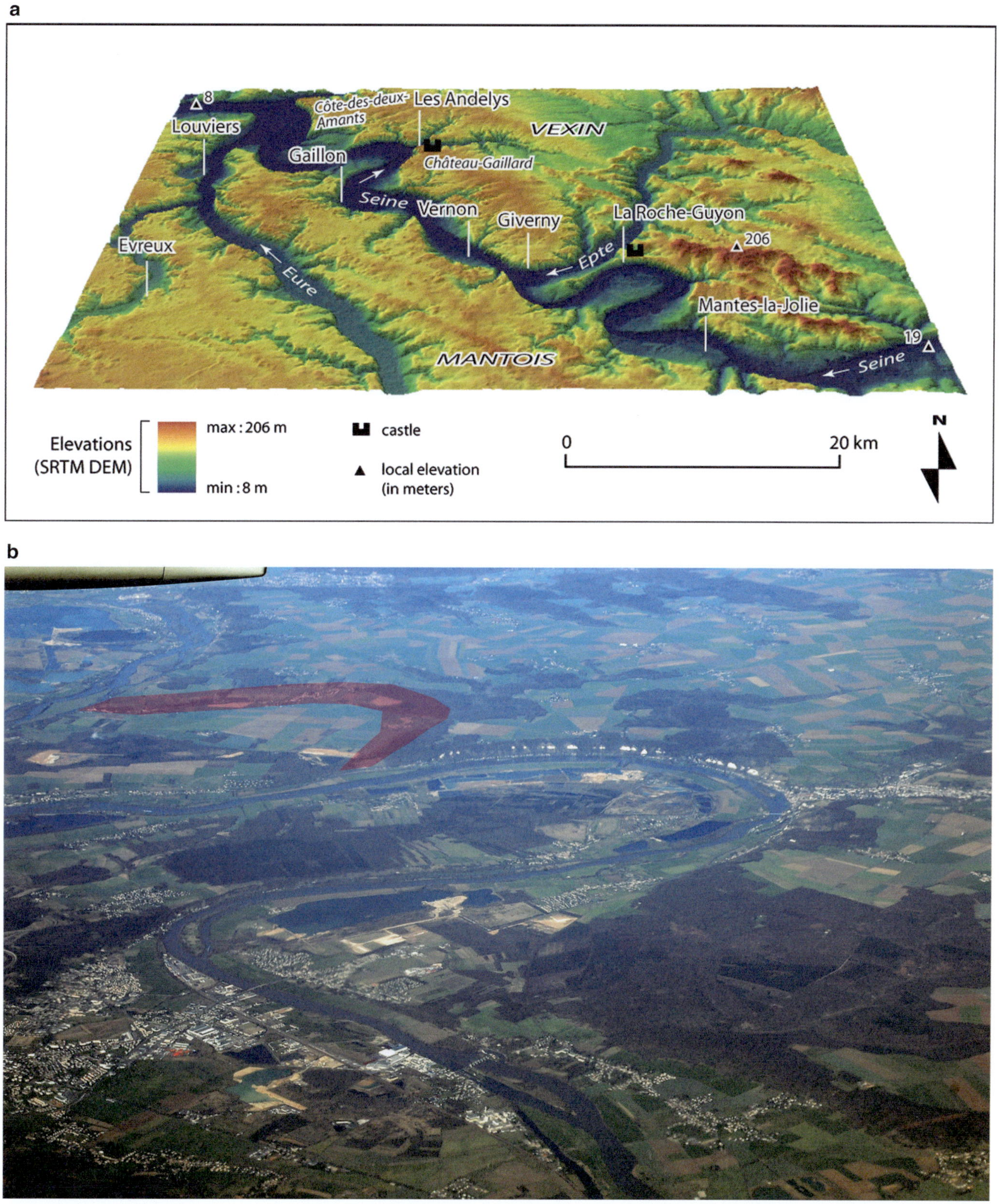

Fig. 3.2 The Seine valley and its surroundings between Mantes-la-Jolie and the Eure-Andelle confluence zone. (**a**) 3D representation, from SRTM DEM (F. Bétard). Note the importance of large enclosed depressions on the valley floor, mainly downstream of Mantes-la-Jolie: active and former sand pits. (**b**) The Gaillon-Les Andelys meander. Note the chalk pinnacles around Les Andelys (the small city on the *right side*), the large sand pits in the alluvial plain and the Port-Mort dam and lock. *Red strip*: the abandoned meander of Daubeuf (Photo J.P. Peulvast)

Fig. 3.3 The concave side of the Les Andelys and Poses meanders at les Andelys, from Château-Gaillard (**a**) and at Amfreville-sous-les-Monts, from the Côte-des-Deux-Amants (**b**) (Photos J.P. Peulvast). Inherited periglacial landforms (Richter slopes, chalk pinnacles) overlooking a narrow "modern" alluvial plain

However, two bigger rivers coming from the humid Pays de Bray, to the north, form wider confluence zones. The lower Epte valley, a well-calibrated fluvial trough, only widens at Giverny, separated from the Seine valley on its last 5 km by an elongated plateau strip reduced to a low crest near La Roche-Guyon. On the contrary, the lower Andelle valley, also turning from SSW to WSW, directly opens on the 3–4 km wide Seine-Eure confluence zone at Pîtres, below a protruding spur named Côte-des-Deux-Amants.

The sinuous parts of the valley, with wavelengths of 4–5 km, classically display low convex and asymmetric lobes gently sloping to the downstream side, opposite to high, abrupt, and partly rocky slopes sharply cut into the concave sides, either directly into the plateau or into the upstream side of the inner lobes. The most of these convex lobes are elongated perpendicularly to the general trend of the valley, over 6–7 km. With altitude decreasing from 80 to 90 m to the floor, they entirely belong to the valley domain, the total width of which is therefore highly variable. Moreover, blunt hillsides with concave outlines, disconnected from the river, draw ancient contours of meanders which later migrated downstream, also contributing to increase the total width of the valley, up to 2 (Vernon, Port-Mort) to 4 km (Buchelay-Mantes-la-Jolie, Gaillon) or even 6 km (abandoned meander of Daubeuf, west of Les Andelys) (Fig. 3.2b).

3.3 Geology

Downstream of Mantes-la-Jolie, picturesque outcrops of Cretaceous chalk form an increasing proportion of the valley-side profiles (Fig. 3.3). Unique among landforms usually covered by soils and superficial deposits, they announce the white vertical cliffs of the Normandy coast and reinforce the scientific interest of landscapes which may be considered as potential geomorphosites (Reynard et al. 2009). They reflect the entrenchment of the Seine valley into the oldest of the layers that form this part of the Paris Basin, to the west and northwest of the Cenozoic plateaus of Île-de-France.

After a long period of emersion since the end of the Jurassic, a marine transgression occurred in the Paris Basin in mid-Cretaceous times (Aptian-Albian). This was the beginning of a long sequence of high sea levels during which carbonate sediments were deposited on vast regions, in warm and shallow water. Hundreds of meters of chalk layers were deposited over large parts of northwestern Europe. Here, the last of them were deposited in the Campanian (83–70 million years), after which the basin progressively emerged and was submitted to erosion. Older chalk layers (Santonian, Coniacian, Turonian) outcrop in the lower parts of the valley slopes, downstream of Bonnières and mainly on the right side. All of them generally contain more or less regularly spaced flint layers and nodules which contribute to increase their relative resistance to weathering and slumping. This resistance allows the formation and preservation of steep cliffs along rivers and shores.

The Cenozoic sediments that overlay the Late Cretaceous chalk are only represented to the southeast (Vexin Français) and to the south, between the Seine and Eure valleys. Paleocene sediments are absent, except in tiny depressions where reefal and peri-reefal limestones are preserved along paleo-fault scarps (Vigny). In the Eocene and Oligocene, the sea invaded periodically the center of the Paris Basin from the English Channel, along a low Seine corridor limited to the north by the NW-SE Bray antiform. After a lacustrine phase in the Ypresian (Sparnacian clays), the last important layer of Lutetian marine limestone ("calcaire grossier") was deposited in a large bay. It presently forms structural surfaces in the Vexin Français and south of the Seine River. Continued by lacustrine and lagoonal sand, marl, and limestone locally interbedded with gypsum and silcretes of Late Eocene age, this series was finally covered by thick littoral sand layers deposited in the old marine corridor and in the Fontainebleau gulf, capped by thin lacustrine silicified clay ("meulière") closing the sedimentary cycle in Chattian-Aquitanian times.

The later erosion of this series was induced by shallow inversion of the basin, accompanied by the progressive formation and incision of a paleo-Loire-Seine drainage axis oriented to the Channel along NW-SE structures (Seine fault, shallow synforms and antiforms). Leading to the formation of the surface that presently forms the main plateau, this phase of denudation resulted in partial erosion of the upper layers of the Cenozoic series. It only preserved the narrow buttes of Oligocene sand capped by hard silcrete ("meulière"), which form the remote frame of the valley in the Vexin Français and Mantois regions (Fig. 3.4). Before the Plio-Quaternary dissection, these buttes were overlooking a large plain where structural surfaces shaped into the Cenozoic limestone layers merged to the west (Vexin Normand) with the exhumed sub-Eocene erosion surface cut into the chalk layers (Dewolf 1982). Reworked until the Pliocene, this surface corresponds to the "flint clay surface" identified in Normandy. Pockets of reddish flint clay are visible under this surface, in quarry walls around Vernon (Fig. 3.5a). They indicate the contribution of subsurface karstic processes to its shaping and the preservation of weathering products in depressions and pits formed by dissolution.

3.4 Landforms and Quaternary Evolution

The landscapes of the Seine valley were formed during the incision phase that followed the last planation stages, the age of which is given by alluvial deposits of the paleo-Loire-Seine preserved on the plateau and known as "sables de Lozère" or granitic sands – i.e., produced by weathering of granitic gravels transported from the Massif Central. The

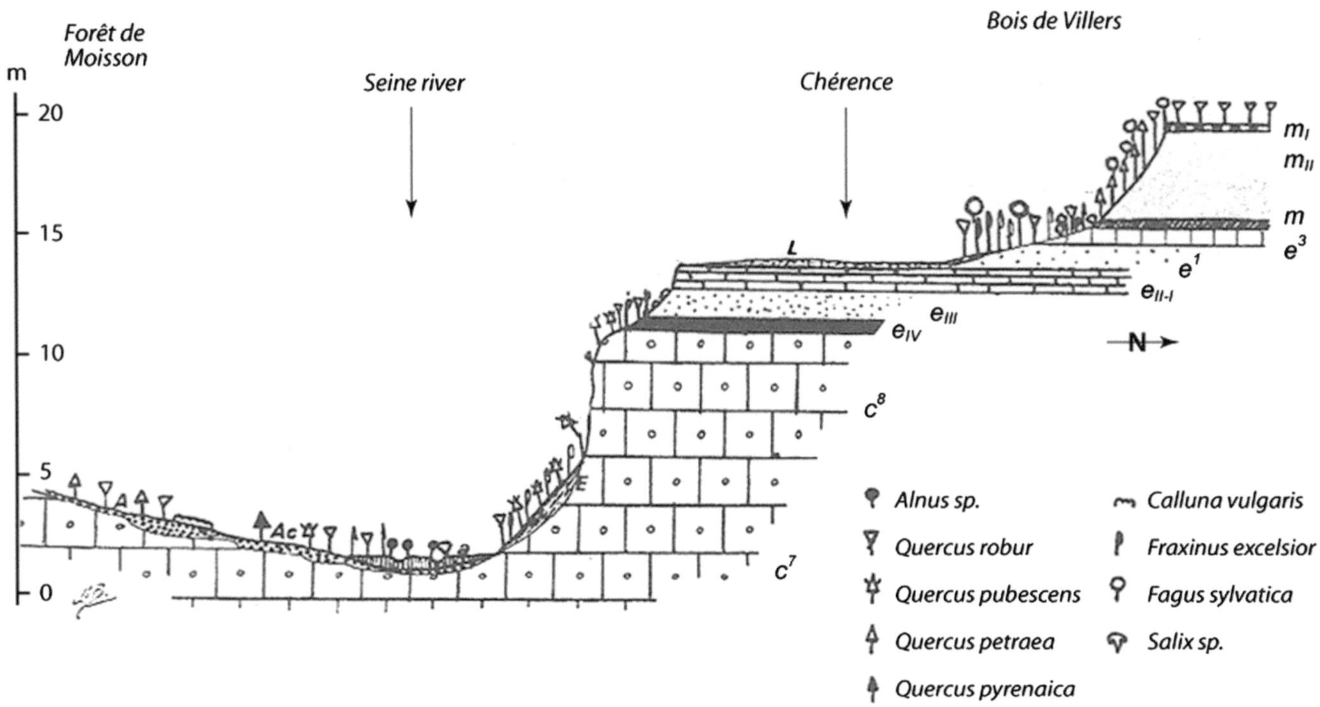

Fig. 3.4 Geological profile of the Seine valley and southern Vexin Français between Vétheuil and La Roche-Guyon (From Pomerol and Feugueur (1974), modified). m_i silcrete ("meulière de Montmorency"), Chattian; m_{ii} Fontainebleau sand, Stampian (Rupelian); *m* marl and green clay, "Sannoisian" (Early Rupelian); e^3 marl, "Ludian" (Priabonian); e^1 Beauchamp sand (Bartonian); $e_{ii}-_i$ limestone ("calcaire grossier"), Lutetian; e_{iii} Cuise sand (Ypresian); e_{iv} plastic clay, "Sparnacian" (early Ypresian); c^8 white chalk, Campanian; c^7 white chalk, Santonian-Coniacian

Pliocene to Early Pleistocene (Gelasian) age of these sands is established from outcrops in the region of Fécamp, close to the English Channel coast, which also registered a regional uplift of 60–100 m since that time (Quesnel et al. 2003). Combined with lowering sea levels and repeated Quaternary episodes of emersion of the Channel, this uplift determined the incision of the lower Seine valley in the former plain. Although deprived of part of its water by the diversion of the Loire River to the Atlantic in the Early Pleistocene, the Seine River maintained its former direction, but its incision in a slowly uplifting region was accompanied by the formation of large meanders progressively entrenched into the plateau.

This history is inseparable of its bioclimatic context, with the alternating glacial-interglacial and stadial-interstadial periods of the last 2.6 million years, succeeding to warmer Pliocene conditions. Like the rest of Western Europe, the region suffered increasingly severe, long, and frequent periods of periglacial climate during the Pleistocene, especially in the last 700,000 years and until the end of the Younger Dryas, 11,500 years ago. In such cold and mostly dry conditions, with continuous permafrost blocking water infiltration, rock disintegration was more efficient than dissolution in carbonate rocks, especially in the chalk, which suffered deep frost shattering. Mass wasting was efficient in steep slopes formed by fluvial sapping and submitted to processes such as rock falls, solifluction, debris flows, or superficial runoff (Fig. 3.5b). Large quantities of waste were provided to the Seine River which, under periglacial climates, carried and deposited thick layers of sand and gravels. These alluvial deposits often contain huge blocks ("blocs démesurés") of sandstone or limestone transported on ice floes during powerful flows triggered by seasonal thawing of snow and fluvial ice.

Also efficient in the affluent valleys, owing to impermeability of the frozen bedrock, these processes interacted with fluvial erosion all along the Pleistocene, leading to the formation, deepening, exaggeration, and migration of the large meanders. In a magisterial study of the Quaternary geology in the middle course of the Seine valley, F. Lécolle (1986) described the alluvial deposits preserved in the form of wide glacis on the lower parts of the convex lobes of the meanders and the original type of evolution that characterized this segment deprived of true fluvial terraces (Fig. 3.6a). The system of meanders evolved both by continuous deepening and by progressive lateral migration of the river, leading to the formation of "conservative meanders."

From a detailed study of the large sand pits opened in the alluvial cover of the convex lobes down to the modern alluvial plain, F. Lécolle evidenced series of 100 m wide stepped terraces shaped into the bedrock and separated from each other by scarps of 1–2 m (Fig. 3.6b). Each step bears oblique and partly embedded sets of deposits defined as "alluvial nappes," the lower and upper limits of which are those of the major erosion phases which shaped the bedrock steps or paleochannels (Fig. 3.6b). In the meanders of Guernes and Moisson, 16 of them were identified, corresponding to Mid- and Late-Pleistocene sequences of Late Elsterian to

Fig. 3.5 Weathering mantle and superficial deposits on the Seine valley sides (Photos J.P. Peulvast). (**a**) Karstic pits and flint clay at the edge of the Vexin Normand plateau above Sainte-Geneviève-lès-Gasny. (**b**) Stratified talus over longitudinal debris flow deposit and frost-shattered chalk and flints at the mouth of a short periglacial gully incised in the lower Epte valley side between Sainte-Geneviève-les-Gasny and Giverny. Legend: *1* Frost-shattered chalk, *2* Longitudinal debris flow deposit, *3* Lateral colluvial deposit

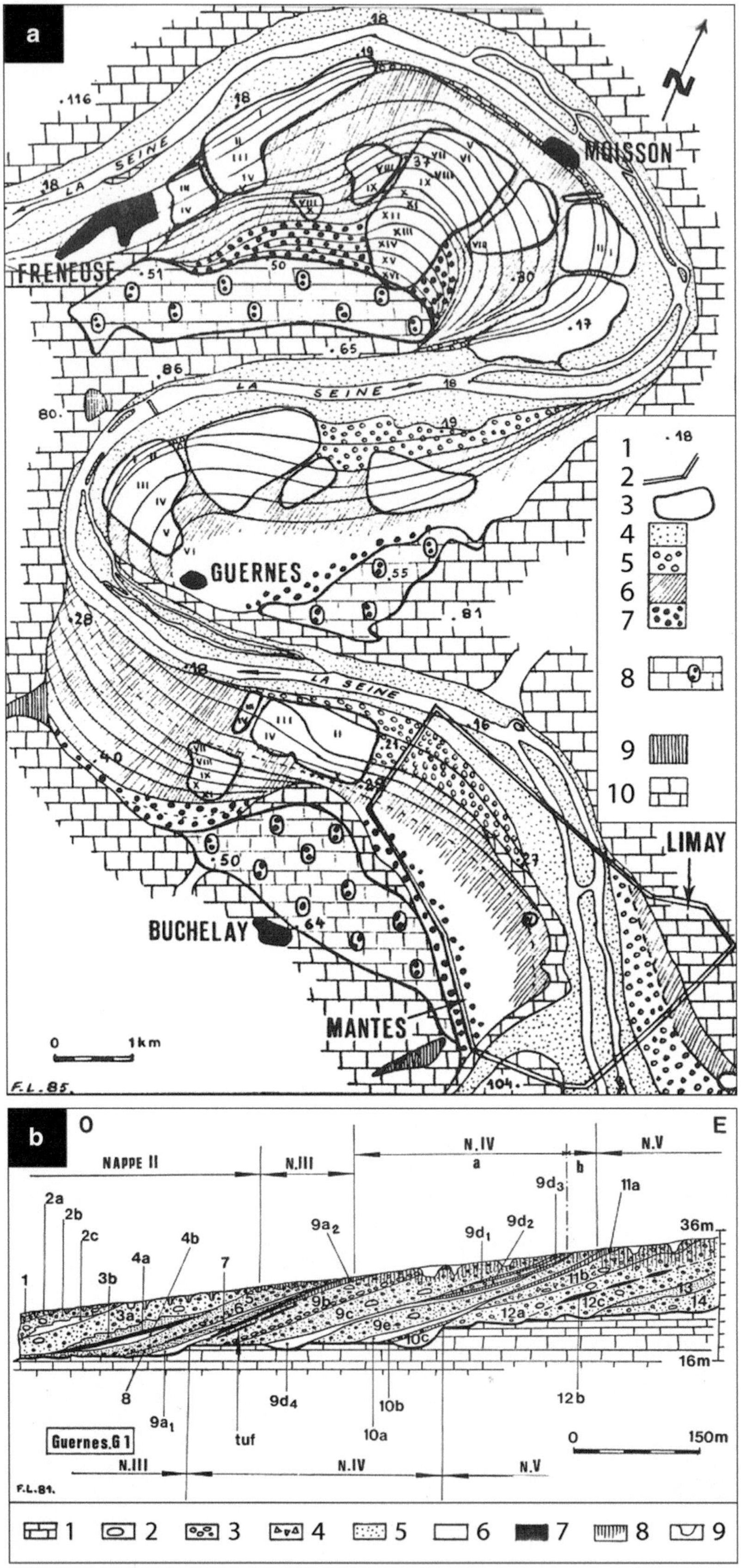

Fig. 3.6 Meanders and alluvial deposits in the Seine valley between Mantes-la-Jolie and Freneuse. Slightly modified from F. Lécolle (1986). (**a**) Map of the alluvial deposits and paleochannels showing the downstream migration of the entrenched meanders of Guernes and Moisson. Legend: *1* Altitude, *2* Limit of the Mantes-la-Jolie agglomeration, *3* Sand or gravel pit, *4* Holocene deposits, *5* Early and mid-Weichselian deposits, *6* Saalian deposits, *7* Elsterian deposits, *8* High level terrace (Elsterian ?), *9* Loess, *10* Pre-Quaternary bedrock. I, II, III, IV: numbers of the alluvial "nappes" (see **b**). (**b**) Profile of the alluvial nappes and paleochannels in the Guernes meander. Note the oblique pattern corresponding to the lateral migration accompanying the progressive incision and the lack of fluvial terraces. Legend: *1* Chalk; *2* Ice-transported block; *3* Pebbles, gravel, sand, *4* Frost-shattered pebbles; *5* Sand; *6* Silt; *7* Fluvial clay, *8* Soil, with pedogenetic reddening; *9* Cryoturbation features

Mid-Weichselian ages, as determined from their degree of weathering and cryoturbation as well as from their fossil fauna and their prehistoric industries. A large part of these deposits reflect catastrophic sedimentation events in cold, periglacial conditions with strong seasonal contrasts. Erosive stages might correspond to the beginning of milder periods with liberation of water giving to the river an increased transport capacity. Their disposition, their superficial truncation by wide glacis, as well as the preservation of stepped paleochannels on the convex side of the valley floor result from simultaneous lateral migration of the river and vertical erosion in the meandering segments. This migration both reflects an exaggeration of the sinuosity and a downstream migration of the valley meanders (Fig. 3.6a). On convex lobes as well as on all concave sides of the meanders, a narrow strip of modern alluvial deposits separates the hill slopes from the river.

Spectacular chalk pinnacles alternating with steep funnels and rectilinear slopes ornate the right side of the valley in the meanders of Moisson, Les Andelys, and Poses. Capped in the southeast (Limay, Vétheuil) by a short cornice of Lutetian limestone which slightly receded by sliding over the Sparnacian clay (Figs. 3.1 and 3.4), these hillsides bear the marks of periglacial processes. Rectilinear profiles (Richter slopes) are well-developed downstream of Les Andelys (Côte-des-Deux-Amants; Fig. 3.3b) and on concave segments of the left side, face to the north (Bonnières-Port-Villez). Their contact with the alluvial plain is sharp, reflecting the scarcity of talus deposits. With their alternate layers of fine gravel and bigger debris, these "grèzes" reflect the variable efficiency of frost shattering in the Late-Pleistocene cold stages and the redistribution of fine-grained material by superficial solifluction. Another type of deposit was observed at the bottom of one of the regularly spaced funnels that separate the pinnacles, east of Giverny. Here, thick layers of heterometric material reflect transport and accumulation processes involving debris flows along the talweg (Fig. 3.5b). This might confirm the suggestion of C. Pomerol (Pomerol and Feugueur 1974) that seasonal thawing of snow patches and of the permafrost active layer might have activated combined processes of gully incision and mass wasting in slopes exposed to SW, S, and SE.

3.5 Landscapes: Human Impact, Cultural Perspectives, and Preservation

The most famous sites of this region result from historic valorization of the steepest hillslopes and pinnacles on the right side of the valley. The medieval castles of La Roche-Guyon and Château-Gaillard were built on rocky promontories of the concave sides of the Moisson and Les Andelys meanders, opposite to large convex lobes, offering excellent views on the river and some of the widest panoramas over the valley (Fig. 3.7). Although largely destroyed after being won by Philippe August in 1203, the Château-Gaillard still presents imposing remains, including an enormous keep, of a fortress which had been built at the top of a narrow chalk promontory, over the confluence zone of the Cambon and Seine Rivers. The natural site itself had been modified in order to make it impregnable (unsuccessfully!), with the excavation of a deep moat isolating the pinnacle from the plateau. The site of La Roche-Guyon was also fortified at the end of the

Fig. 3.7 Medieval castles in the Seine valley. A. Château Gaillard, Les Andelys (Photo C. Giusti)

Fig. 3.8 The human impact on the landscapes: Haute-Isle (E of La Roche-Guyon). (**a**) The Assumption troglodyte church, seventeenth century (Photo C. Giusti) (**b**) Haute-Isle: the chalk pinnacles, toward Vétheuil. Postcard, ca 1900, coll. J.P. Peulvast. Note the importance of the cultivated area, even on steep slopes (orchards); compare with Fig. 3.1. Upslope, the Lutetian limestone cornice is excavated by multiple quarries, now abandoned

Fig. 3.9 The Seine River at Lavacourt, painting by Claude Monet, 1880. Dallas Museum of Art, Munger Fund (Public Domain). Note the partly vegetated sand banks in the middle of the shallow channel

twelfth century, in order to watch both the Seine and Epte valley from the top of a 38 m high keep built on a crest between two steep funnels. Reinforced by two surrounding walls in the thirteenth century, it later lost 22 m of its height. Subterranean stairs connect it to the modern castle built in the sixteenth to eighteenth centuries in the partly artificial cliff shaped below the old keep. Like the castle, the surrounding village comprises many troglodyte dwellings, called "boves." Natural caves are scarce and rather small in the porous chalk (Rodet 1991), but many cavities perforate the pinnacles. Most of them were hollowed out as premises for breeding pigeons or making wine. They are already mentioned in the seventeenth century by the poet Boileau, the nephew of whom ordered the excavation and construction of a troglodyte church in 1670, at Haute-Isle (Fig. 3.8a).

The concave sides of the meanders are also known for their vegetation adapted to the high lime content of their thin rendzina soils. Although rich in protected species of Mediterranean-alpine affinities representing heritages of warmer Holocene times, the grass formations typical of the steepest slopes exposed to the south between the funnels tend to recede owing to the abandonment of the former pastoral activity. Among those that are progressively occupying these slopes and closing the landscapes, many tree species derive from the fruit trees which formerly were grown here together with vine. Photos and postcards dating to the beginning of the twentieth century show an entirely cultivated landscape, with narrow plots elongated in the slopes, even in their steepest parts (Fig. 3.8b). This may explain the erosion of the already thin soils. Currently, actions of protection (such as the classification of the "coteaux de La Roche-Guyon" as ZNIEFF or zone of faunistic and floristic interest) hardly succeed in maintaining the calcicole grasslands in spite of their botanic value (various species of orchids, in particular). However, grasslands are maintained around Château-Gaillard, owing to the Natura 2000 program which helped restoring sheep breeding (Dumont-Fillon 2002).

These spectacular landscapes face convex lobes where vast sand pits and other shallow depressions reflecting former sand and gravel extraction only leave scattered rural or wooded areas, around isolated villages and farms. Some of them were reforested or used as golf links (Moisson), whereas others, excavated below the water table, were transformed in lakes dedicated to nautical leisure (Sandrancourt, Lavacourt, Poses) or bird protection and reproduction (Guernes, Moisson, Bernières, Poses). The river itself was strongly modified: its shallow mobile channels and islands, still visible on Monet's paintings (around 1880) (Fig. 3.9), were replaced by a large fluvial way artificially deepened through dredging and construction of dams and locks (Méricourt, Port-Mort, Amfreville; Fig. 3.2b). Only big winter floods resuscitate for a while the former systems of anastomosing channels in the modern alluvial plain (Fig. 3.1).

These heavily transformed landscapes are hardly reminiscent of those represented by Corot, Maugendre, Monet (Skeggs 1988), and the other artists who had visited the region or settled there. However, not only the changing parts of the landscapes were painted but also the fields, crops, curtains of willows, alders and poplars along the Seine River, and natural or artificial water stretches reflecting variable skies. Permanent elements were also immortalized such as Château-Gaillard and the chalk pinnacles at Les Andelys, by Joseph M. W. Turner, Paul Signac, or Félix Vallotton, and the hillsides of Vétheuil, Rolleboise, and Giverny by Claude Monet, Theodore Robinson, or Maximilien Luce. Even the twentieth century comics art valorized some of them, since the site and castle of La Roche-Guyon were used by Edgar P. Jacobs (2007) as the scenery (between the Cretaceous and an apocalyptic future!) of "Le Piège diabolique," one of the adventures of Blake and Mortimer. Thus, in spite of the extractive activity, still ongoing in parts of the valley floor, the harmony of the landscapes is preserved forever through these famous works which contribute to give this region its unique combination of natural and cultural heritages, making it a high place of "cultural geomorphology" (Panizza and Piacente 2003).

References

Dewolf Y (1982) Le contact Île de France/Basse Normandie, évolution géodynamique. Thèse, Mémoires et Documents de Géographie, CNRS, Paris

Dumont-Fillon N (2002) Les politiques publiques de paysage et de patrimoine: un outil de gestion des territoires. Le cas du Marais Vernier (Eure) et des coteaux de La Roche-Guyon (Val-d'Oise). Thèse de Doctorat, ENGREF, Paris

Jacobs EP (2007) Le piège diabolique. Editions Blake et Mortimer, Bruxelles

Lécolle F (1986) Le cours moyen de la Seine au Pléistocène moyen et supérieur. Géologie et préhistoire. Thèse de Doctorat d'Etat, Université Pierre et Marie Curie, Paris VI

Panizza M, Piacente S (2003) Geomorfologia Culturale. Pitagora Editrice, Bologna

Pomerol C, Feugueur LL (1974) Bassin de Paris. Ile-de-France. Pays de Bray. Guides géologiques régionaux. Masson, Paris

Quesnel F, Catt JA, Laignel B, Bourdillon C, Meyer R (2003) The Neogene and Quaternary Clay-with-flint north and south of the Channel: comparisons of distribution, age, genetic processes and geodynamics. J Quat Sci 18(3–4):283–294

Reynard E, Coratza P, Regolini-Bissig G (2009) Geomorphosites. Pfeil, München

Rodet J (1991) Les karsts de la craie. Etude comparative. Thèse de Doctorat d'Etat, Université Paris-Sorbonne, Paris IV

Skeggs D (1988) Monet et la Seine. Impressions d'un fleuve. Albin Michel, Paris

4 The High Normandy Chalk Cliffs: An Inspiring Geomorphosite for Painters and Novelists

Stéphane Costa

Abstract

The High Normandy coast specificity relies on the existence of remarkable landforms, such as white chalk cliffs and pebble beaches. Due to the low mechanical strength of chalk, these cliffs are very susceptible to erosion, and the rapid cliff retreat (~20 cm/year) threatens homes located inconveniently at the top of the cliff. Because of cumulative factors (exposure to strong west wind, beach sedimentary crisis, low altimetry of valleys), this study area of the eastern English Channel is also particularly sensitive to storm flooding. Moreover, the High Normandy cliffs, with their imposing verticality, their whiteness varying in tone with the ever-changing light and tide and their ghostly shapes, exerted also a powerful attraction to painters and novelists. One of the most famous chalk cliffs known in High Normandy, and probably in France, is Etretat. This site is remarkable because of its geological features and its very low rates of cliff erosion at a historical scale (few cm/year). The touristic site of Etretat, which has been affected by the fashion of sea bathing of the nineteenth century, has inspired painters. French writers, such as Guy de Maupassant and André Gide, also fell in love with the site, not to mention Maurice Leblanc, father of the "gentleman burglar", Arsène Lupin, who hid the treasure of the kings of France in the "hollow stack".

Keywords

Chalk cliff • Coastal erosion • High Normandy • Etretat • Geomorphosite

4.1 Introduction

The High Normandy coast is characterised by imposing chalk cliffs with flint bands, 70 m high on average, overlooking gravel beaches and shore platforms, themselves derived from the cliff retreat. These cliffs are interspersed by deep valleys either drained or dried (*valleuses*), which are the only links between the sea and the inland. This is why coastal communities and their activities are preferentially located in these low areas, though liable to flooding. Due to its morphostructural characteristics (chalk cliffs highly susceptible to weathering), its orientation to the disturbed atmospheric westerly flux and important urbanisation at elevation often lower than high water level of spring tides, the High Normandy coast is particularly vulnerable to natural hazards, namely, storm surge and coastline erosion. The cliff retreat is fast (average 20 cm/year on average between 1966 and 1995), though not uniform. These scarps, located at the north-west end of the Paris Basin, are carved in the Upper Cretaceous chalk, the homogeneity of which is only apparent. Due to tectonic deformations perpendicular to the coastline, various geological strata are outcropping. These lithostratigraphic characteristics determine differences in cliff morphology with contrasted erosion rates.

The cliffs of Normandy, with their imposing verticality, their whiteness varying in tone with the ever-changing light

S. Costa (✉)
Laboratory of Physical Geography and Environment (UMR-CNRS 6554 LETG-GEOPHEN), University of Caen Basse-Normandie, Esplanade de la Paix, Caen 14000, France
e-mail: stephane.costa@unicaen.fr

M. Fort and M.-F. André (eds.), *Landscapes and Landforms of France*, World Geomorphological Landscapes,
DOI 10.1007/978-94-007-7022-5_4, © Springer Science+Business Media Dordrecht 2014

Fig. 4.1 Chalk cliff, shore platform, groyne, gravel beach and valley of Scie (Pourville-sur-Mer) (Photo S. Costa)

and tide and their ghostly shapes, exerted a powerful attraction for painters of the nineteenth century. From Camille Corot to Gustave Courbet, with a special mention for the impressionist painter Claude Monet, they have given us representations of Yport, Dieppe and Etretat. As this site, although daily harassed by the sea, the "Aval Arch" (compared by Maupassant to an elephant dipping its trunk into the sea) has not been eroded and reduced since Claude Monet's time (painter of a "series" that represented it under various lighting conditions at different times of day).

Writers have been also inspired by these white walls and pebble/gravel beaches. Thus, Alphonse Karr (1808–1890) said, "if I had to show a friend the sea for the first time, it is Etretat I would choose". Similarly, Guy de Maupassant presented one of the arches of Etretat, the Manneporte, "like a vault through which a ship could pass".

Here below, we present the morphological and landscape characteristics of this coastline that are related to the regional geology. These geological characteristics also explain the terms and rates of shore evolution which undergoes much erosion. Finally, we will focus on the exceptional site of Etretat that attracted geomorphologists, painters, writers and tourists fascinated by the geology and landscape of its remarkable arches, cave and stack.

4.2 Geographical and Geological Setting

Over 120 km of shoreline, between Cap d'Antifer and Le Tréport, we can observe an original coast, characterised by chalk cliffs and gravel beaches (Figs. 4.1 and 4.2). Averaging approximately 70 m in height, they do not exceed 40 m at Cap d'Ailly (Fig. 4.4, cf. infra), but 105 m at Antifer and 101 m at Le Tréport.

Geologically, this area corresponds to the north-west termination of the Paris Basin (Fig. 4.3a). The plateau of High Normandy and Picardy, and therefore its cliff, is formed of Upper Cretaceous chalk (Cenomanian to Campanian) more or less rich in flints (Cavelier et al. 1979; Mégnien and Mégnien 1980; Mortimore and Pomerol 1987; Juignet

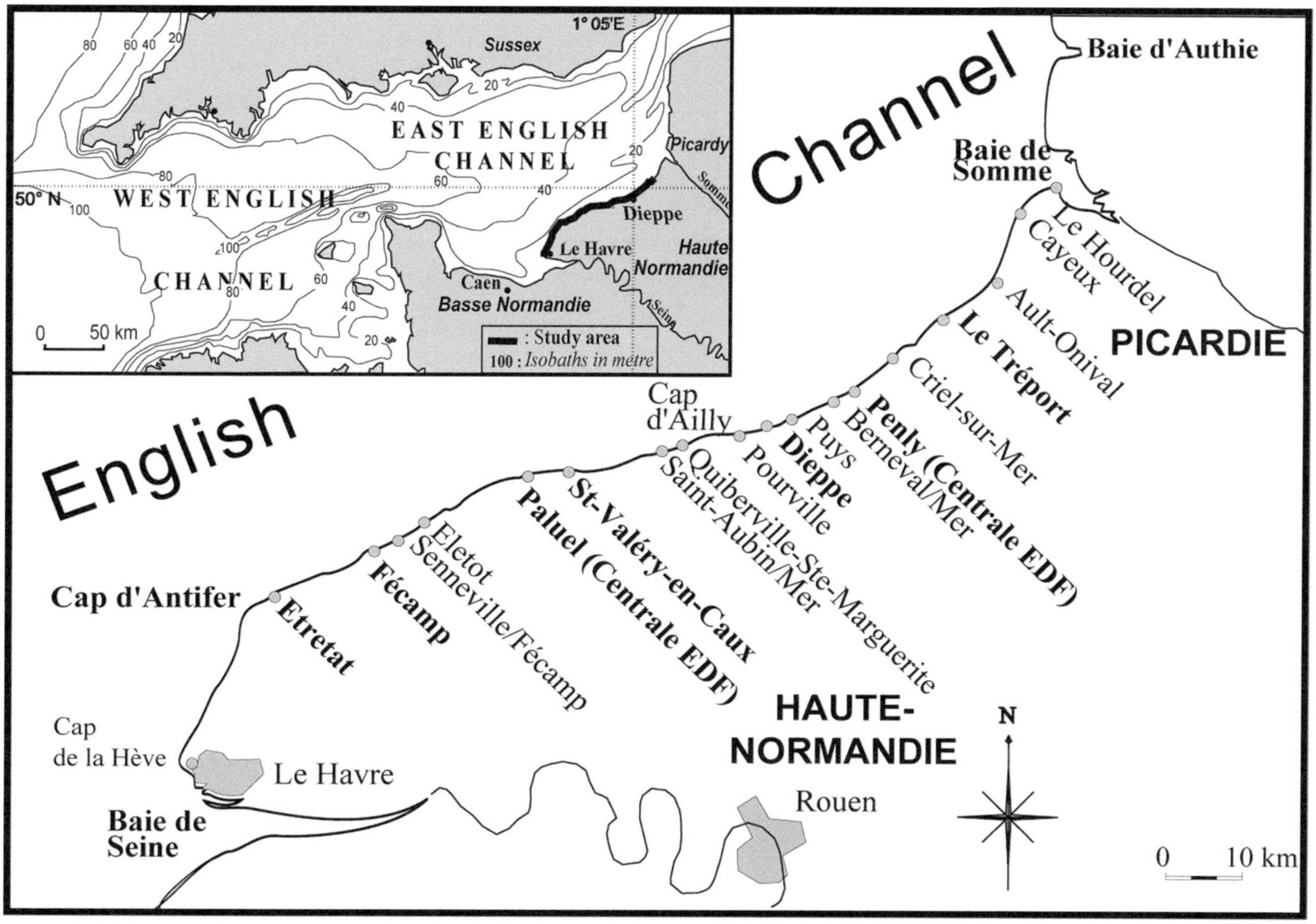

Fig. 4.2 Study area

and Breton 1992; Laignel 1997). Residual formations with flint and Quaternary loess are also deposited on this karstified chalk.

During the Upper Cretaceous (97–70 Ma), an important sea level rise caused flooding of the entire Paris Basin by the "sea chalk", which deposited the imposing chalky layers. It was a shallow, epicontinental sea, in which were deposited large amounts of microorganisms producing calcareous tests that gave rise to thick chalk strata. This lime-rich sedimentary rock is white or greyish, soft and friable and porous hence permeable. Its main component is calcium carbonate (90 % $CaCO_3$). It is formed mainly by an aggregate of coccoliths (skeletal remains of coccolithophores, 1–10 μm in diameter) and often contains planktonic foraminifera that swarmed in the Cretaceous seas. Other microorganisms have concentrated silica in their skeleton (diatoms, radiolarians, sponges, etc.) to generate interbedded bands of flint. The silica content in seawater originally comes from the weathering of silicate minerals constituting the continental crust and was fed by rivers or by wind (dust). This silica can also come from oceanic volcanoes (lithogenic silica) or phytoliths (mainland biogenic silica). Oxygenated and cooling waters are major factors in the formation of flint. The hypothesis most often used to explain the rhythmicity of flint bands in the cliffs is based on climatic variations due to astronomical forcing (Milankovitch cycles): the precipitation of silica would have taken place during a cold period, with the development of upwelling currents being felt up to the edge of the continental shelf and the surrounding epicontinental sea.

Major tectonic deformations in NW-SE directions result in the outcrop of various strata from the Upper Cretaceous (Fig. 4.3b). The majority of cliffs are formed by Senonian white chalk rich in flints (Coniacian, Santonian and locally Campanian). However, this apparent homogeneity hides more complex details. Turonian chalk comprising clay, greyish to whitish, with little or no flint, protrudes near Antifer, Etretat, Fécamp and Le Tréport where it reaches its maximum extension at Penly. Locally, Cenomanian chalk protrudes too. These are heterogeneous, sometimes rich in detrital components (clay and quartz), and can be glauconitic or nodular. A 10 m thick cover of sandy, clayish sediment of Palaeogene age can be found on the cliff top at Cap d'Ailly, at Sotteville and at Bois de Cise (Bignot 1962). In detail, the lithostratigraphy of chalk is more complex (Mortimore 2001; Hoyez 2008).

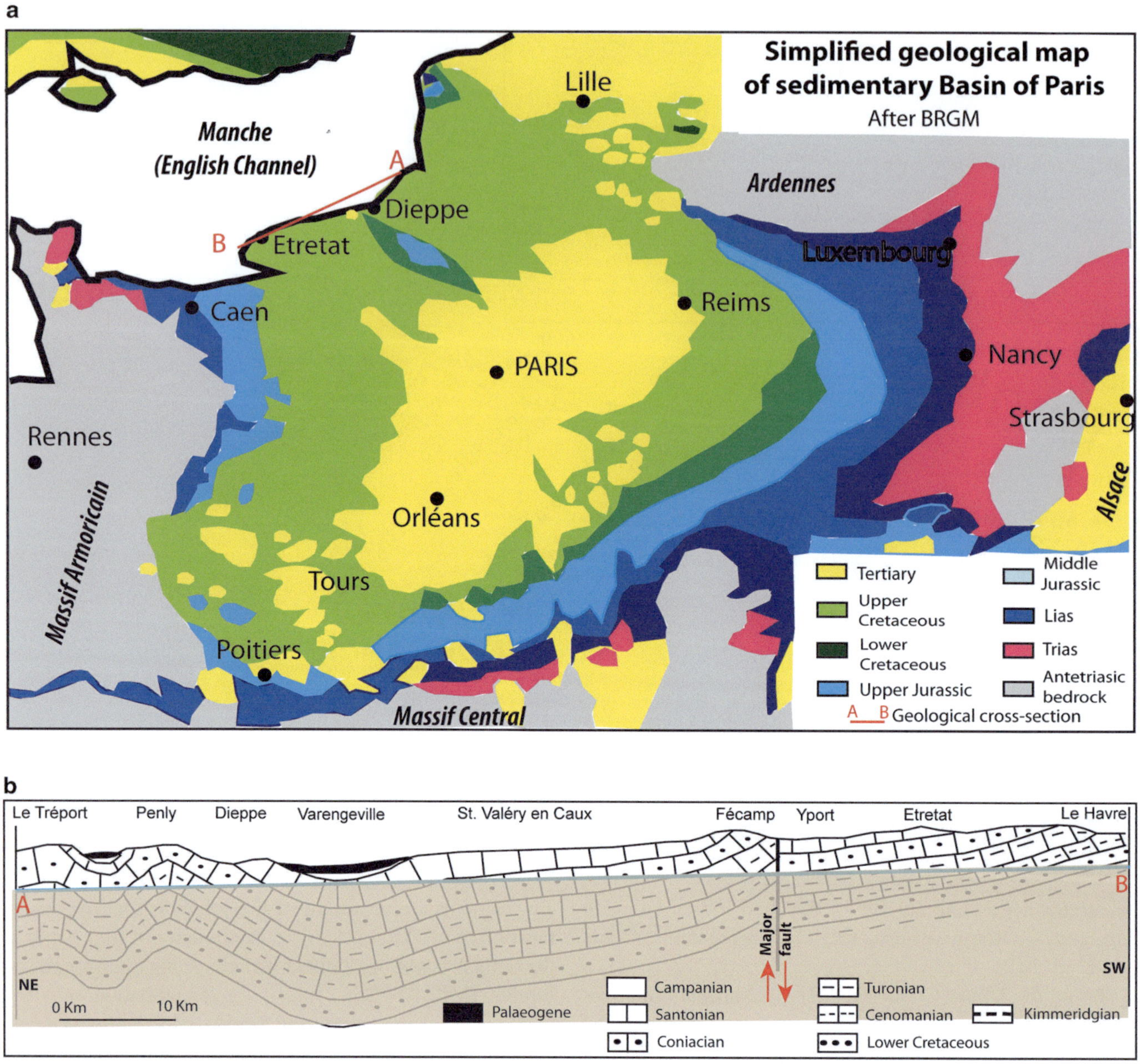

Fig. 4.3 (**a**) Simplified geological map of sedimentary basin of Paris (after BRGM); (**b**) geological cross section of the coast between Le Havre and Le Tréport (Costa 1997)

Because of their different structural characteristics, weathering of chalk layers does not proceed uniformly, and, hence, the various chalk strata correspond with different types of cliff morphology with contrasting rates of evolution. Three main types of cliff are recognised (Fig. 4.4): "simple cliffs" cut in Senonian chalks are stretching for approximately 70 km around the coastline; "resistant pedestal" cliffs correspond to a resistant outcrop, generally the Turonian layer; and lastly, the "complex cliffs", in particular around Cap d'Ailly, mainly consist of sand and clay layers—the cliff foot is cut in the Santonian chalks, which are then covered by the non-cohesive Tertiary sediments.

A shore platform is cut into the chalk at the foot of the cliffs. The seaward slope is gentle (1° maximum at Antifer). This wide shore platform (150–300 m) is covered on its landward margin by a thin gravel beach.

During the Miocene, the uplift of the High Normandy plateau has been initiated, as a consequence of the Alpine orogeny (marine sands (from Redonian transgression) trapped in karst formations (at Valmont) above 120 m; Cavelier et al. 1979; Mégnien and Mégnien 1980). This uplift of about 150 m led to the formation of coastal cliffs, to progressive entrenching of streams in connection to jointing and the individualisation of the Seine course from

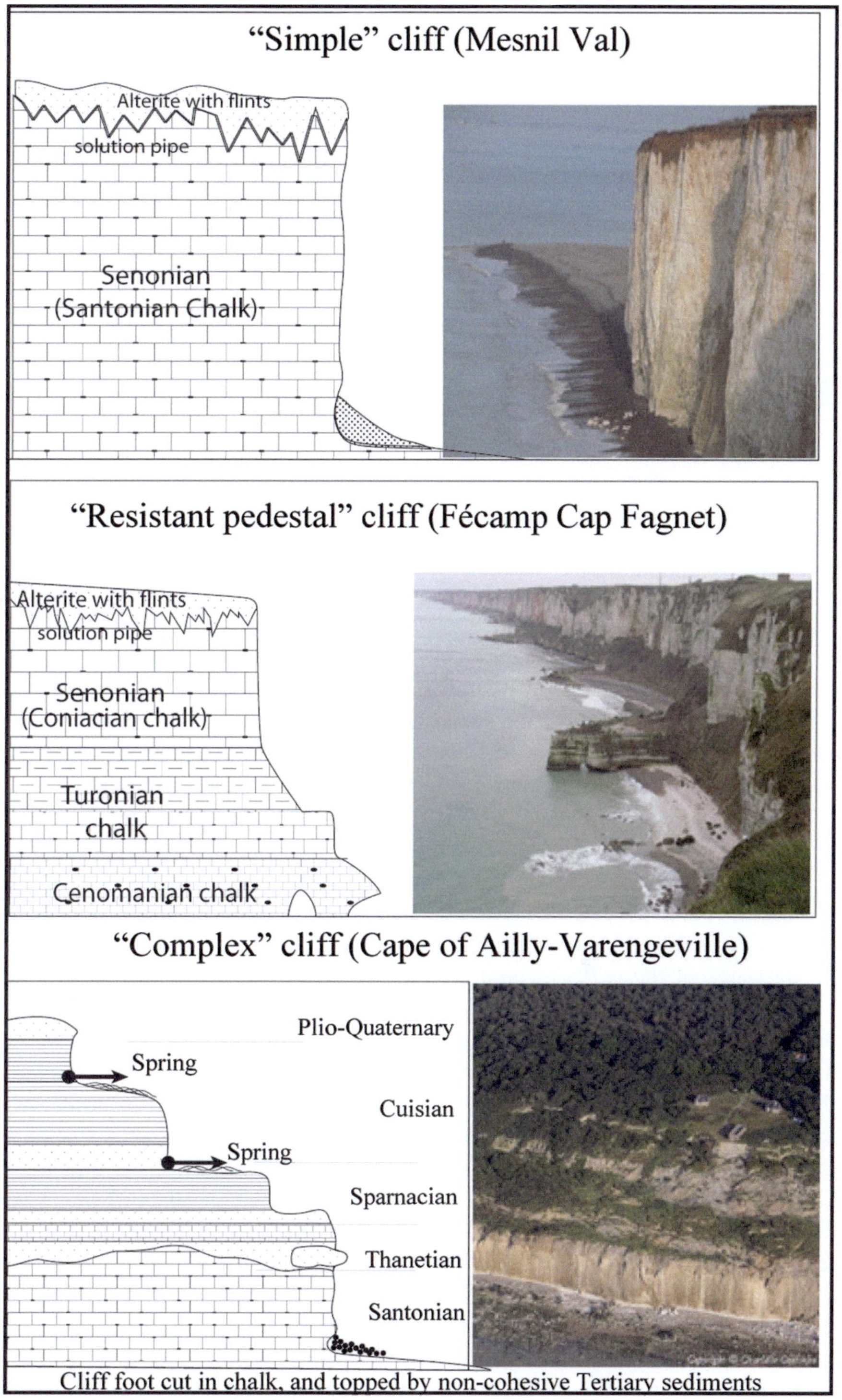

Fig. 4.4 Types of High Normandy cliffs according to stratigraphy (Costa 1997)

that of the Loire. At the same time, the water table fall increased the dislocation of chalk and led to karst formation. Glacial periods caused the lowering of sea level, and during the last one, the Würm glaciation (−30,000 years), the ice sheet reached the middle of British Isles and the northern of Europe. In this climatic context, the channel dried up completely and crossed by large rivers flowing into the Atlantic far beyond the present-day coastline (Paleo-Seine, Somme, Thames, Meuse, Scheldt, Rhine, etc.). During this low sea level, the cliffs were not the vertical and white walls that we know today, but the landscape looked like the banks of the Seine today, namely, sides covered by head deposits and framed by deeply incised valleys.

Fig. 4.5 Three hanging valleys above shoreline (valleuses): Puys (with houses), Camp de César and Parfond-Val (Photo S. Costa)

After the end of the last glaciation (−18,000 years), the sea level rose up to nearly 100 m (Flandrian transgression) to reach the foot of the coastal side (5–6,000 years ago) and remove the limbs and loose debris. At this time, the chalk cliffs were reactivated, vertical arches were created and hanging valleys overlooking the shoreline (*valleuses*) were formed. These are dry valleys perched atop the cliffs. They result from the competition between a slow digging of the river (which has seeped into the ground) and the rapid retreat of cliffs (Fig. 4.5).

Among the best known *valleuses* are those near Varengeville-sur-Mer, which host the park of Moutiers. The presence of deposits of sand, clay and sandstone of Tertiary age in the *valleuses* and their sheltered position from the disturbed westerly flow favour biodiversity as testified by the renowned flora park of the *Bois des Moutiers*. This park, which displays luxuriant rhododendrons, azaleas and camellias growing on acid soils, was inspired from the eighteenth-century British gardens, with the contribution of the famous British landscape designer Gertrude Jekyll.

Another characteristic of the Varengeville area is the existence of dwellings (*Manoir d'Ango*) and a church, built in the seventeenth century in local materials, such as red and white clay bricks, flints and sandstones. This church, threatened by cliff erosion, has a choir bathed in blue light reflected by the stained glass windows, partly created by the artist Georges Braque, whose grave is now located in the "sailor cemetery" adjoining the church.

The numerous dry and drained valleys perpendicular to the shoreline are protected by a gravel beach that is often thick (due to the presence of jetties and groynes) and 30–100 m wide. These valleys, the only connection between sea and inland, represent also the lowest parts along the cliffs: their elevation is not higher than the highest high tide level and makes these zones very fragile. Yet, it is in these low coastal areas that are affected by storm surge that the population and their activities are concentrated (Fig. 4.6).

4.3 Chalk Cliffs Affected by Rapid Retreat

Photogrammetric analyses were carried out to quantify the retreat of chalk cliffs. The 1966 and 1995 vertical aerial photographic surveys (1:10,000) from the French National Geographic Institute have been used (Costa et al. 2001, 2004). The mean retreat rate of the entire shoreline under study is approximately 6 m between 1966 and 1995, which yields a rate of 0.21 m/year. This average rate, however, masks a very high spatial variability of cliff retreat (Fig. 4.7). In fact, the analysis allows three distinctive areas to be distinguished: (1) an area of low retreat rate (0.8–0.13 m/year) between Antifer and Fécamp, (2) an area of moderate retreat rate (approximately 0.19 m/year) between Fécamp and Saint-Valéry-en-Caux and between Dieppe and Le Tréport and (3) an area of rapid retreat (0.21–0.28 m/year) between Saint-Valéry-en-Caux and Dieppe (Costa et al. 2001, 2004).

Even within each of these coastal sections, important variations do exist. These sharp variations are linked with the influence of cliff collapses or anthropogenic obstacles, such as harbour arms or major groynes that disrupt the gravel transit from the south-west to the north-east. Then down current from these obstacles, the cliff retreat can be multiplied by 2 or 3. These observations are confirmed by the analysis of the retreat at 50 m intervals (Fig. 4.10).

The identification of three sectors with distinctive retreat rates raises questions about the causes of this spatial distribution of the cliff retreat rates and especially its relation to the outcropping of different chalk strata (Costa 1997; Costa et al. 2004). The sectors with "low"

Fig. 4.6 Storm surges at Etretat in 27–28/02/1990 (Photo S. Costa)

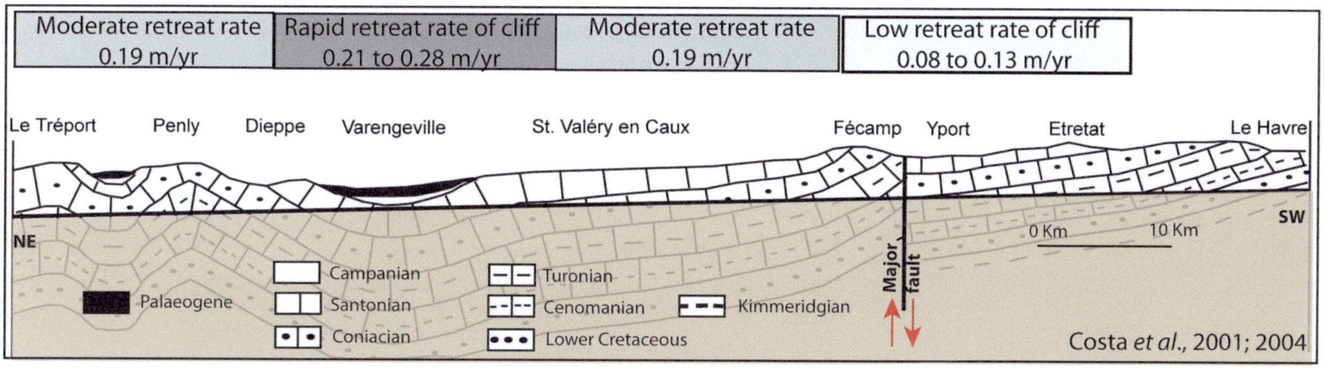

Fig. 4.7 Relationship between chalk rates, cliff retreat and lithology (Costa et al. 2001, 2004)

and "moderate" retreat are affected by infrequent but voluminous rockfalls and correspond to cliff-foot cut into Turonian, Cenomanian and even Coniacian outcrops (Antifer/Fécamp; Fécamp/Saint-Valéry-en-Caux; Dieppe/Le Tréport). By contrast, the rapidly retreating sectors, affected by frequent but less voluminous rockfalls, correspond to Santonian and Campanian outcrops (Saint-Valéry-en-Caux/Dieppe).

4.4 Etretat as a Specific Area: Cave, Arch and Stack, Affected by Very Slow Retreat

The site of Etretat is one of the most famous coastal chalk cliffs in France. The town is settled in a dry valley perpendicular to the shoreline, which is framed by specific cliffs, marked by stacks, caves and arches (Fig. 4.8). This site comprises four principal capes and arches: *la Porte d'Amont* (upstream door), la *Porte d'Aval* (downstream door) with his stack, *la Manneporte* and *la Pointe de la Courtine* (tip of La Courtine).

The Etretat cliffs are chiefly cut in the Coniacian chalk, but the upper Turonian appears locally. Morphological characteristics of the site of Etretat result from specific geological phenomena.

Chalk stratification is affected by pseudo-folds, which suggest that the bedding planes were originally wavy and therefore the seafloor during the Turonian-Coniacian was very irregular. Etretat is one of the most remarkable examples of this phenomenon. In fact, these undulations were induced by ocean currents that formed on the seabed antidunes and megaripples, hectometric wavelength and decametric amplitude on the surface of sludge chalk, hence created the so-called dips that are not in relation with tectonics. This area is also affected by synsedimentary slumping, and hard-ground development is also found, particularly in the Coniacian chalk (Fig. 4.9). These bioturbated nodular chalks, pluricentimetric to metric in thickness, were formed during interruptions of sedimentation (Lasseur 2007). These hard grounds are mechanically resistant, due to their induration/dolomitisation, and they block the water infiltration into the cliff and allow the development of endokarst.

Fig. 4.8 Cave, arch and stacks of Etretat (Photo S. Costa)

Lastly, the flint vertical development locally reaches up to 5 m in thickness. The inverted pear shape, called *paramoudra*, is the most classic; it is a secondary silicification around an axis whose origin is variously interpreted (possibly worm burrow). All these structural characteristics make the cliffs of Etretat more resistant to the development and preservation of endokarstic and exokarstic features such as solution pipes, which are often filled by alterites with flints and Tertiary deposits.

For many authors, the alignment of the arches and caves would be the result of fracturing parallel to the shore that would have allowed the installation of a karst pipe hollowed out by the sea after the last Holocene marine transgression. The natural evolution of this site, by regressive erosion of the cliffs, will lead gradually to expanding caves, which will form an arch, the top of which will collapse to form a stack. Structural features also explain the low rates of cliff retreat at Etretat that did not seem to have evolved on a historical timescale (Fig. 4.10). The coastal evolution calculated by photogrammetric analysis is about 3–4 cm/year between 1966 and 1995 (of the same order of magnitude as the error margin of the method!) but increases (10–15 cm/year) to the north of la *Porte d'Amont*. This is the lowest retreat rate of the High Normandy chalk coast.

Fig. 4.9 Hard-ground outcrop at the bottom of the cliff at Etretat, foreground right (Photo S. Costa)

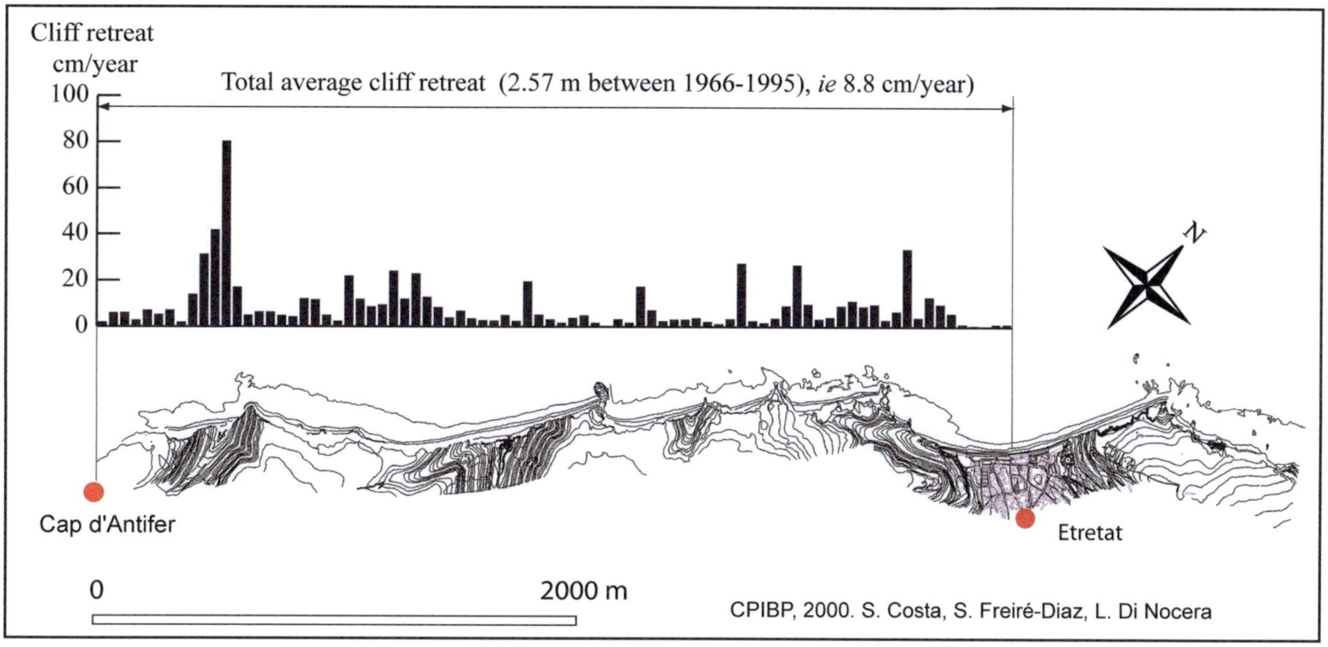

Fig. 4.10 Rates of cliff retreat at Etretat between 1996 and 1995 (Costa et al. 2001, 2004)

4.5 Etretat: A Geomorphosite Exerting Powerful Attraction to Nineteenth-Century Painters and Writers

The cliffs of Normandy, with their imposing verticality, their whiteness varying in tone with the ever-changing light and tide and their ghostly shapes, exerted a powerful attraction to painters of this period. From Gustave Courbet (1819–1877) to Claude Monet (1840–1926), they have given us representations of Yport, Dieppe and Etretat reflecting their fascination with geology. The high hygrometric content of the air explains the colour shades and the quality of the light. The "quest of light" has been the aim of the impressionist painters.

Since 1824, Dieppe and Brighton were connected by paddle steamer, facilitating a meeting point of the arts midway between London and Paris. Whereas the story has often focused on two central personalities, Walter Sickert and Jacques-Emile Blanche, these represented only the nucleus of a much wider artistic community, starting with Bonington and Turner in the earlier 1820s and ending a century later with the important relationship between Ben Nicolson and Georges Braque, who met regularly in Varen-geville. Coastal Normandy played a prominent role in the development of Impressionism during this period: by the late 1860s and early 1870s, the beach was the central character in the scenes depicted by Manet, Monet and Boudin. This new playground for the fashionable holidaymakers provided both subject matters and clients along the *côte d'Albâtre*.

Etretat beach, called "the beach of artists and writers", was the nineteenth-century resort frequented by the "upper class". In 1852, the casino opened, followed by bathing establishments, regatta clubs, tennis and golf. The last decade of the Second Empire (1860–1870) was probably the most brilliant period of this worldly Etretat. The villas made of bricks and flints with slate roofs located on the hillsides are still today witnessing this great time. The above-mentioned low rates of erosion are supported by a variety of paintings on the site of Etretat. They show that the landscape has changed very little since the nineteenth century. In general, the channel and its seashores have been in the middle of an important sequence of the universal history of painting, especially *landscape painting* during the nineteenth century and the first half of the twentieth.

Affected by the fashion of sea bathing of the nineteenth century, the touristic site of Etretat has given us works of painters as Camille Corot (1796–1875) and Gustave Courbet (1819–1877), or Claude Monet (1840–1926) who found inspiration there and contributed to its fame (Fig. 4.11). There were also famous musicians like Jacques Offenbach (1819–1880), who built his house in Etretat and called it "Orpheus in the Underworld", the name of his opera buffa. Similarly, French writers such as Guy de Maupassant (1850–1893) and André Gide (1869–1951) fell in love with the place, not to mention Maurice Leblanc (1864–1941), father of the "gentleman burglar", Arsène Lupin, who hid his treasure in the "hollow stack", 70 m high (Fig. 4.12). In his novel, Maurice Leblanc built the plot around the mystery of the *Aiguille Creuse* (hollow stack) which was assumed to hide the most fabulous treasure ever imagined, transmitted by the kings of France since the time of Julius Caesar.

Fig. 4.11 The cliffs at Etretat Claude Monet 1885. Williamstown. Clark Art Institute, USA (Public Domain)

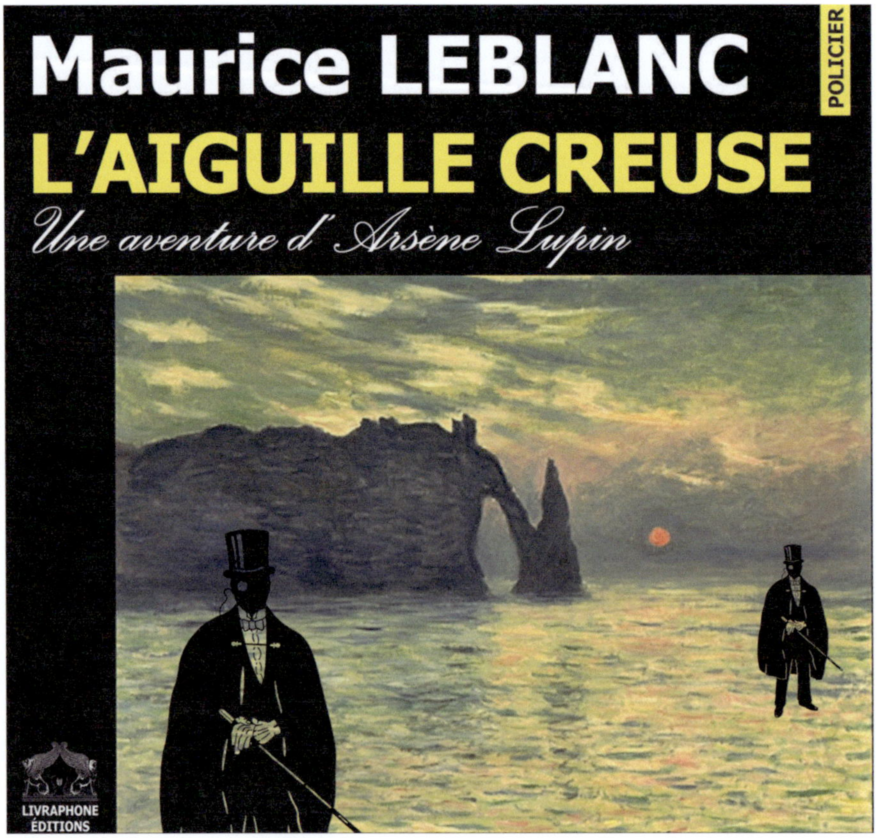

Fig. 4.12 Cover of L'Aiguille creuse (Hollow stack) by Maurice Leblanc, one of the most famous adventures of Arsène Lupin, the gentleman burglar (© Livraphone Edition, with permission)

4.6 Conclusion

Together with the English coasts of Sussex and Yorkshire, the High Normandy coast is a rare example in Europe of chalk cliffs and cobbly silex beaches. Such coastlines help us to understand the geological history of the sites, such as the modes of chalk deposition (including hard-ground formation during interruptions of sedimentation), the role of ocean currents forming the formation of pseudo-folds on seabed chalks and the influence of the Pyrenean and Alpine orogenies in the tectonic undulations of chalk strata and the preservation of Tertiary sandy clays of Tertiary age in synclinal structures. These various outcrops of differential resistance to erosion explain the existence of cliff profiles and thus the variety of retreat rates. The overall low resistance of chalks explains the rapid erosion that threatens at places coastal communities and their activities. This great wall of chalk also fascinated famous nineteenth-century painters and novelists, who played their part in making these remarkable landscapes famous worldwide. Nowadays, local and regional communities are working together to have the High Normandy coastline added to the World Heritage UNESCO List, in order to make this heritage recognised for its geological, cultural and aesthetical outstanding qualities.

References

Bignot G (1962) Etude sédimentologique et micropaléontologique de l'Eocène du Cap d'Ailly (près de Dieppe-Seine-Maritime). Thèse 3ème cycle, Faculté des Sciences, Université de Paris

Cavelier C, Mégnien C, Pomerol C, Rat P (1979) Le Bassin de Paris. Bulletin d'Information des Géologues du Bassin de Paris 16(4):2–52

Costa S (1997) Dynamique littorale et risques naturels : L'impact des aménagements, des variations du niveau marin et des modifications climatiques entre la Baie de Seine et la Baie de Somme. PhD thesis, University of Paris I – Panthéon Sorbonne, Paris

Costa S, Freiré-Diaz S, Di-Nocera L (2001) Le littoral haut normand et picard: une gestion concertée. Annales de Géographie 618:117–135

Costa S, Delahaye D, Freiré-Diaz S, Davidson R, Di-Nocera L, Plessis E (2004) Quantification by photogrammetric analysis of the Normandy and Picardy rocky coast dynamic (Normandy, France). In: Mortimore RN, Duperret A (eds) Coastal chalk cliff instability. Geological Society, vol 20. Engineering Geology Special Publications, London, pp 139–148

Hoyez B (2008) Falaises du pays de Caux: lithostratigraphie des craies turono-campaniennes. Publications des universités de Rouen et du Havre, Mont-Saint-Aignan

Juignet P, Breton G (1992) Mid-Cretaceous sequence stratigraphy and sedimentary cyclicity in the western Paris Basin. Paleogeography, Paleoclimatology, Paleoecology 91:197–218

Laignel B (1997) Les altérites à silex de l'ouest du Bassin de Paris: caractérisation lithologique, genèse et utilisation potentielle comme granulats. PhD thesis, University of Rouen, Edit. BRGM, Orléans

Lasseur E (2007) La Craie du Bassin de Paris (Cénomanien-Campanien, Crétacé supérieur). Sédimentologie de faciès, stratigraphie séquentielle et géométrie 3D. PhD thesis, University of Rennes 1, Rennes

Mégnien C, Mégnien F (1980) Synthèse géologique du Bassin de Paris. Mémoire du Bureau de Recherches Géologiques et Minière, 3 vol, n° 101, 102, 103

Mortimore RN (2001) Chalk: a stratigraphy for all reasons. Keynote paper. The Scott-Simpson Lecture Proceedings of the Ussher Society 10:105–122

Mortimore RN, Pomerol B (1987) Correlation of the Upper Cretaceous (Upper Cenomanian to Campanian) white chalk in the Anglo-Paris Basin. Proceedings of the Geologist's Association 98:97–143

The Mont-Saint-Michel Bay: An Exceptional Megatidal Environment Influenced by Natural Evolution and Man-Made Modifications

5

Chantal Bonnot-Courtois, Anne-Véronique Walter-Simonnet, and Agnès Baltzer

Abstract

Due to its exceptional tidal range and the immensity of its tidal flat, the Mont-Saint-Michel bay is a favored model for research on coastal hydro-sedimentary processes, reconstruction of the depositional wedge during the Holocene, and landscape evolution. The western tidal flat consists of a large muddy to sandy embayment, whereas the eastern intertidal domain consists of estuarine deposits characterized by finely laminated muddy sand, enriched in biogenic carbonate. The present-day morphology and sedimentary dynamics are the result of a complex balance between tide-dominated sediment supply and erosive processes under wave action and tidal channels migration. The bay has been filled for thousands of years to the rhythm of sea level oscillations, leading to the gradual progradation of the shoreline. In addition to this natural evolution, human activities have developed in the bay, both on the western tidal flat (fixed fisheries and shellfish farming) and on the eastern estuarine side (polder formation, dikes, and dam construction), and have progressively changed the landscapes. In 1884, Victor Hugo issued an appeal for the protection of the Mont-Saint-Michel in these terms: "*The Mont-Saint-Michel is to France what the Great Pyramid is to Egypt. It must be protected from any damage. The Mont-Saint-Michel must remain an island. This combined work of nature and art must be conserved at all costs.*" Since that time, salt marshes have progressed, the land reclamation and man-made developments to facilitate the access to the mount also continued, so the maritime character of Mont is still endangered.

Keywords

Mont-Saint-Michel bay • Sedimentary dynamics • Man-made modifications • Recent evolution

5.1 Introduction

Located on the western coast of France in the English Channel, the Mont-Saint-Michel bay is a wide (500 km^2) and shallow (below 20 m) depression extending between the Cotentin peninsula (Normandy) and the Brittany coast. The bay is known all over the world through its highly touristic "Mont-Saint-Michel" (Fig. 5.1), site inscribed on the UNESCO World Heritage in 1979 and also through its tidal range that reaches about 15 m at highest spring tides. The huge intertidal area covers 250 km^2 between the "Pointe du Grouin" at Cancale in the northwest and Champeaux cliff in

C. Bonnot-Courtois (✉)
CNRS, UMR 8586 PRODIG, Laboratoire de Géomorphologie et Environnement littoral, Dinard, France
e-mail: chantalbonnot0577@orange.fr

A.-V. Walter-Simonnet
UMR 6249 Chrono-Environnement, Université de Franche Comté, Besançon, France
e-mail: Anne-Veronique.Walter@univ-fcomte.fr

A. Baltzer
Géosciences Marines, Ifremer Brest, Plouzané, France
e-mail: Agnes.Baltzer@ifremer.fr

M. Fort and M.-F. André (eds.), *Landscapes and Landforms of France*, World Geomorphological Landscapes, DOI 10.1007/978-94-007-7022-5_5, © Springer Science+Business Media Dordrecht 2014

Fig. 5.1 The abbey of Mont-Saint-Michel, a monument of legend for 708 years. Rising up from sand flat and tide flooding, Mont-Saint-Michel appears like a human challenge to the nature, architecture, and time (Photo P. Bretel, 2011)

the northeast. The morpho-sedimentary organization of the bay is controlled by tidal currents, wave, and fluvial regimes (Larsonneur et al. 1989; Migniot 1998; L'Homer et al. 1999; Bonnot-Courtois et al. 2002). Recent scientific programs have provided an opportunity to study the present-day intertidal sedimentary dynamics in finer detail. These studies were conducted to address environmental issues related to the insularity of the mount and the development of shellfish aquaculture activities. A new morpho-sedimentary map of the bay has been produced (Bonnot-Courtois et al. 2009) and coupled with a LIDAR topographical survey of the entire bay, highlighting the levels of submergence of the tidal flat. In addition, a chronologically arranged set of aerial photographs dating from 1952 allowed a quantitative assessment of the dynamics of the main morpho-sedimentary tidal flat structures such as shell banks and salt marshes. Finally, the evolution of the landscapes of the bay was reconstructed from old documents. This made it possible to trace the impact of successive modifications made to the entire tidal flat over the centuries (Bonnot-Courtois and L'Homer 2012).

5.2 Geological Setting, Hydrodynamic Context, and Sediments Distribution

An ancient basement of metamorphic formations (570 million years old) and of granitic bodies (540 and 525 million years old) forms the bedrock and the reliefs around Mont-Saint-Michel bay (L'Homer et al. 1999). The hills surrounding the southern part of the Mont-Saint-Michel bay are composed of Brioverian shales locally intruded by granodioritic batholith (*Saint-Broladre, Avranches, and Carolles*). Late intrusions of leucogranite are responsible for Mont Dol, Mont-Saint-Michel, and Tombelaine residual hills (Fig. 5.2). In the bay, there remains no proof of Paleozoic to Cenozoic formations because intense erosion and weathering resulted in the formation of a peneplain throughout the region. During Quaternary falls in sea level and lowstands, rivers incised the shale substrate and shaped depressions between granitic reliefs. Because regional subsidence is negligible, only the last postglacial transgression has been recorded in the sediment infill of the bay (L'Homer et al. 1999). The current configuration of the bay results

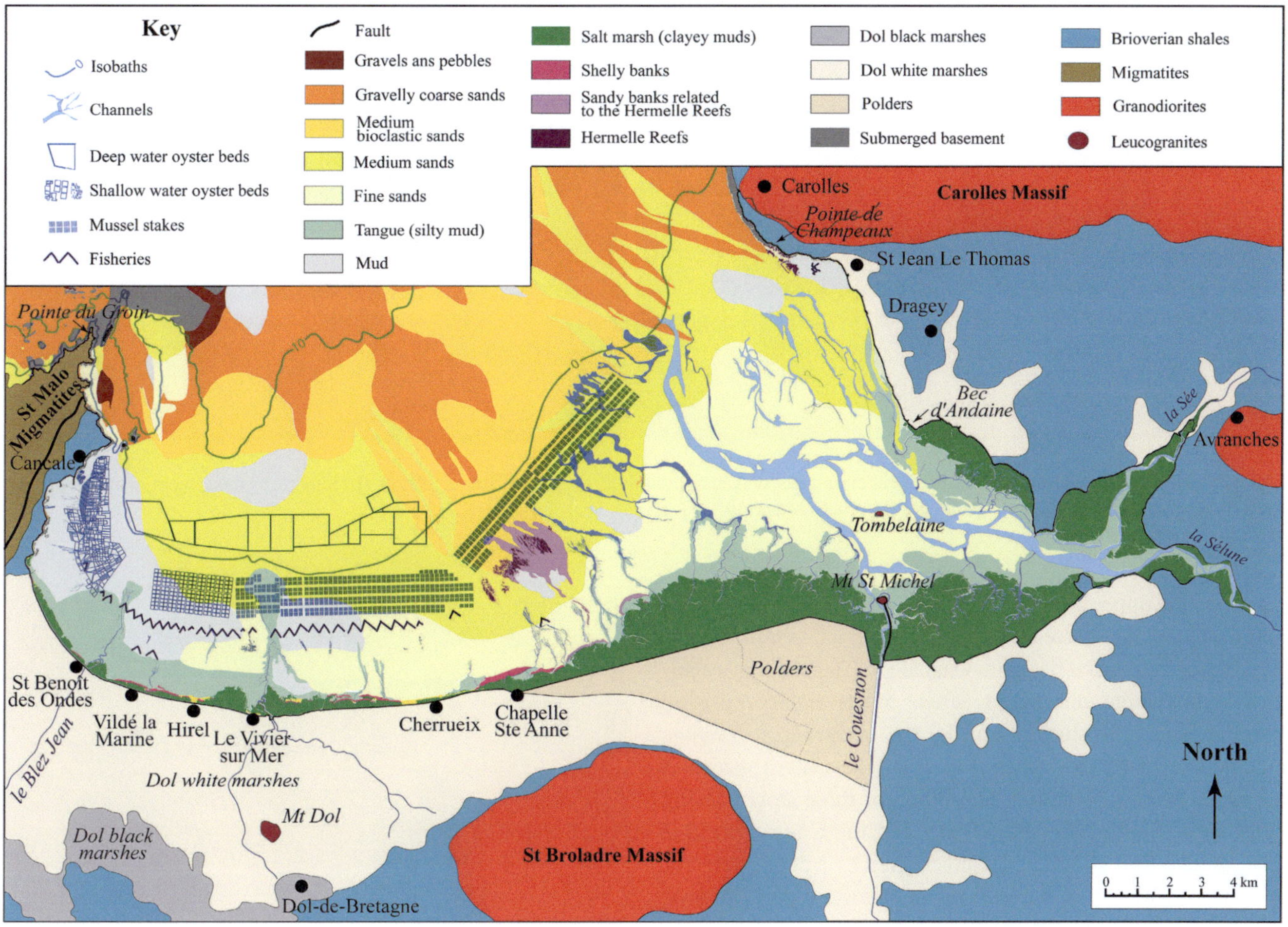

Fig. 5.2 Geological units and superficial sediments distribution in the Mont-Saint-Michel bay (After L'Homer et al. 1999; Bonnot–Courtois et al. 2009)

from the sea level rise following the last deglaciation (18,000 years BP) when the sea level was 120 m below the present-day sea level.

The Holocene transgression included an initial phase marked by a rapid sea level rise, from −120 to −10 m below the present-day sea level from 18,000 years BP to 6,000 years BP (10–6 mm/year from Lambeck 1997), followed by a slower second phase with the sea level rising at an average rate of 1.5 mm/year (Fairbanks 1989 in Billeaud 2007). This overall sea level rise did not take place steadily but through a sequence of oscillations causing the shoreline to advance and retreat (Morzadec-Kerfourn 2002). As from 8,000 years BP, the sea flooded the bay and the estuaries and the embayments, i.e., the *Dol* marsh, were gradually filled with 15 m thick sedimentary deposits. Since that time, the coastal environments have evolved under the influence of hydrodynamics, sediment supply, and man-made modifications that make up the current landscape of the bay.

Two tidal current regimes coexist in the bay. The currents alternate off the Pointe du Grouin and to the northeast on the median line of the bay; the flood tide flowing toward the southeast is more intense than the ebb tide flowing toward the northwest. At the entrance to the bay, the currents can reach over 1 m/s, before slowing down to about 0.6 m/s on the tidal flat, except in the estuarine channels where they can exceed 2 m/s. To the southwest, the currents rotate. The flood tide still predominates over the ebb tide but at a highly reduced speed in the Cancale embayment.

At the entrance to the bay, the NW to W swells are diffracted around the Pointe du Groin such that the Cancale embayment is completely protected from swell action. Moving along the median line of the bay, the wave height decreases almost completely due to friction on the broad tidal flat in the estuarine zone. The northeastern sandy littoral of the bay is exposed to prevailing NW swells that induce beach erosion at Saint-Jean-le-Thomas. The eroded sands are transported southward by the littoral drift and form two important sand spits at Dragey and the Bec d'Andaine (Fig. 5.3a) (Tessier et al. 2006). At the center of the bay, the diffracted northern swells cause the accumulation of shell

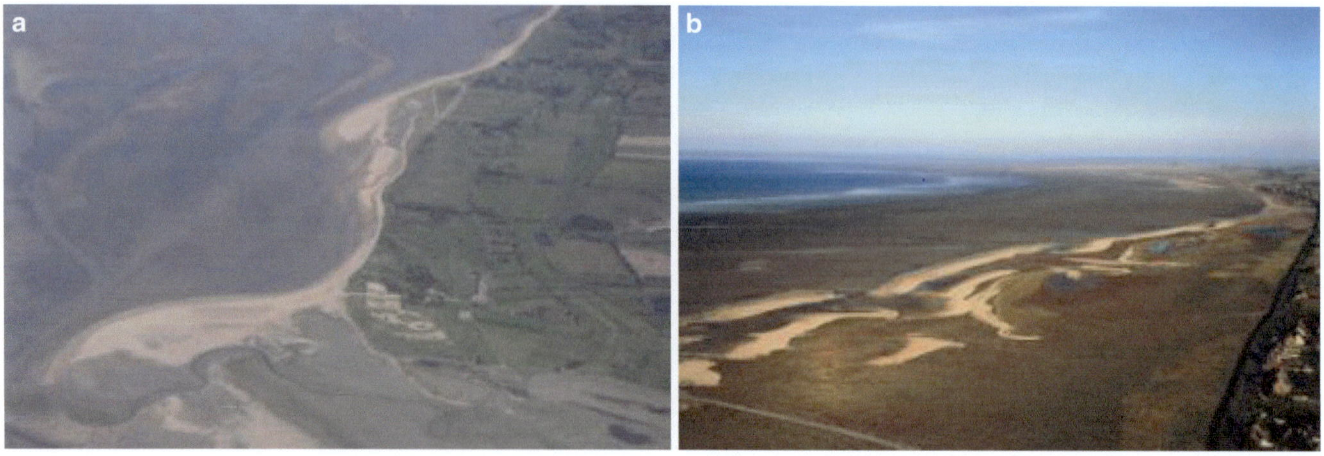

Fig. 5.3 (**a**) Northeastern sandy coastline: Dragey and Bec d'Andaine sand spits (Photo P. Bretel, 2011). (**b**) Southern embayment: wave-generated shell banks migrating on the upper tidal flat (Photo A. Mauxion, 2007)

banks on the upper tidal flat between Saint-Benoît-des-Ondes and La Chapelle Sainte-Anne (Fig. 5.3b).

The southern tidal flat from Cancale to Cherrueix is a sheltered embayment uncovered across 5 km and comprising homogeneous silts cut through by the outflows of drainage channels from the Dol marsh (Fig. 5.2). The eastern section is a huge estuarine domain at the mouth of three slow-flowing coastal rivers: the Sée, the Sélune, and the Couesnon. This eastern tidal flat, 10 km wide and crossed by a dense network of tidal channels, is subject to an intense hydrodynamic regime which removes the sand beds. The boundary between the western bay and the eastern estuarine zone is marked by an annelid reef (*Sabellaria alveolata*) which developed on the lower tidal flat, forming the "Hermelles" bank (Larsonneur & coll. 1989; L'Homer et al. 1999; Bonnot-Courtois et al. 2002).

5.3 Current Sedimentary Dynamics

5.3.1 Shell Bank Dynamic

The western upper tidal flat of Mont-Saint-Michel bay between Saint-Benoît-des-Ondes and La Chapelle Sainte-Anne is characterized by large shell banks aligned parallel to the coastline. These bioclastic accumulations, made of shell fragments, are related to wave action and present different types of morphology according to their morpho-sedimentary characteristics: grain size, bioclastic content, amplitude, position on the tidal flat, etc. Analysis of a chronological set of aerial photographs between 1947 and 2006 shows that the rate of migration of the banks from the lower tidal flats to the coastline varies from several dozen meters per year for the banks on the middle tidal flat to less than a few meters per year for those located at the edge of the salt marsh (Fig. 5.4).The mobilization of bioclastic compo-

Fig. 5.4 Shell bank alignment on the upper tidal flat at Vildé-Hirel. Located at the edge of the salt marsh, these banks isolate a lagoon-like depression behind them. The road was constructed over a dike, itself built on an ancient coastal barrier (Photo B. Caline, 1980)

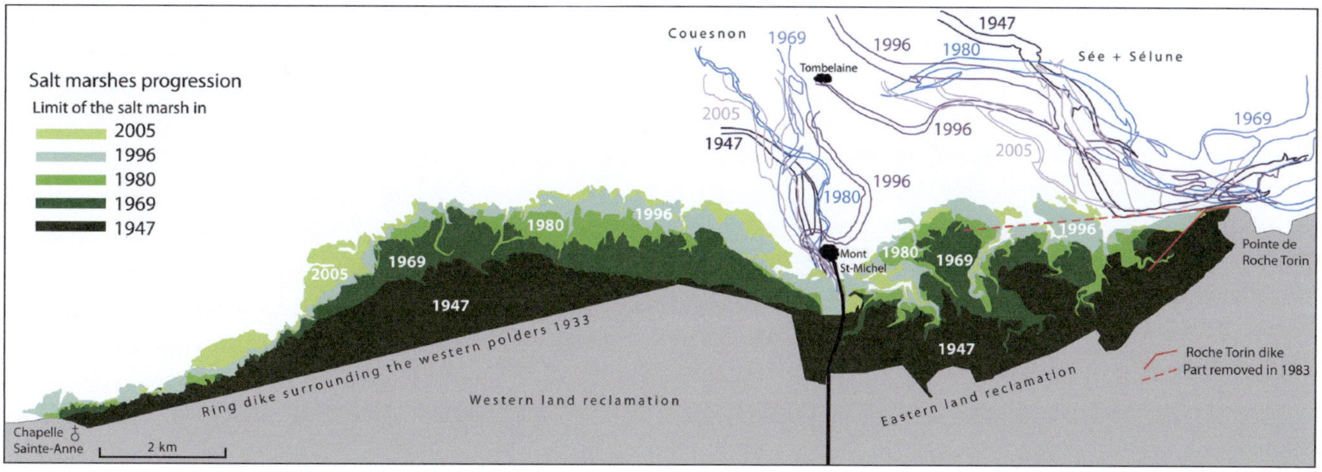

Fig. 5.5 Salt marshes advance toward the upper tidal flat and channels shifting around the Mont-Saint-Michel between 1947 and 2005

nents by the swell depends on submergence, and carbonate bioclasts are easily set in motion by a light swell on the lower and middle tidal flat (Weill 2010). When the shell banks reach the upper tidal flat, they accumulate and thicken sufficiently to form a coastal barrier which stabilizes and protects the coastline. Conversely, a strong swell combined with high water spring tides may destroy the structure of the coastal barrier.

Two successive barriers isolated the Dol marsh from the coastal domain during the Middle Ages. The oldest, internal bar was breached in many places, hence attesting there were major seawater incursions into the marsh. The buildup of this first bar dates back to the sixth century and continued until the eighth century (L'Homer et al. 1995). The second bar formed during the ninth and tenth centuries, when numerous marine submergences appear to have occurred. The Duchesse Anne dike, built during the eleventh century on this second line of ancient shell bars (Fig. 5.4), definitely fixed the coastline and isolated the Dol marsh from marine incursions.

5.3.2 Salt Marshes Evolution

When the sediment level is high enough, salt-tolerant plant species colonize the upper tidal flat and settle according to their tolerance to salinity and consequently to submergence (Bonnot-Courtois and Levasseur 2000). The different plant species that make up the salt marshes take over from each other in space and time, and their distribution is a good indicator of the topographic evolution of the tidal flat. In the eastern part of the bay, the salt marshes are most developed beyond La Chapelle Sainte-Anne and extend over almost 4 km on the upper tidal flat around Mont-Saint-Michel itself (Lefeuvre et al. 2009). The sedimentary deposition at the salt marsh front mainly depends on the duration of submergence and hence on the elevation of the mud flat and on the morpho-dynamic context of the upper tidal flat. In calm conditions, fine particles are deposited at a rate of 1–2 cm/year at the boundary between the lower salt marsh and the upper mud flat. The more exposed salt marsh edges, located in dynamic conditions, may be subject to erosion, often linked to the proximity of a channel. Sedimentary accretion diminishes from the upper mud flat to the middle and upper salt marsh, where it stands at only a few mm/year.

The natural settling and extension of the salt marsh around the Mont-Saint-Michel were analyzed between 1947 and 2005 (Fig. 5.5) and show that the area covered by salt marshes to the west of Mont-Saint-Michel has doubled in 50 years (654 ha in 1947 and 1,366 ha in 1996), while the progradation is slower to the east (634 ha in 1947 and 1,049 ha in 1996). The salt marshes currently cover almost 2,500 ha between La Chapelle Sainte-Anne and the Pointe de Roche Torin, spreading over the upper tidal flat with an average rate of 15–20 ha/year (Fig. 5.6). The progression of the salt marshes around the Mont causes the feeling of a less maritime environment in the vicinity of the Mont-Saint-Michel.

5.4 Man-Made Development Impact on Landscape Evolution

The most ancient wooden remains of a fish dam was discovered during a major erosion of the beach at Saint-Jean-le-Thomas and dates over 3,450 years B.P. Archeological findings in different sedimentary contexts constitute markers of human occupation in the Dol marsh, going back to the end of the Gallic period. As early as the eighth century, human activities such as salt production in the Sée and Sélune estuaries and marine activities (fisheries, ports) marked the evolution of the landscapes of the bay. The reconstruction of the shorelines during the Roman times and the Middle Ages shows the formation of coastal sandbars that border the salt

Fig. 5.6 Wide extension of the salt marshes in front of the polders around the Mont-Saint-Michel. The reduction of the salt marshes spreading is one of the main objectives of the project of rehabilitation of the Mont-Saint-Michel's maritime character (Photo P. Bretel, 2011)

meadows in the Dol marsh and the development of marine activities (ports, fisheries) and establishment of many salt works in the estuarine area (L'Homer et al. 2009; Bonnot-Courtois and L'Homer 2012). Between the tenth and the twelfth centuries, coastal villages such as Hirel and Cherrueix were founded on the sandbar partially cutting off the southwestern part of the Dol marsh. To the east, the shifting course of the Couesnon prevents sustainable buildup of the coastal sandbars, and a number of rural establishments were destroyed by the combined action of tides and the shifting Couesnon channel. The relative stability of the tidal flats to the southwest and northeast of the bay led to the installation of many fisheries, and more and more salt works were established in the estuarine bay and contributed to change the shoreline by huge sandy accumulation on the upper tidal flat (L'Homer and Piquois 2002; Bonnot-Courtois and L'Homer 2012).

5.4.1 Western Embayment: Fisheries and Shellfish Farming

On the middle tidal sand flat of the southwest of the bay, the first fixed fisheries date back to earlier than the eleventh century. The almost continuous line of wooden fisheries became denser during the seventeenth century, and the number of fisheries reached about 40 across the entire middle tidal flat (Fig. 5.7). Expansion to the east was limited by strong tidal currents and the shifting of the Couesnon channel which, at the time, flowed toward the northwest.

The main stages of shellfish-farming (oysters and mussels) development in the western tidal flat of the Mont-Saint-Michel bay are as follows:

- Eighteenth century: Exploitation of natural common oyster beds by dredging
- Nineteenth century: Farming of common oysters at Cancale
- 1906: First intertidal oyster leases in uncovered areas on the middle and lower tidal flats at Cancale (Fig. 5.7)
- 1970–1996: New deep-water oyster leases

The main impact of farming oysters in bags at Cancale embayment was to slow down the currents and raise the level of fine silt under the installations. During the shellfish-farming restructuring process of 2004, the most silt-laden oyster farms were transferred eastward in the place of the mussel beds opposite Vildé.

Fig. 5.7 Wooden fisheries in the middle tidal flat opposite Vildé (Photo A. Mauxion, 2007) and intertidal oyster farming at Cancale (Photo A. Humbert, 2001)

Mussel farming was introduced in the bay in 1954 on either side of the Vivier channel. From 1960 onward, the lines of mussel stockades spread extensively between Saint-Benoît-des-Ondes and the annelid reef ("*Hermelles*" bank). Between 1975 and 1980, new mussel leases were created off this reef, and in 2004, the Vildé mussel stockades, replaced by oyster trestles, were transferred to the northeast of the "*Hermelles*" bank in the shift zone of the estuarine channels.

5.4.2 Eastern Estuarine Area: Land Reclamation Around the Mont-Saint-Michel

At the end of the eighteenth century, the mouths of the three main rivers of the bay were located to the east: the Couesnon, the Sée, and the Sélune, which flow in a northwesterly direction. Before the modifications around Mont-Saint-Michel have been made, the tide flooded into a large estuarine system formed by the joint channel of the Sée and Sélune and by that of the Couesnon (Bonnot-Courtois et al. 2002). The mouth of the Couesnon was located in Moidrey cove, and the main channel flowed in a northwesterly direction alongside the dike of the marshes located below the Saint-Broladre massif. Periodically destabilized by tidal currents and sediment deposition, the Couesnon channel and the tidal channels broke the dikes in many places, destroying all the earlier enclosures, both on the Breton side and on the Norman side (Bonnot-Courtois 2012) (Fig. 5.8).

In 1856, a company named Mosselmann and Donon reached an agreement with the State to reclaim 4,350 ha around Mont-Saint-Michel as polders, with the company undertaking to channel the Couesnon. The main phases of land reclamation took place as follows (Fig. 5.9):

- *1858*: Diversion of the Couesnon by Mosselmann by building an earth dam known as the "Le Pas au Bœuf" at Moidrey (Fig. 5.8).
- *1863*: Channeling of 3,800 m of the Couesnon, starting from the Moidrey cove (Fig. 5.8).
- *1857–1861*: Construction of the Roche Torin dike impinging on the "Pointe de Roche Torin" at the mouth of the Sélune river (Fig. 5.5), initially planned to extend up to Mont-Saint-Michel itself. This dike was built to divert the joint channel of the Sée and the Sélune northward in order to enable land reclamation to the east of Mont-Saint-Michel.
- *1868*: The sea broke through the dikes and flooded the first polders to the west. Following these storms, a recommendation was put forward to rock-bank the ring dike.
- *1878–1879*: Construction of the insubmersible dike access road to Mont-Saint-Michel.
- *1906, 1928 and 1947–49*: Land reclamation to the east of Mont-Saint-Michel.
- *1914*: Permission for southward displacement of the western polder concession limit, originally planned to extend up to Mont-Saint-Michel itself, was obtained by the "Beaux-Arts."
- *1933–1934*: Construction of the ring dike surrounding the polders to the west in alignment with Mont-Saint-Michel.
- *1969*: Construction of the first Caserne dam on the Couesnon.

The land around Mont-Saint-Michel was therefore not reclaimed from the sea (Fig. 5.9) until successful channeling of the Couesnon in 1863 meant that polders could be built on its old estuary. Successive polder construction, implemented in parallel with the advance of the salt marshes, is marked by a network of dikes built thanks to the hard labor of the Mosselmann workers.

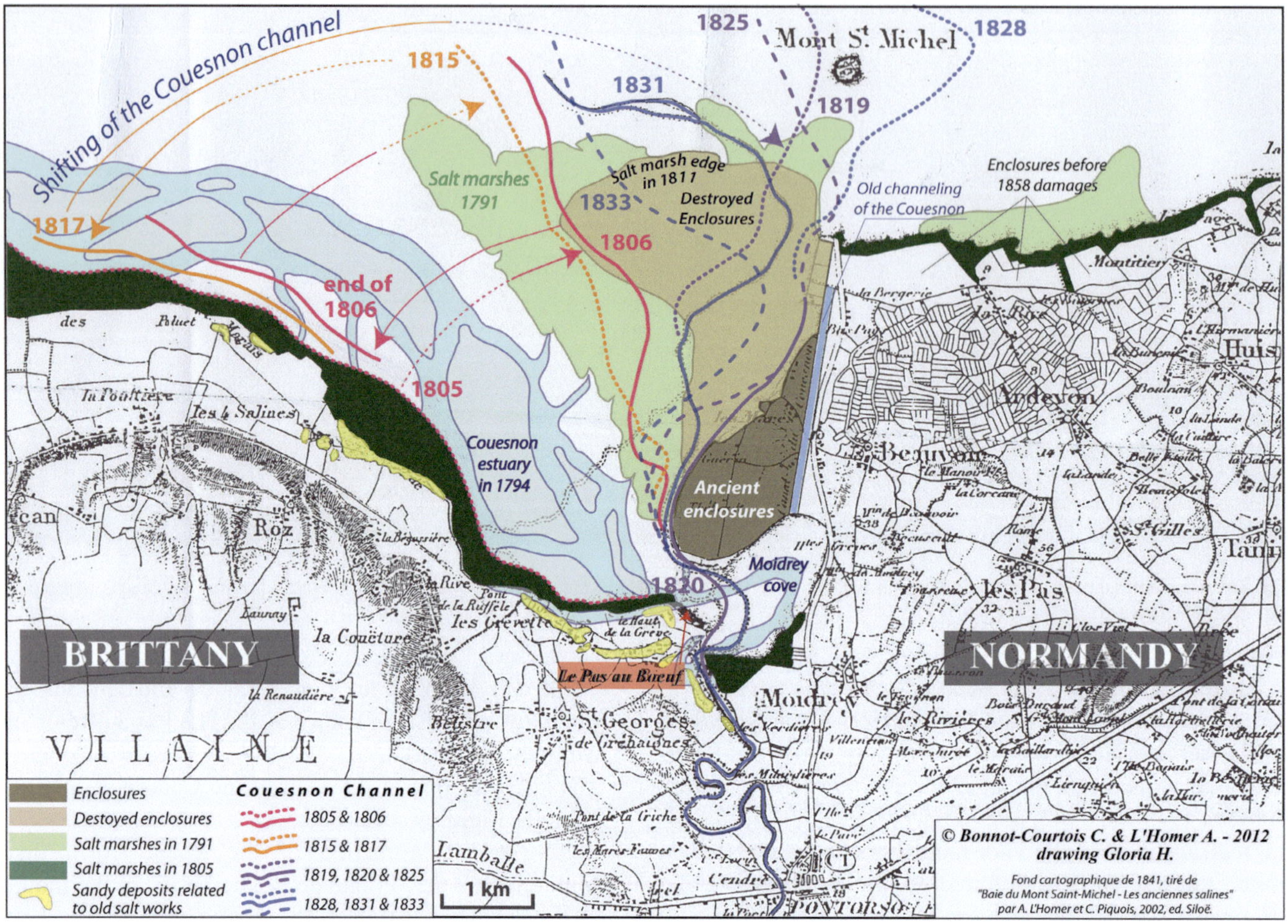

Fig. 5.8 The Couesnon channel migration at the beginning of the nineteenth century (From 1805 to 1833), before the river was channeled

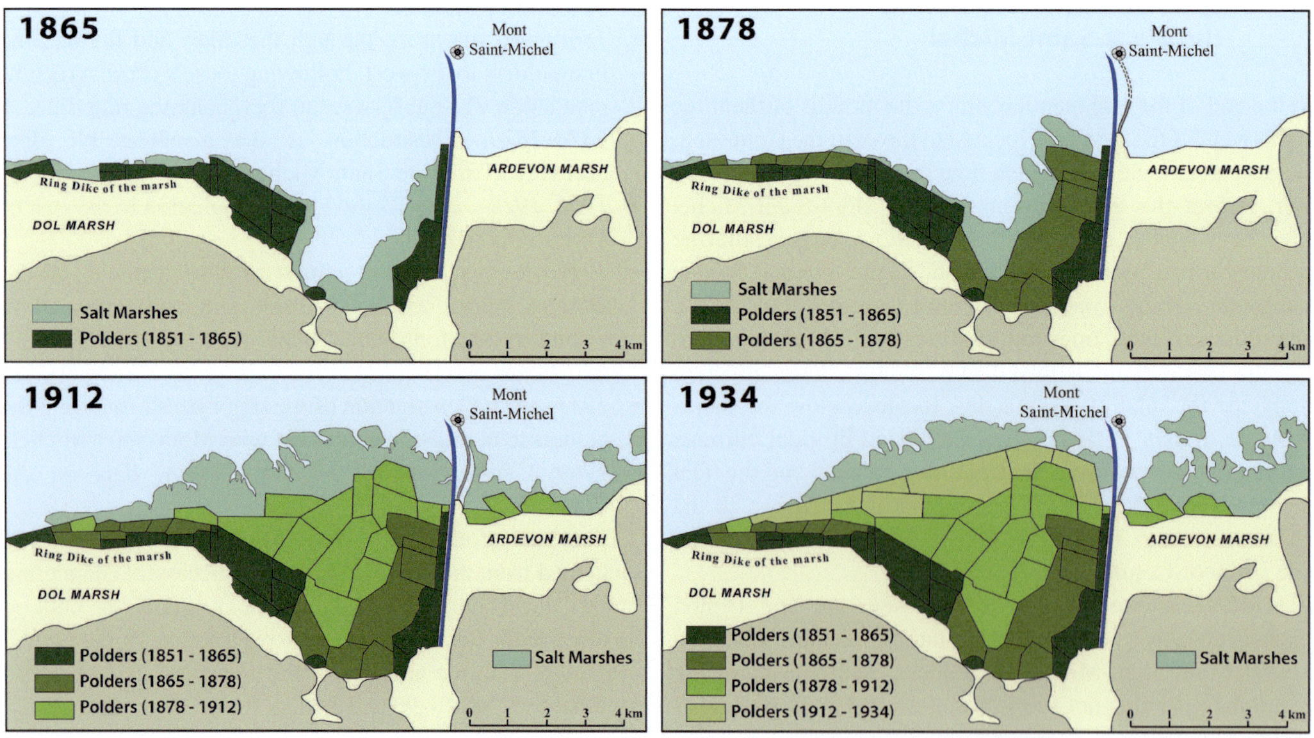

Fig. 5.9 The main stages of land reclamation around the Mont-Saint-Michel

5.5 Tourism and Restoration of the Mont-Saint-Michel's Maritime Character

The history of the bay from the eighth to the tenth century is closely linked to that of Mont-Saint-Michel Abbey, initially a simple oratory founded in 708 by Aubert, the Bishop of Avranches at the time. The Mont-Saint-Michel, with around three millions visitors per year, is one of the major French touristic sites, either for the village and its gothic abbey settled on top of the granitic mount or for the exceptional size of the tidal flat (Tessier et al. 2012). However, due to the natural infilling, enhanced by land reclamation, rapid sediment accretion and salt marsh progradation occur around the Mont-Saint-Michel so that the attractive spectacle of incoming tidal bore is becoming very rare (Fig. 5.10).

In order to restore the maritime character of the Mont-Saint-Michel, a research project was initiated in 1995, including hydrodynamic, sedimentological, and environmental studies and numerical and physical models. All data and information about the project as well as on the progress of the restoration works that began in 2005 are described in detail on the following web site: http://www.projetmont-saintmichel.fr (official website of the "Syndicat Mixte Baie du Mont-Saint-Michel" for the rehabilitation of the Mont-Saint-Michel's maritime character). The achievements at the project halfway point take place as follows:

- *2009*: Completion of the construction of the dam over the Couesnon which is the cornerstone of the project's hydraulic aspect. The hydraulic flushing through this dam began to remove sediments around the Mont-Saint-Michel in May 2009 (Desguée 2012).
- *2010–2011*: Beginning of the reception work (car park, reception, and service buildings) and access work for the Mont (pedestrian footbridge and causeway) enabling a completely new approach of the rock.
- *2012*: Completion of the new car park and the public transport shuttles to bring visitors to the Mont.
- *2014*: Opening of the pedestrian footbridge to visitors, pedestrians, and shuttles and to the Mont's permanent security services.
- *2015*: More symbolically, the operation will be completed with the destruction of the causeway, which is over 100 years old (1879), and took visitors from the continent to the Mont. The works to restore the maritime character around the Mont-Saint-Michel will be then achieved.

Fig. 5.10 General view of the salt marshes, the causeway and the Mont one hour after the high water of an exceptional spring tide (Photo C. Bonnot, 2001)

5.6 Conclusion

The present-day landscapes of Mont-Saint-Michel bay have been formed firstly by its natural evolution following the Holocene sedimentary infill linked to the last phase of the sea level rise and secondly from developments which have marked the human occupation of this coastal fringe. The gradual clogging of the bay meant that large areas of marine marshland could be reclaimed. These were subsequently protected from marine incursions and from the shifting channels by dikes which demarcated the coastline. The exceptional extent of the tidal flat favored the installation of fixed fisheries in the western part, and numerous salt works were established, mainly in the estuarine zone. The channeling of the Couesnon, land reclamation, and the building of dikes in the nineteenth and twentieth centuries fundamentally changed the shoreline around Mont-Saint-Michel. Shellfish farming, developed in the twentieth century, currently occupies most of the western lower tidal flat up to the network of estuarine channels. Present-day sedimentary dynamics are dominated by tidal processes, which in turn are superimposed by wave action which is setting in motion once again the coastal barriers formed by the shell banks or the sand spits of the northeast coast.

The fact that sedimentation takes precedence over erosion and the steady advance of the salt marshes show that the natural accretion process in the bay is continuing. Developments undertaken to re-establish the marine character of Mont-Saint-Michel aim to restore sufficient hydraulic power to exceed the cohesion of the sediments, whether overgrown or not, in order to reduce sand deposits in the immediate vicinity of Mont-Saint-Michel itself. The redistribution of eroded sediments on either side of Mont-Saint-Michel will result in lateral extension of the salt marshes, but will have no significant effect on the sediment budget with respect to the entire Mont-Saint-Michel bay.

Acknowledgments The authors are very grateful to B. Caline (Total), B. Tessier (Caen University), and R. Desguée (Syndicat Mixte Baie du Mont-Saint-Michel) for their work, their help, and scientific discussions in the bay over the years. They also wish to thank P. Bretel (Caen University), A. Mauxion, and A. Humbert for the aerial photographs of the bay.

References (Non-comprehensive List)

Billeaud I (2007) Dynamique de construction d'un prisme sédimentaire littoral en régime mégatidal (la Baie du Mont Saint-Michel). Thesis, University of Caen, pp 239

Bonnot-Courtois C (2012) A tribute to A. L'homer(†), Dynamique sédimentaire intertidale en baie du Mont Saint-Michel entre évolution naturelle et aménagements. XIIèmes Journées Nationales Génie Civil Génie Côtier, Cherbourg, pp 187–222. doi:10.5150/jngcgc.2012.021-B, http://www.paralia.fr

Bonnot-Courtois C, Levasseur JE (2000) Etudes en Environnement. Volume 5. Contribution à la connaissance de la dynamique des herbus. Final Report of the Joint Initiative for Restoring the Marine Character of Mont-Saint-Michel. Mission Mont-Saint-Michel, Manche Department Infrastructure Directorate (DDE). Univ. Rennes 1, CRI 8586 Prodig CNRS, 196 p

Bonnot-Courtois C, Caline B, L'homer A, Le Vot M (ed) (2002) La baie du Mont-Saint-Michel et l'estuaire de la Rance. Environnements sédimentaires, aménagements et évolution récente. The bay of Mont-Saint-Michel and the Rance estuary. Recent development and evolution of depositional environments. Bull. Cent. Rech. Elf-Explor. Prod., Mémoire 26, Total Fina Elf-CNRS-EPHE, pp 164–176

Bonnot-Courtois C, Maze JP, Le Vot M, Augris C, Ehrhold A, Simplet L, Blanchard M (2009) Carte morpho-sédimentaire de la baie du Mont Saint-Michel (Ille-et-Vilaine & Manche) échelle 1:25,000, ed. Quae – Ifremer

Desguée R (2012) Technical notes. Suivi de l'efficacité du Barrage. Syndicat Mixte Baie du Mont Saint Michel. La Baie 29:6–7

Fairbancks RG (1989) A 17,000 year glacio-eustatic sea-level record influence of glacial melting rates on the Younger Dryas event and deep-ocean circulation. Nature 342:637–642

L'homer A, Piquois C (2002) Mont Saint-Michel bay Les anciennes salines. Siloë, Laval, 175 p

L'homer A, Caline B, Bonnot-Courtois C (1995) Intérêt archéologique des bancs coquilliers du secteur occidental de la baie du Mont Saint-Michel. In: Baie du Mont Saint-Michel et Marais de Dol. Milieux naturels et peuplements dans le passé. Langouet & Morzadec-Kerfourn Dir. Les Dossiers du Ce.R.A.A. Suppl. 1995 n° R, pp 47–65

L'homer A, Courbouleix S, Chantraine J, Deroin JP (1999) With the cooperation of Bonnot-Courtois C, Caline B, Ehrhold A, Lautridou J-P, Morzadec-Kerfourn MT (1999) Carte géologique de la France (1:50,000), feuille Baie du Mont-Saint-Michel, (208), Orléans, BRGM. Notice explicative, 183 p

L'homer A, Bonnot-Courtois C, Casset M, Levalet D, Nicolas-Mery D, Syndicat Mixte Baie du Mont-Saint-Michel (2009) Evolution de la baie autour du Mont depuis 708. Exposition pour le 13ème centenaire du Mont-Saint-Michel. 8 Panneaux. Syndicat Mixte, Ardevon

Lambeck K (1997) Sea-level change along the French Atlantic and Channel coasts since the time of the Last Glacial Maximum. Palaeogeogr Palaeoclimatol Palaeoecol 129:1–22

Larsonneur C & coll. (1989) La baie du Mont-Saint-Michel: un modèle de sédimentation en zone tempérée. Bull Inst Geol. Bassin Aquitaine 46:1–75

Lefeuvre JC, Mouton JP, Mauxion A (2009) L'histoire de la baie du Mont Saint-Michel et de son abbaye. Editions Ouest-France, 287 p

Migniot C (1998) Rétablissement du caractère maritime du Mont-Saint-Michel. Synthèse des connaissances hydro-sédimentaires de la baie. Mission Mont-Saint-Michel, Manche Department Infrastucture Directorate (DDE), 111 p

Morzadec-Kerfourn MT (2002) La sédimentation holocène dans la partie occidentale de la baie du Mont-Saint-Michel: l'évolution du marais de Dol de Bretagne. In Bonnot-Courtois C, Caline B, L'Homer A, Le Vot M (dir) La baie du Mont- Saint-Michel et l'estuaire de la Rance. Environnements sédimentaires, aménagements et évolution récente. Bull Cent Rech Elf-Explor Prod, Mémoire 26, Total Fina Elf-CNRS-EPHE, pp 153–163

Syndicat Mixte Baie du Mont-Saint-Michel. Rétablissement du Caractère Maritime. http://www.projetmontsaintmichel.fr

Tessier B, Billeaud I, Lesueur P (2006) The Bay of Mont-Saint-Michel northeastern littoral: an illustrative case of coastal sedimentary body evolution and stratigraphic organization in a transgressive/highstand context. BullSoc Geol Fr, t. 177(2):71–78

Tessier B, Bonnot-Courtois C, Billeaud I, Weill P, Caline B, Furgerot L (2012) The Mont Saint-Michel ba (NW France). Facies, sequences and evolution of a macrotidal embayment and estuarine environment. TIDALITES 2012 8th international conference on tidal environments. Field Trip Book Publ. ASF, n 72, 46p

Weill P (2010) Formation et évolution de cheniers en contexte macrotidal. Approches expérimentales et in situ. Thesis, University of Caen, p 230 + Appendices

The "Pink Granite" Coast (Northern Brittany)

6

Yannick Lageat

Abstract

The "Pink Granite" coast, in northern Brittany, some 30 km long, is more familiar to tourists than to geomorphologists. The following paper highlights the limitations of the explanations propounded to the 700,000 visitors who annually come to admire the fantastic shapes into which the coarse-grained granites, which are reddish in fact, have been eroded. The occurrence of assemblages of boulder piles involves their subsurface initiation by differential weathering and the stripping of the saprolite by marine processes. Following their exhumation, basins and flutes have developed on the granite outcrops under subaerial conditions. At all scales of analysis, the "Pink Granite" coast exhibits an assemblage of landforms, which are not specific of coastal environments, illustrating the concept of convergence or equifinality, that is, the production of similar landforms by different processes.

Keywords

Granite • Weathering • Boulders • Platforms • Denudation coast

6.1 Introduction

The "Pink Granite Coast" is a stretch of coastline, some 30 km long, from Ploumanac'h to Trébeurden, in northern Brittany. Paradoxically, however, only the red granites, which outcrop on the two extreme parts of the shoreline, have attracted attention thanks to the fantastic shapes into which they have been eroded (Fig. 6.1). Since the end of the nineteenth century, the tourism has been closely linked to the singularity of this scenery, but, for unknown reasons, its promotion has been associated with the pink colour of some featureless outcrops. One century after its "creation", this morphosite remains to be scientifically valued, as this coast is more familiar to tourists than to geomorphologists, except for a short development by Migon (2006).

Y. Lageat (✉)
Géomer-LETG, UMR 6554-CNRS, University of Western Brittany, Place Nicolas Copernic, Plouzané 29200, France
e-mail: yannick.lageat@univ-brest.fr

6.2 Geographical and Geological Setting

The granitic intrusion of Ploumanac'h is elliptic in shape and covers a surface of 35 km^2, not including the tidal platform whose remarkable extension is due to the conjunction of the strong amplitude of the tide and its progressive dive into the Channel.

The age of this pluton is Stephanian (293 ± 15 Million years), therefore late Carboniferous. It results from the association of three main units appearing on Fig. 6.2 which, according to Barrière (1977), chronologically and from the outside to the inside of the pluton are formed as the following:

- Coarse-grained red granites forcefully emplaced in a gneissic basement
- Medium-grained pink granites injected along ring fractures
- Fine-grained grey granites intruded through cauldron subsidence

The complex has been excavated below the Tregor surface, which is preserved at an elevation of about 100 m. It has been subject to a differential weathering in a continental

M. Fort and M.-F. André (eds.), *Landscapes and Landforms of France*, World Geomorphological Landscapes,
DOI 10.1007/978-94-007-7022-5_6, © Springer Science+Business Media Dordrecht 2014

Fig. 6.1 The natural scenery of Ploumanac'h at the beginning of the twentieth century (Personal collection)

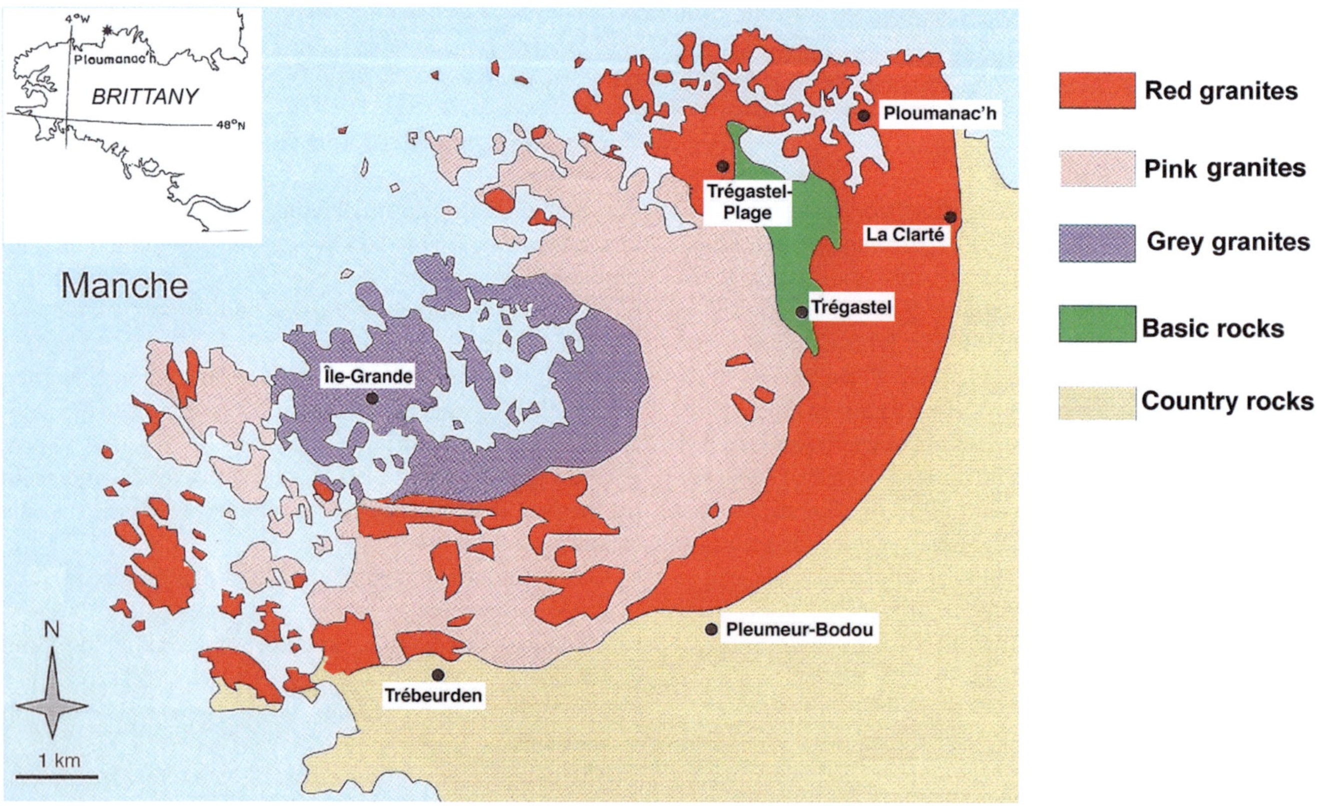

Fig. 6.2 Geological map of the Ploumanac'h complex (Adapted from Barrière 1977)

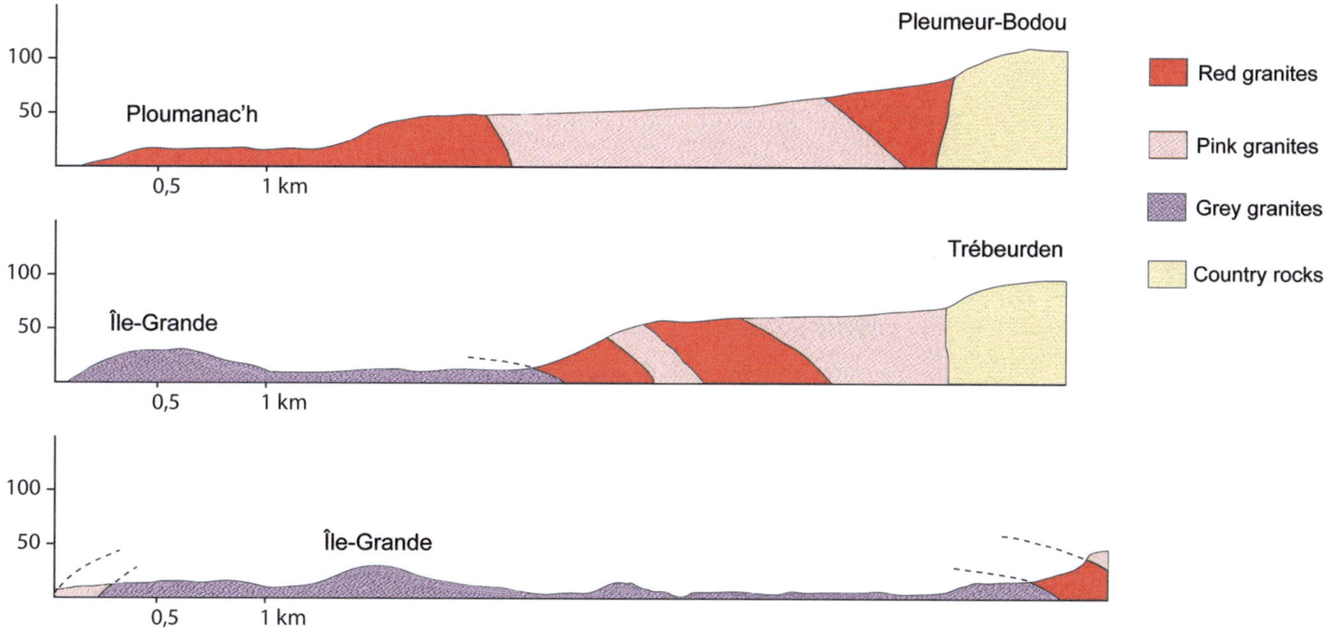

Fig. 6.3 Sections across the Ploumanac'h complex (Adapted from Lageat 1981)

context, and, as a result, an arcuate scarp follows strictly the contact between the plutonic intrusion and its country rocks (Lageat 1981).

At the foot of this scarp, stepped erosion benches have been bevelled through the granitic intrusion down the rolling upland plateau (Fig. 6.3).

- An intermediate platform, between 3 and 5 km large, stands between 70 and 30 m high and cuts across the various rocky materials of the first two units.
- The coastal bench, below 20 m, clearly corresponds spatially to the merged part of the so-called platform with skerries (*plate-forme à écueils*), which was described in the Tregor region by Battistini and Martin in 1956 and which does not relate to the differences of lithology within the plutonic suite (Fig. 6.4).

An apparent anomaly deals with the so-called depressed position of the central fine-grained granites, which are supposed to be more resistant (Rondeau 1959, 1978). However, they outcrop as the original roof of the inner part of the intrusion exhumed from an onion-shaped arrangement.

6.3 Landforms

The most unusual sceneries are linked to the external coarse-grained granite when it outcrops close to the coast in the three different sites of Ploumanac'h, Trégastel and Trébeurden.

The former site has been described by the novelist Thomas A. Janvier in 1904: "Huge strange-shaped boulders of red granite are scattered singly or are heaped up in wonderfully balanced mounds. The mere geologist will tell you that these rounded rocks have been dropped by a glacier. Any Breton, of course, knows that this explanation is all nonsense (…) In every one of these scattered boulders a Pagan soul is imprisoned until the Day of Judgment". In fact, Brittany was never glaciated, and Ploumanac'h is not "the site of a terminal moraine", as suggested by Janvier, but is the result of a mechanical stripping of the incoherent products of chemical action. So corestones are left stacked in different, often precarious, positions through the erosion of the loose weathered granite.

The touristic attractivity of the most external sites has been favoured by the spacing of the discontinuities across the red granites. It has therefore allowed the shaping of the picturesque landforms where marine processes have operated upon a pre-existing weathering mantle.

After their exhumation the exposed boulders have gone on evolving through subaerial weathering. Their flat summits have been hollowed out by circular to elliptical pits and their vertical faces scoured by parallel flutes. The present-day development of these microforms can easily be checked since a single friction is sufficient to remove a crystal grain. As a matter of fact, freeze-thaw has nothing to do with this granular disintegration in the climatic context of the Brittany coast. By contrast, salt crystallization proves to be efficient in the process of weathering as it is demonstrated on faces exposed to the input of aerosols from the sea. The deepening of "gnammas" and "karrens" is favoured by the presence of water, either when it stagnates or when it flows down temporarily.

The peculiar position of these microforms may inform about the tipping of some boulders, one of them having even been overturned during a storm in March 1920 on the beach of Coz-Pors in Trégastel (Fig. 6.5).

Fig. 6.4 Contact between the intermediate and the lower platforms near the Grève Blanche in Trégastel (Photo Y. Lageat)

Fig. 6.5 The beach of Coz Pors, Trégastel, before and after March 1920 (Personal collection)

Fig. 6.6 The chaos of the Little Traouïero, Ploumanac'h (Photo Y. Lageat)

However the term "chaos", frequently referred to, should only be mentioned when boulders accumulate on valley floors after having been stripped from the weathering mantles by incision. This term could apply to the boulder stream of the Traouïero Valley carved below the intermediate platform (Fig. 6.6).

The rate of the granular disintegration can be measured in only one place, at the Saint-Uzec menhir in Pleumeur-Bodou which is more than 5 m high (Fig. 6.7). One of its faces has been scored by nine flutes since 5,000 BP, at a rate amounting to 30 mm ka^{-1}. That the slab was horizontal is demonstrated by the presence on this face of two pre-megalithic gnammas which could only have developed on an exposed flattish rock surface (Lageat et al. 1994).

It proves that the rock basins on the site of Ploumanac'h could not have been carved out during the sole postglacial times because of their average dimensions (640×440×220 mm), based on measurements of 250 basins. As the boulders have been lastly exhumed thanks to the Eemian transgression, these microforms might have partially been deepened by frost shattering during the Weichselian period, before their possible fossilization by aeolian loess during the Pleniglacial.

6.4 Morphological Evolution

The first authors to question the origin of the intermediate platform cut across the complex were De Martonne and Robert (1904) who mentioned that marine deposits might be present.

Fig. 6.7 The Saint-Uzec menhir (Pleumeur-Bodou) showing a pre-megalithic rock basin and postmegalithic flutes. It stands on the intermediate platform and the marginal scarp is visible in the background (Photo Y. Lageat)

Fig. 6.8 Cliff retreat in 5-m-thick loess deposits in Ploumanac'h (Photo Y. Lageat)

Actually, there still is a remnant of an abrasion surface at 30 m a.s.l. in Île-Grande, and perfectly rounded pebbles confirm this statement. That bevelling might date back to the Piacenzian sea level (3.3–2.9 million years) according to Miller et al. (2012). By contrast with this bare surface, 10–15 m thick weathering mantles are preserved on the intermediate bench, and, in order to open the red granite quarries at La Clarté, the crypto-alteration front has to be cleared so that huge cubic blocks can then be extracted. On one hand, they would have been washed away by the waves if the sea level had gone up to elevations above 30 m. On the other hand, this saprolite could hardly have been formed since the Pliocene time.

The intrusion, which is excavated below the country rocks, surprisingly juts out in the sea (Godard 1977). Although the metamorphic basement is edged by high cliffs, whose base bears the marks of marine abrasion, the granitic shoreline corresponds to a denudation coast. The irregular weathering front has been revealed by the wave attack during the interglacial sea levels. The sea has swept away not only the in situ grus but also the surficial cold deposits, either periglacial "head" deposits (consisting of angular fragments embedded in a finer matrix) or aeolian loess. The work of the sea is still in progress as shown by the retreat of soft low cliffs, which deepens bays between rocky headlands (Fig. 6.8).

Along this typically "contraposed" coast, the sandy grus and the Quaternary deposits are scattered away according to their granulometry, and they form mudflats in sheltered sites, sandy beaches with variable slope gradients too and even gravel beaches according to wave energy conditions.

6.5 Cultural Value and Touristic Promotion

Long before the first geological map in 1909 has singled out the red granite inside the complex, the picturesque features of Ploumanac'h have caught the imagination of many painters belonging to different trends, from the Nabies to the Symbolists even through the Cubists.

They probably were the first to name by French terms some of the most curious landforms, the Palette of course (Fig. 6.9), but also different words taken from the bestiary – the camel, the dog, the eagle, the rabbit, the tortoise (Fig. 6.10) and even the imperial three-cornered hat (Fig. 6.11): "Is Napoleon's hat still in Perros-Guirec?"; this message broadcasted by the BBC on 3 August 1944 called for liberating insurrection against the occupying forces in Brittany.

It was a writer, the future academician Charles Le Goffic (1920), who was the first to warn against the damages done by the "ignorant" quarry workers and villa builders. Since 1901, an artistic association has aimed at the protection of the site of Ploumanac'h. This association intended to preserve those sites from being degraded. Nevertheless, this preservation took quite a long time until the "coastal protection agency" (*Conservatoire du Littoral*) became the owner of 86.5 ha in 1991.

Nowadays, the site has become the victim of its fame as 700.000 visitors come to admire it each year. A conservation plan has been established to protect the site against the degradation, and the visitors are made to walk on dedicated footpaths among the granitic boulders. Eighty-eight percent of them, however, claim they are not well-enough informed about the origin of the granitic "chaos" (Lageat and Nicolazo 2009). How could they possibly be well informed when such explanations as follows are given to them? "The granit (sic) which composes these boulders is more than 300 million years old. All along the years, the blocks have been eroded through unceasing aggressions by water and frost and have come to form these spectacular accumulations worth being admired".

Fig. 6.9 The painter's palette, Renote Island, Trégastel (Photo Y. Lageat)

One of the main problems geomorphology is facing is to make visitors aware of the different time scales ranging from the crystallization of the huge granitic body to the sculpting out of the tiny microforms. The erosion concept is too often advocated by non-specialists, but it is too vague and confusing in so far as it mixes up too many processes, times and resultant forms. Much is to be done to increase the level of understanding.

6.6 Conclusion

As mentioned by Migon (2006), "detailed studies of granite coastal geomorphology are surprisingly few". As a matter of fact, at all scales of analysis, the "Pink Granite" coast exhibits an assemblage of landforms, which could be quite as much typical of continental environments:

- Gentle footslopes surfaces across preserving thick saprolites
- Boulders fields depending on primary joint spacing
- Large weathering pits and incipient flutes occurring on exposed rocky surfaces

So despite the coastal setting, very few "classic" coastal landforms are present along this shoreline, with the only exception of narrow wave-cut platforms developed on granites. Actually the intervention of marine agents has evidently

Fig. 6.10 The "Tortoise", Ploumanac'h (Photo Y. Lageat)

Fig. 6.11 Napoleon's hat, Ploumanac'h (Personal collection)

favoured the stripping of the grus cover (by wave attack) and enhanced the development of microforms (by sea spray). However, in the light of such facts, a simple acceleration of the efficiency of some processes remains consistent with the concept of convergence or equifinality as the development of identical landforms could have been also achieved in inland environments, particularly in the French Massif Central (Godard et al. 2001).

Acknowledgements I thank the guest editors Professor Marie-Françoise André (University Blaise Pascal, Clermont-Ferrand 2) and Professor Monique Fort (University Denis Diderot, Paris 7) for their invitation to prepare this paper. I am especially grateful to my sister Maryvonne Le Rouzic who improved the form of the original manuscript, to Laurence David who revised the Figs. 6.2 and 6.3 and to an anonymous referee who provided helpful reviews, and I am deeply indebted to Professor Dominique Sellier (University of Nantes) for a stimulating dialogue over the years.

References

Barrière M (1977) Le complexe de Ploumanac'h (Massif armoricain). Essai sur la mise en place et l'évolution pétrologique d'une association plutonique subalcaline tardi-orogénique. Thesis of geology, University of Brest

Battistini R, Martin S (1956) La "plate-forme à écueils" du nord-ouest de la Bretagne. Norois 10:147–161

Godard A (1977) Pays et paysages du granite. Introduction à une géographie des domaines granitiques. P.U.F, Paris

Godard A, Lagasquie JJ, Lageat Y (2001) Basement regions. Springer, Berlin

Janvier TA (1904) A Breton shrine. Harper's Monthly Magazine, 109, 650:276–287

Lageat Y (1981) Contribution à la connaissance de la porosité des roches cristallines. Physio-Géo 1:3–33

Lageat Y, Nicolazo J (2009) L'invention de la Côte de Granite Rose (Bretagne) et les étapes de la valorisation d'un géomorphosite. Bull Assoc Géogr Franç 2:124–135

Lageat Y, Sellier D, Twidale CR (1994) Mégalithes et météorisation des granites en Bretagne littorale. France du Nord-Ouest. Géogr Physique et Quaternaire 48:107–113

Le Goffic C (1920) Alerte pour Ploumanac'h. Un des plus renommés paysages de Bretagne est menacé. Bull Soc Protection des Paysages de France 86:12–14

Martonne E de, Robert E (1904) Excursion géographique en Basse-Bretagne (Monts d'Arrée-Trégorrois). Bull Soc Scient Et Méd de l'Ouest XIII:293–334

Migon P (2006) Granite landscapes of the world. Oxford University Press, Oxford

Miller KG, Wright JD, Browning JV, Kulpeez A, Kominz M, Naish TR, Cramer BS, Rosenthal Y, Peltier WR, Sosdian S (2012) High tide of the warm Pliocene: implications of global sea level for Antarctic deglaciation. Geology 40:407–410

Rondeau A (1959) Corse et Bretagne: à propos du modelé granitique de la région de Trégastel (Côtes-du-Nord). Norois 20:451–457

Rondeau A (1978) Sur le comportement des leucogranites (ex-"granulites") en milieu littoral. Bull Assoc Géogr Franç 449:47–49

Another Loire: The Armorican Loire

7

Nathalie Carcaud and Hervé Davodeau

Abstract

Over part of its course, the river Loire today presents an archetypal landscape. This is the royal river, with vineyards and châteaux listed as UNESCO World Heritage Sites since 2000. In this chapter it is demonstrated that the river presents other landscapes of interest, notably those of the Armorican Loire. This downstream section of the river made of river landscapes is characterised by the geomorphological constraints of the morphostructural system of the Armorican Massif. The valley, quite narrow, is bordered by hillsides, which sometimes forms steep slopes. The hydrosystem consists of a patchwork of geomorphic compartments (valleys, islands, multiple branches of the river, etc.). The promontories, which punctuate the river Loire, allow for remarkable oblique viewpoints which are still little known. The landscape of the Armorican Loire is also a lived space with diverse social uses around the river, which has progressively acquired heritage recognition. It is nevertheless important to avoid the phenomenon of heritage face.

Keywords

Loire Valley • River patterns • Lived space • Cultural heritage

7.1 Introduction

The inscription of the Loire Valley on the UNESCO World Heritage List in November 2000 contributed to shape its perception as the "last wild river in Europe" around landscape patterns which are recognised worldwide: a majestic and slow-moving river lined with royal castles behind which lie the gentle contours of famous vineyards (Fig. 7.1).

The construction of this image has induced a touristic boom around this World Heritage Site, i.e. along the 250 km of the Loire Valley located between the towns of Sully and Chalonnes.

> The Loire Valley is an outstanding cultural landscape of great beauty, containing historic towns and villages, great architectural monuments (the châteaux) and cultivated lands formed by many centuries of interaction between their population and the physical environment, primarily the river Loire itself.
>
> Criterion (i): The Loire Valley is noteworthy for the quality of its architectural heritage, in its historic towns such as Blois, Chinon, Orléans, Saumur and Tours, but in particular in its world-famous castles, such as the Château de Chambord.

N. Carcaud (✉)
AGROCAMPUS-OUEST, Centre d'Angers, 2 Rue A. Le Nôtre, Angers Cedex 01 40045, France
e-mail: nathalie.carcaud@agrocampus-ouest.fr

H. Davodeau
Laboratoire: UMR CNRS 6590 ESO (espaces et sociétés), AGROCAMPUS-OUEST, Centre d'Angers, 2 Rue A. Le Nôtre, Angers Cedex 01 40045, France
e-mail: herve.davodeau@agrocampus-ouest.fr

M. Fort and M.-F. André (eds.), *Landscapes and Landforms of France*, World Geomorphological Landscapes,
DOI 10.1007/978-94-007-7022-5_7, © Springer Science+Business Media Dordrecht 2014

Fig. 7.1 Saumur castle (Credit: H. Davodeau)

Criterion (ii): The Loire Valley is an outstanding cultural landscape along a major river which bears witness to an interchange of human values and to a harmonious development of interactions between human beings and their environment over two millennia.

Criterion (iii): The landscape of the Loire Valley, and more particularly its many cultural monuments, illustrate to an exceptional degree the ideals of the Renaissance and the Age of the Enlightenment on western European thought and design.

Source : http://whc.unesco.org/en/list/933

In this article, the authors intend to take a wider perspective in order to demonstrate that there are also other Loire landscapes which are less renowned because they belong to another section of the valley, more specifically the stretch of the Armorican Loire which lies between Angers and Nantes (Fig. 7.2).

7.2 Geographical and Geological Setting of the Loire Valley

To understand the Armorican Loire, one should consider this section of the Loire Valley as part of its drainage basin, notably regarding its morphostructural and hydrogeomorphological dimensions.

As shown in Fig. 7.3, the Loire river system can be subdivided into three functional parts (Schulé 2000): the Upper Loire (*haut bassin*), the Central Loire (*Loire moyenne*) and the Lower Loire (*Loire océane*).

7.2.1 The Three Loire

Upstream from the "bec de Nevers" (the confluence of the Loire and the Allier), the two rivers of the high drainage basin run parallel. Their sources both lie in the Cévennes (medium-altitude Mediterranean mountains) and their flows are irregular, being strongly influenced by the climate and landscape of the Cévennes, but also practically equivalent in terms of discharge (150 m^3/s for the Allier, 182 m^3/s for the Loire at the confluence). The term "cévenol regime" is used to describe the regimes of both these rivers.

Downstream from Nevers, between the Nivernais and Touraine areas, the Loire leaves the Massif Central and enters the Paris Basin, where it opens out into the wider floodplain of the Loire Valley. The river landscapes of both sections are subject to similar flows and river regime to those described for Nevers. In Orleans, for instance, the discharge is only 350 m^3/s.

Downstream from Tours, the shape of the river is changing. The term *Loire océane* ("Atlantic Loire"), suggested by Schulé (2000), refers to the arrival of the river in a region exposed to a maritime oceanic climate. Schulé also suggests that the Atlantic Loire may in turn be subdivided into four functional sectors: the Touraine Loire, the Saumur Loire, the Armorican Loire and the Loire Estuary.

Fig. 7.2 The île Batailleuse (an island near Saint-Florent-le-Vieil) (Credit: N. Carcaud, 12/2012)

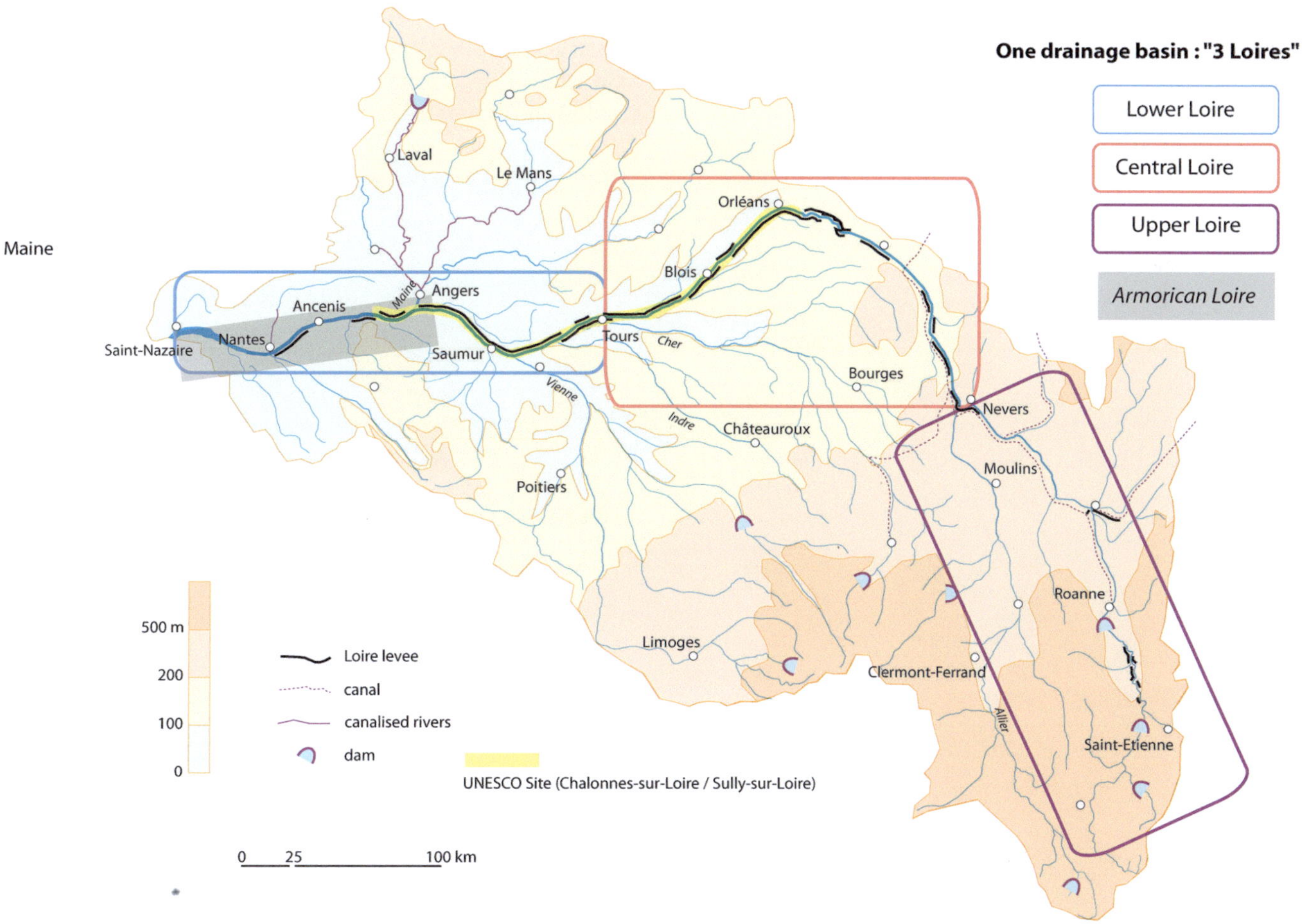

Fig. 7.3 The lower (Atlantic) Loire in the drainage basin of the river Loire (Credit: N. Monnier and N. Carcaud)

The Touraine Loire is characterised by several confluences. Three tributaries join the Loire over a distance of less than 60 km: the Cher (discharge 90 m^3/s), the Indre (discharge 20 m^3/s) and above all the Vienne (discharge 200 m^3/s). The flow of the Loire is thus almost doubled and strongly influenced by water from the Limousin Region, notably from the Millevaches Plateau, and therefore by medium-sized maritime mountains. There are no major tributaries along the Saumur Loire (the discharge at Saumur is 698 m^3/s), but this section remains influenced by the increased discharge of the Touraine section of the Loire thus allowing the river to open up a wide floodplain in the Authion Valley. It is at the end of this valley that the river leaves the Paris Basin and enters the Armorican section of the Loire. The last major confluence (the Maine) is to be found downstream from Angers. The river Maine leaves a drainage basin, which is quite distinct from the previous ones since its waters come from the crystalline plateaus of the Perche and the sedimentary plateaus of Normandy, with typically maritime landscapes and climates and an average discharge of 120 m^3/s. The Maine and its drainage basin impose a different hydrological rhythm and new constraints on local populations living along this section of the Loire. The Loire Estuary is even more radically different. The limit of the estuary has varied over time. Today it is located at Ancenis, which marks the limit of tidal influence. This part of the Loire is subject to numerous threats.

7.3 The Armorican Loire and Its Specific Landscapes

7.3.1 The Armorican Massif: A Source of Constraints

As the Loire runs through the Armorican Massif, it crosses a rigid morphostructural system, which is reflected in the shape and course of the river valley. The Armorican Massif was affected by the Cadomian (about 600 Ma) and Hercynian deformations (about 300 Ma). It is a polygenetic mountain range with Appalachian morphology. Outcrops reveal its basic metamorphic terrains with granite injections and Palaeozoic "syncline folds". The Loire crosses a series of eroded structures which today correspond, in most instances, to sedimentary basins separated by large-scale-oriented structures which have reinforced the separation of these basins (Fig. 7.4) (Alcaydé 1990; Sellier 1986).

The broken line of the Armorican Loire Valley is a result of the underlying tectonic framework. Similarly, the contrasting lithology created by these tectonic movements provides a fundamental explanation for the shape of the valley. After Ponts-de-Cé (the point at which the river enters the Armorican Massif), the valley narrows to a width of just 1.5 miles. Downstream, apart from the two main narrow points at Denée (1 mile) and Champtoceaux (0.8 mile), the morphological fea-

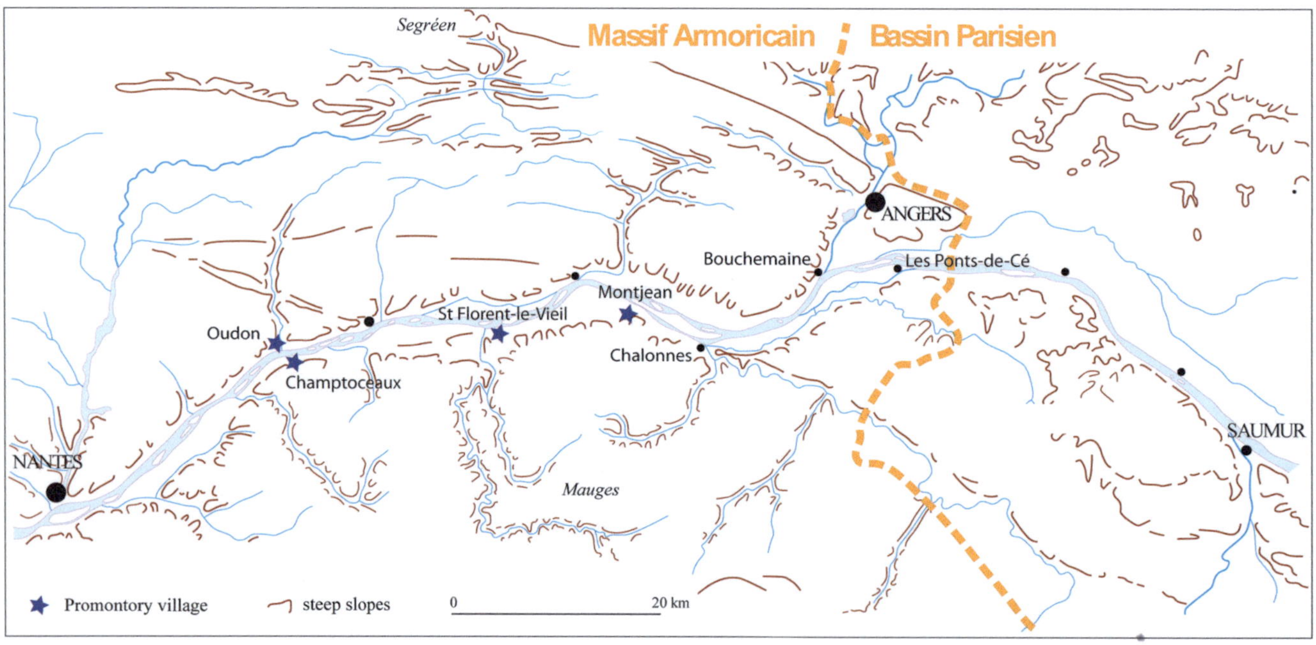

Fig. 7.4 The Armorican Loire and its specific features: islands, valleys and promontories (Credit: N. Monnier and N. Carcaud)

tures of the valley remain more or less homogeneous. In the Segréen and Châteaubriant plateaus towards the North and in the Mauges plateau towards the South, the river has generally cut convex hillsides, which locally form steep slopes and cliffs. This is notably the case on the left bank of the river where it narrows at Ponts-de-Cé (Mûrs-Erigné cliffs), and also on the Bouchemaine/Chalonnes-sur-Loire section of the river (Corniche Angevine and the Coteau de Savennières hillsides), or at the Champtoceaux river gorge (Fig. 7.5a–c).

7.3.2 A River Regime Marked by Temporal Irregularity

As is the case for all "maritime" rivers, the Loire merely appears more stable from a statistical point of view. The river still presents considerable behavioural instability with alternating periods of water abundance and shortages. The water monitoring station at Montjean-sur-Loire provides an accurate picture of the Armorican Loire. The river regime (measured from 1900 onwards) reveals an average yearly discharge of 875 m^3/s. The maximum annual flow occurred in 1910 (1,960 m^3/s), the same year as the greatest flood (6,330 m^3/s); the lowest discharge occurred in 1949 (282 m^3/s) with the lowest water level of the century. The annual variation in flow is thus within the range of 1–6.95. Floods have also periodically occupied the Loire landscape. On average there is a flow of 4,500 m^3/s once every 5 years. The highest floods per decade are over 5,300 m^3/s. The flood levels of December 1910 and December 1982 return on average at least every 50 years.

7.3.3 A Mosaic of Fluvial Landforms

Regarding its hillside morphology, the *Atlas of Maine and Loire Landscapes* and *the Atlas of Loire Atlantique Landscapes* describe the Armorican section as the "Loire of promontories" (Conseil Général de Maine-et-Loire 2003). Its hydrosystem consists of a patchwork of interdependent and dynamic geomorphic compartments formed at different temporal and spatial scales, which offer a wide range of possibilities for human life (Fig. 7.6).

In the floodplain, we find the "*Val de Loire*" as defined by Dion (1961). Between Nevers and Nantes, it covers a considerable surface area of 133,750 ha, a large part of which is today equipped with levees which substantially reduce the floodplain. Between the *Bec d'Allier* and Nantes, over a distance of 487 km, there is linear protection of 484 km, covering 72 % of the floodable area. A reconstruction of the landscape of the Loire Valley before the building of the levees would reveal two large morphodynamic units. The riverbanks are protected from annual floods. Of variable width, these levees are regularly interrupted by breaches created by the Loire corridors. Lateral depressions of various widths are formed by overflow areas, which are naturally flooded during the annual breaching of the banks. These depressions called "*vals*" are currently occupied by a Loire tributary whose connection with the main channel is obstructed by the bank ridges. Smaller ridges, formed by the remains of a very low Quaternary terrace, sometimes rise several metres above the lateral depression (Arnaud-Fassetta et al. 2010).

In the Armorican Loire River, the "vals" lose some of the size they reach in the Saumur section of the river. Their surface area is around 1,000–2,000 ha as against 34,000 ha in the Authion Valley. The latter were the last valley sections to be equipped with "unsinkable" levees during the second half of the nineteenth century.

Between the levees and the valley sides, the low-water channel is often straight with an average width of 1,000 m. This is a physical river landscape dominated by a moving bed load, a type of hydrosystem which is stable during historical times but subject to annual channel adjustment limiting the river transport (Babonaux 1970). There are two main elements at work in this phenomenon: the islands and the multiple branches of the river. The islands occupy a lateral position and disperse the flow of the river towards branches in which there is a succession of pools and riffles, which can attain a considerable size (several hundred metres). Some of the branches are called *boires* or oxbow lakes (Fig. 7.7). During low-water periods, only the main branch is drained. It was in this multiple-channel system that engineers built the Armorican fluvial groins at the beginning of the twentieth century in order to facilitate the deepening of the main channel and thus allow river transport as far as Angers.

In the low-water channel of the Armorican Loire River, the submerged groins are certainly the most amazing landscape element. They consist of rock blocks placed perpendicular to the river axis, which emerge during low-water flows. These structures were built at the beginning of the twentieth century in order to create a navigable channel for large boats. Between Angers and Nantes, 700 groins border the low-water channel and still allow the river, at least downstream from Bouchemaine, to be recorded among navigable French watercourses. However, these groins did not allow the river to compete with the railways, which began to reach the Loire Valley towards the mid-nineteenth century. Despite fluvial transport continued until the 1980s with the last oil tankers, the groins were already obsolete when they were built.

The project of rendering the Loire navigable is now viewed as a nineteenth-century engineering utopia, and the groins and other hydraulic works are generally considered to have had a negative impact on the natural balance of the river environment. By favouring the digging of a navigation channel, the groins (along with the river sand mining industry which is today banned) decreased the low-water mark by 2 m on average and thus contributed to disconnecting secondary

Fig. 7.5 The promontories of the Armorican Loire: (**a**) Montjean-sur-Loire (Credit H. Davodeau 06/2012), (**b**) Saint-Florent-le-Vieil (Credit H. Davodeau 06/2012), (**c**) Champtoceaux (http://www.champtoceaux.fr)

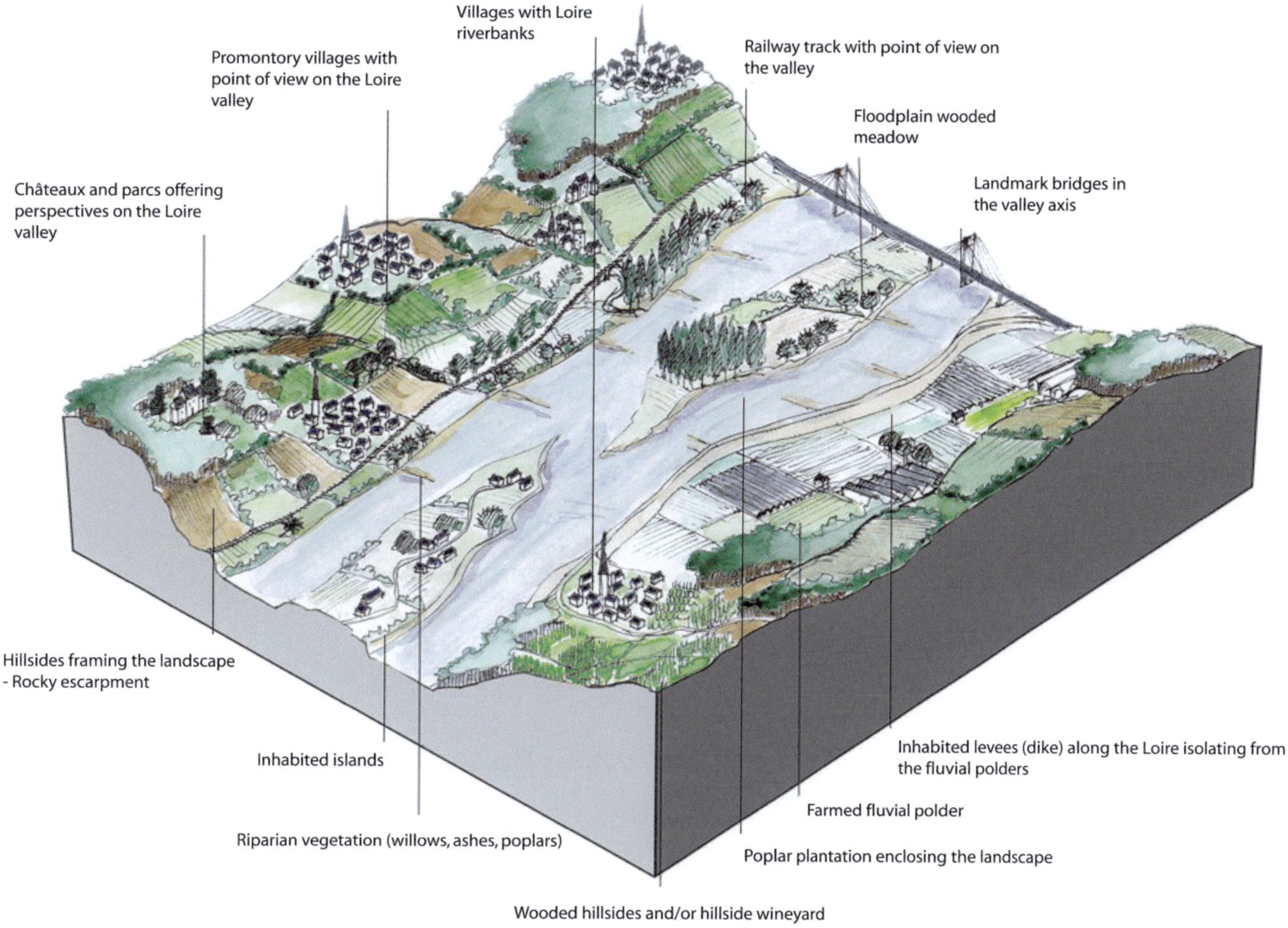

Fig. 7.6 3D diagram. The Loire of the Promontoires (Atlas des paysages de Loire-Atlantique 2012, with permission: http://www.paysages.loire-atlantique.gouv.fr)

branches from the main course of the river. The groins are thus today increasingly seen as a symbol of an era during which the river was outrageously altered and they are therefore the target of environmentally motivated political decisions: in 2009–2010 this led to experimental remodelling of the groins (they were shortened and lowered) which may be a first step to their complete removal. Whereas being criticised by politicians and more particularly by environmentalists, the groins are nevertheless considered useful by riverside residents of the Armorican Loire: not as navigational works but, for example, as observation points (they allow perpendicular access to the river), fishing swims (downstream swims are sheltered from the current) and as bathing spots (e.g. between two groins). The diverse uses of the groins raise the issue of local residents' and river users' attachment to the groins. Recently conducted surveys demonstrate that local residents, although they do not explicitly recognise any heritage value, nevertheless express a certain attachment to these hydraulic works, which surprising as it may seem are not considered as being contrary to the established image of the "wild river" (Davodeau et al. 2013).

The singular nature of the Armorican Loire is established through a series of contrasts with the Saumur and Touraine sections of the Loire. The former is darker (due to the schist), while the other two are brighter (due to the *tuffeau*, a whitish soft limestone); it has steeper valley sides with promontories in contrast with the wider valleys and horizons of the latter two sections. The resulting dichotomy is that of the "wild river" v. the controlled river (with its navigational groins), the Loire Chateaux and vineyards v. the popular industrial river, and thus, as a result, the Saumur and Touraine sections are viewed as a more touristic part of the Loire (the World Heritage Site, the Loire Anjou Touraine Regional Park) v. a more "ordinary" Armorican Loire.

7.4 The Advantages and Potential Risks of Heritage Status

The promontories, which punctuate the valley, allow for remarkable oblique viewpoints, which are much less frequent upstream. Historically, the rocky outcrops of the promontories

Fig. 7.7 Oxbow lake (Credit: N. Carcaud, 09/2008)

favoured the development of villages and towns located in the immediate vicinity of the river and yet protected from its vicissitudes. Situated outside the perimeter of the UNESCO Heritage Site and relatively distant from the two main urban centres of Angers and Nantes, these small rural communities are not subject to intensive tourism. The promontories have therefore not been promoted in accordance with the touristic potential of their panoramic viewpoints. Montjean, Oudon and Saint-Florent-le-Vieil (the birthplace of the famous French writer Julien Gracq) certainly offer scenic viewpoints but they are mostly visited by locals.

7.4.1 A Living Loire

The distant and panoramic views from the promontories are less representative of local habits than the close-knit links at the foot of the valley sides between local residents and their river. The Loire is not so much a spectacle appreciated from above but rather a constant "presence" which locals seek from below, near the riverbanks. This contact with the river is facilitated by the perpendicular groins, which allow access to the centre of the river. The diverse social uses on and around the groins, the "flood culture" which has developed on the large inhabitable islands and specific activities like playing bowls in the river sand (much less well known than *boule de fort*, which is an indoor version of the game, typical to the Anjou region further upstream) testify to a certain Loire lifestyle which may be observed along the whole length of the river but with greater intensity and authenticity along the Armorican Loire (Fig. 7.8).

7.4.2 The Current Heritage Process

The international recognition of the heritage value of the Saumur Loire has perhaps provoked confusion, jealousy and a feeling of injustice among some local politicians who would like to have seen their communities benefit from the same Heritage Status. The heritage area would have greater geomorphological consistency if it had excluded the 15 km stretch of the *Corniche Angevine*, which is already protected under a 1930 French statute relating to natural sites (2003) and which is today the largest protected area (2,400 ha) along the Armorican Loire between Angers and Nantes. Outside this stretch, the regulations pertaining to historical monuments or the protection perimeter for urban and rural sites allows for the protection of the greatest monuments or architectural ensembles. Despite being outside the World Heritage Site, the landscapes of the Armorican Loire have nevertheless progressively acquired heritage recognition. Apart from its buildings, the natural landscape of the Armorican Loire enjoys indirect protection aimed at protecting biodiversity (Natura 2000) or benefits from the banning of any construction work in areas prone to flooding (through a French national plan for the prevention of flooding). Such regulations reinforce the local power of state services and ensure more effective protection than any extension of the Heritage Status perimeter requested by some local politicians, since the Heritage Status provides no specific regulation. They also provide more effective protection than any regional park, which might be created along the Armorican Loire (along the lines of the Loire Anjou Touraine Regional Park along the Saumur stretch of the Loire).

Fig. 7.8 Another function for the groins: an access to the river (Credit: N. Carcaud, 02/2012)

7.4.3 The Risks of Heritage Status

The need for recognition of the Armorican Loire could result in a "heritage race" which we believe presents certain risks: firstly, because the existing regulations are already very demanding, and, secondly, because it could be detrimental to the intimate relationship with the river which we consider to be one of the virtues of this very lively section of the Loire Valley. The "race for Heritage Status" does not appear to be the best response to a legitimate need for recognition because it could result in the opposite effect due to the risk of erasing the specific lifestyles of this section of the valley in order to conform to dominant representations of the river and thus lead, above all, to a reduction in the liveliness of these places. *When I walk down the streets of Saint-Florent or when I walk along the banks of the Loire I am surprised by the absence of games and children shouting in those places where, at the age of eight or ten, my gang spent their lives on the quayside or in the bushes which lined the river banks. At that time the submerged groins of the navigable Loire were being built: big piles of chestnut hurdles lay on the slipway: one of our favourite games was to climb up and then jump down from as high as possible onto the cobblestones of the quay* (Gracq 1967).

7.5 Conclusion

The geomorphological particularities of the Armorican Loire have given rise to landscapes, which do not quite correspond to the world-famous image of the Loire Valley. It would be futile to try to deploy public measures in order to reduce the gap between the reality of life along the river and the dominant representations of the river, since, beyond their physical appearance, these specific landscapes, which possess their own qualities, express, above all, a certain lifestyle along the river. It is this relationship, which must be preserved and placed at the heart of public policies, which today takes as its reference point all Loire landscapes whether Armorican or non-Armorican.

Acknowledgement We would like to dedicate this chapter to our cartographer N. Monnier, who passed away recently (2011).

References

Alcaydé G (1990) Val de Loire, Anjou, Touraine, Orléanais, Berry. Guides géologiques régionaux. Masson, Paris, 199 p

Arnaud-Fassetta G, Carcaud N, Salvador P-G, Castanet C (2010) Fluviatile palaeoenvironments in archaeological context: geographical position, methodological approach and global change – hydrological risk issues. Quat Int 216:93–117

Atlas des paysages de Loire-Atlantique (2012) DREAL Pays de la Loire. http://www.paysages.loire-atlantique.gouv.fr

Babonaux Y (1970) Le lit de la Loire : étude d'hydrodynamique fluviale. Comité des Travaux historiques et scientifiques. Bibliothèque Nationale, Paris, 252 p

Conseil Général de Maine-et-Loire (2003) Atlas des paysages du Maine-et-Loire. Le Polygraphe, Angers, 205 p

Davodeau H, Barraud R, Carcaud N, Montembault D, Pordoy C (2013) Rivers groins along the Armoricain Loire: river responses and local resident reactions. A sociogeographical landscape reading as a complement for ecological engineering. In: Arnaud-Fassetta G, Masson E, Reynard E (eds) Proceedings of the international conference.

Continental Hydrosystems and European Territories faced with Different Water Laws, July 11–13, Sion/Verlag Friedrich Pfeil, Switzerland/München, pp 205–220
Dion R (1961) Histoire des levées de la Loire. Author's Personal Publ, Paris, 312 p
Gracq J (1967) Lettrines Editions José Corti, 256 p
Schulé C-A (2000) Les crues inondantes en Anjou du XIXe au XXe siècles. Archives d'Anjou 4:247–253
Sellier D (1986) Les versants du pays nantais, étude géomorphologique. PhD thesis, IGARUN, Université de Nantes, 506 p

The Arcachon Bay Estuary: A "Collage" of Landscapes

8

Frédéric Bertrand

Abstract

The Arcachon Bay is a typical semi-sheltered lagoon located on the Atlantic coast of France, facing the wave-dominated shelf of the Bay of Biscay. Its long-term evolution was controlled by the development of the Cap Ferret barrier spit and the establishment of a tide ebb-dominant barrier system leading to a spatial split of the bay in two parts: (a) the inner lagoon mostly composed of tidal flats and (b) the outer lagoon with higher hydrological capacities where a large-scale tidal inlet is bordered by the largest coastal dune in Europe. The identification of major interactive sedimentary environments with reference to their representative and complementary characteristics leads to a deductive selection of five landforms supporting oyster cultivation and various recreational (sportfishing, sailing, ecotourism) activities of both economic and landscape value.

Keywords

Biogeomorphology • Barrier system • Salt marsh • Coastal dune • Land reclamation • Coastal management

8.1 Introduction

Arcachon Bay is a large (174 km^2) mesotidal shallow lagoon located on the southwestern Atlantic coast of France (44°40′ N, 1°10′ W; Fig. 8.1) and the only major embayment along the remarkably drift-aligned coast of the Landes of Gascony. It is connected to the Atlantic Ocean by a 2–3 km wide and about 12 km long channel called "the passes du Bassin" that enables important seawater exchanges and therefore the development of an important oyster farming industry. Likewise, various recreational activities have been developed since the early creation of Arcachon seaside resort in the late 1850s – with the extension of the rail line – and early 1860s, in relation with local (Bordeaux) and foreigner (England) upper middle-class interest for sporting activities and beach pleasures. From a geomorphological point of view, the Arcachon area is more well known for the dunes that surround the lagoon, especially the impressive Great Dune of Pyla than for its inland low-lying and yet varied landscapes. As a matter of fact, Arcachon Bay occurs within a patchwork of interactive sedimentary environments (i.e., mainland, backbarrier, lagoon, inlet, barrier, barrier platform, and shoreface) nested at different scales, each with distinct morphology, sediment, and habitat patterns but associated in a complete and unique barrier island geosystem.

8.2 General Setting

8.2.1 Long-Term Morphological Development

Arcachon Bay as such is one of these six interactive sedimentary environments, bounded on the seaward side by the Cap Ferret barrier spit and on the mainland side, both to east and to the south, by a very crenulated shoreline as the lagoonal water intersects small drowned river valleys (*crastes*) and artificial

F. Bertrand (✉)
Department of Geography, Paris-Sorbonne University, Institut de Géographie, 191, rue Saint-Jacques 75005, Paris, France
e-mail: frdric.bertrand@gmail.com

M. Fort and M.-F. André (eds.), *Landscapes and Landforms of France*, World Geomorphological Landscapes,
DOI 10.1007/978-94-007-7022-5_8,

Fig. 8.1 True color composite image (R3G2B1) of Arcachon Bay showing the *triangular-shaped* shallow lagoon within a pine forest environment partly cultivated (cornfields), facing growing urban and recreational land-use pressures. Sedimentary environments, selected geomorphological landscapes of high value (*polygons*), and stratigraphic cross section (*broken line*, Fig. 8.2) are indicated (U.S. Geological Survey, Global Land Survey, 2001, Landsat ETM+, 30 m scene ETM+, L71200029_02920050712, USGS, Sioux Falls, South Dakota)

headlands due to salt-marsh embankments for reclamation. The total barrier-lagoon system is a three-dimensional feature (Fig. 8.2) including a sequence of sediments up to 30–35 m thick that have developed over a time span of 14,000 years from 18,000 years B.P. to 4,000 years B.P. (Féniès and Lericolais 2005). The processes forming sand barriers lead to the development during the postglacial sea-level rise of a transgressive barrier system with the overlapping of coarse-grained sands and sandy gravels with marine shell debris – typical of tidal channel deposits – over Neogene to lower Pleistocene alluvial deposits truncated by the extensive Aeolian coversand of the Landes region (13,000 km^2).

When the sea level reached a stillstand, around 4,000 years BP, the Leyre estuary started being transformed into an open lagoon with a main channel oriented to the northwest according to the structural axis of the northwest-southwest trending Leyre normal fault (Faugères et al. 1986). The present-day triangular shape of the bay is a consequence of the formation of the Cap Ferret littoral spit between 2,200 years B.P. and 1,900 years B.P. in response to a wave power increase following the subboreal/sub-Atlantic climate instability or a decrease in tide power related to the lagoon sediment infill (Allard et al. 2009). The unconsolidated, easily eroded underlying Pleistocene sands permitted an extensive (30 km) shift of the inlet to the south. Since that period continued sediment supply from the foreshore, stable sea level, low subsidence, and natural fixation of the primary generation of coastal dunes (Tastet and Pontee 1998) prevent the Cap Ferret spit from submerging or migrating landward. Instead, the barrier continued to migrate laterally into an open-end barrier inlet, inducing a morphodynamic response by all the elements of the barrier system (Fig. 8.3).

8.2.2 Hydrological and Hydrodynamic Characteristics

The most obvious characteristic of the landscape of the backbarrier lagoon is surface-area variability related to ebbing and

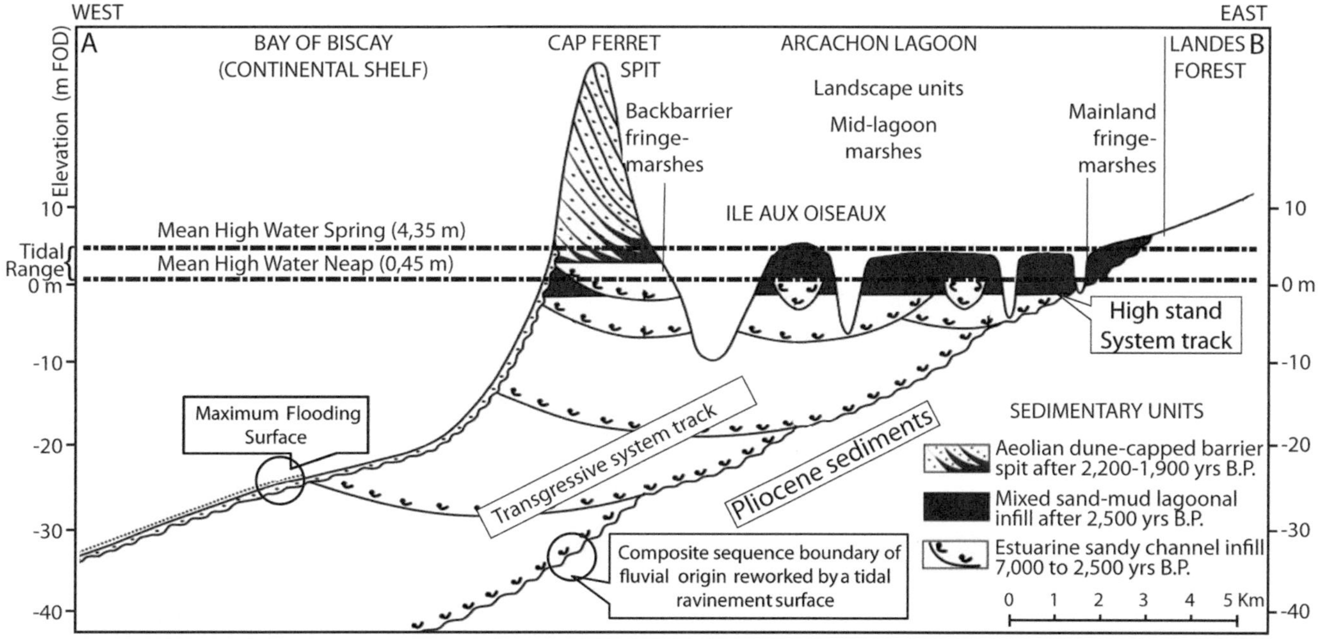

Fig. 8.2 Stratigraphic cross section revealing the Holocene internal architecture of the incised valley of the Leyre and the distribution of subenvironments in the Arcachon barrier lagoon. The cross section is located on Fig. 8.1 (Redrawn and modified with the permission of Gilles Lericolais after Fig. 8.3 in H. Feniès and G. Lericollais, 2005)

flooding tides and resulting from emergence of extensive intertidal features (oyster reefs, tidal flats, bars) during low-water stages of the tide (Fig. 8.1). At millennial scale this open-water backbarrier lagoon must be related to the original pattern of the basin fill – a relatively deep engulfed shoreface – and a low rate of sediment input with respect to the basin capacity. Thus, the tidal prism (384 millions of cubic meters) is directly related to a mesotidal range (varying from 2.1 to 4.8 m), and water surface area increases greater than 200 % between spring low (40 km^2) and high tides (172 km^2). Annual freshwater inputs average 1.25×10^6 m^3 with a major part (4/5) coming from the Leyre River located in the southeastern part of the bay (Plus et al. 2009). Since the tidal prism increases seaward, a funnel-shaped morphology typifies all tidal flats and salt-marsh channels. As a result of an asymmetrical current pattern causing the development of flood-dominated and ebb-dominated channels, the lagoon surface exhibits an intricate network of anastomosing channels (57 km^2) isolating large tidal flat areas (117 km^2) made of permeable muddy sand.

In contrast, the external part of the bay is dominated by subtidal open water with a typical mixed-energy environment. Flow velocities can reach 2 m/s southward the Cap Ferret sand spit (Cayocca 2001), whereas high energy North Atlantic of swells traveling mainly from the W–NW sector generate a strong southerly longshore drift of approximately 700,000 m^3 of sediment per year (Michel and Howa 1997). Nevertheless, current profiles and water-level measurements recorded during the severe Joachim storm in December 2011 show that under severe wave conditions (Hs > 6.0 m) and despite high incidence of waves from the NW (nearly 10°), the longshore drift is weaker (1 < 1 m/s) than under "usual" storm conditions associated to Hs = 4–5 m (Sénéchal et al. 2013).

8.3 Geomorphological Landscapes

8.3.1 The Ebb Tidal Delta Including the Arguin Bank

On the seaward side of the inlet shoreline, dynamics is dependent on the configuration of a typical ebb tidal delta where strong littoral drift implies significant channel and sandbank migration (Lafon et al. 2002). Michel and Howa (1997) proposed a one- to two-channel 80-year cyclical morphological evolution under the control of external hydrodynamic forcing. When there is a single ebb channel, longshore drift on the updrift side of the inlet accumulates sediment into a spit, while sand bypassing delivers sediment to the downdrift shoreline, which is subsequently caught in littoral drift reversal and causes a progradation of the shoreline. Conversely, when the two-channel morphology becomes dominant, more sediment is trapped on the ebb delta, which leads to a deficit on the downdrift shoreline, and subsequently erosion occurs. Although the overall cyclic behavior of Arcachon tidal inlet is not yet clearly defined, it is important to acknowledge the cyclical change in ebb tidal volume that was associated during the second half of the nineteenth century with the Arguin bank migration and the consequences for the adjacent shoreline and coastal dune.

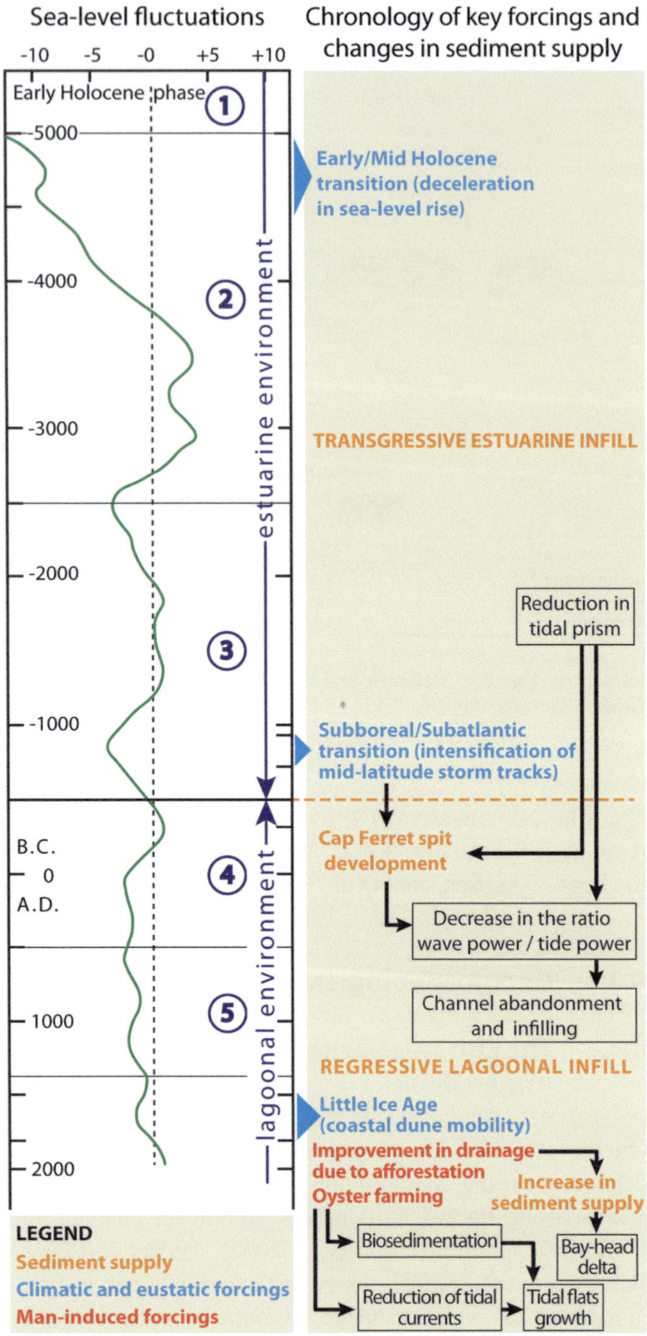

Fig. 8.3 Millenary evolution from the Leyre estuary to the Arcachon lagoon revealing the main parameters forcing their infill. The estimate ages of the five filling successive stages are indicated (After Froidefond and Legigan 1985; Féniès and Lericolais 2005; Allard 2008)

8.3.2 The Pyla Perched Dune

The adjacent beach south of the Arcachon tidal inlet is crowned by the highest (108 m) coastal dune in Europe and appears from above as a somehow unlikely simple feature of the Arcachon Bay landscape (Fig. 8.4).

Dark lines corresponding to old soils (paleosols) appear on the landward side of the dune although partly buried when crowd shuffling is greater. They make it possible to reconstruct a polyphase vertical accretion of that enormous accumulation (approximately 60×10^6 m^3) of sand (Froidefond and Legigan 1985). From the base of the dune to summit, four main paleosols have been distinguished and dated by radiocarbon ages, palynologic analysis, and historic dating. The base of the dune is underlined by an old podzol dated at 1650 BC and exposed by the eroding shoreline. It represents the remnant of a sandy soil covered by a large forest when the sea was 6–8 km further west. Paleosol 2 located between 2 and 5 m above sea level and paleosol 3 (Fig. 8.5) with altitude from 20 to 40 m correspond to the upper part of two major transgressive dunefields composed of compound parabolic dunes built by the reworking of the coversand of the Landes and fixed by the natural development of pine forest until the early 1700s. The latter undulating level dated at 1600 AD marks a major discontinuity in dune development with an evolution of the system from parabolic to crescentic in response to climatic changes (i.e., during the "Little Ice Age"), channel migration, and large-scale management strategies. Thus, the following part of the accumulation limited by paleosol 4 represents the backslope of a crescentic dune that invaded the parabolic system and buried paleosol 3 below 50 m of sand reaching 80 m in height.

Since this dune called "Dune de la Grave" was moving very quickly, an attempt was begun to stabilize sands by planting pines. It failed because of the migration of about 500 m of the southern tidal channel to the foot of dune. As a consequence vegetation began to be stripped away. Part of the sands nourished the tidal banks, but another part was transported by the wind toward the summit and deposited on the young pines (paleosol 4) resulting in the present Pilat dune construction as well as on the lee side increasing eastward progression of the overall sand accumulation.

Presently Aeolian activity and dune mobility are clearly demonstrated by wind-borne accumulations around the crest of the dune as well as avalanche of sands and sand encroachment burying surrounding roads and houses. On the seaward side of the dune, the landward displacement of both the dune crest and shoreline (dune toe) can be attributed to the recent construction of a seawall in 1994 along the "Corniche road" to the Pyla beach generating diffracted but still energetic waves instead of deflecting the wave energy away from the coast (Fig. 8.5). In order to reduce sand losses in the surf zone and to prevent eastward inlet channel migration from taking more of this strongly frequented beach, sand nourishment has been reinforced since 2003 and may temporarily enlarge the Aeolian sand transport rate to the perched Pyla dune.

Fig. 8.4 Great Transverse dune of Pyla (108 m) facing the south tidal channel of Arcachon lagoon inlet and the Arguin sandbank. Oyster farms have established since the 1990s on shoals sheltered from wave action but not from surge-tide floods. Arcachon Bay in the background (With the kind permission of Michel Boudigues, publication director of three Axes Services Loisirs)

Fig. 8.5 Pyla retrograding beach and steep cliff-like scarp at the back (Photo F. Bertrand 2010). "Paleosol 3" represents the surface of the parabolic dunes fixed by the natural development of pine forest until the early 1700s. It is overlain by the backslope of the "Dune de la Grave" that invaded the parabolic system and buried it under 50 m of sand. In the nineteenth century, an attempt was made to stabilize these Aeolian sands by planting a pine forest. The rocky revetment placed at the toe of the slope to protect the Corniche road has caused erosion and shoreline retreat more than 30 m south of the breakwater

8.3.3 Tidal Flats and Seagrass Meadows

Since the lagoon surface is not heavily aggrading, the shallow floor behind Cap Ferret spit has maintained an extensive system of tidal flats mainly made of muddy sand that encloses the largest (70 km^2) seagrass meadow of Zostera noltii in Europe (Auby and Labourg 1996).

Although not a spectacular landscape, these tidal flats are of major ecological and socioeconomic value for the Arcachon Bay since they exhibit substantial rates of both primary and secondary productivity, support diverse epiphyte and faunal communities, and are important nursery environments for many fish species. Moreover, the seagrass dense leaf canopy and root/rhizome structure provide a high potential for exporting the large anthropogenic N-inputs to the lagoon – due to forest areas in the catchment being replaced by intensive maize farming – as well as reducing internal N-loading. These morphological attributes also help to stabilize bottom sediments and trap fine particles and sediment from the water column (Ganthy et al. 2010). Thus, the changing character of the accretional surfaces that maintain elevations suitable for Zostera spp. growth is likely to play an important role in determining marsh distribution patterns and evolution within a sediment-deficit basin where suitable sites for marsh plants colonization are confined to the margins of the basin.

8.3.4 Leyre Bayhead Delta and Marshes

Although they represent less than 10 % of the total surface area of the barrier lagoon (including reclaimed areas), the Arcachon Bay salt marshes exhibit a wide range of morphological variations and landscapes. This is due to the variety of relict antecedent surfaces that intercept the upper part of the tidal zone and to a complex history of alterations by human activities through enclosure for salt production, fish production, and agriculture as well as urban infill since the three last centuries. Two distinct geomorphic settings for marsh development can be distinguished: valleys and headlands (or interfluves) between the valleys.

The extensive Leyre valley marshes are the most distinctive marsh unit surrounded by the mainland on each coast side of the south (Leyre du Teich) and north (Leyre de Malprat) distributaries of a typical fluvial-sourced bayhead delta. The reconstruction of the chronostratigraphic framework of the Leyre incised valley shows that this original setting was made possible by progradation from about 2,500 B.P. to present day of a very thin depositional system (a few meters thick) of fluvial origin over a well-developed (30–35 m thick) retrogradational system deposited during early Holocene phase (before 7,000 B.P., unit 1) until well after the slowdown of sea-level rise. Thus, the fluvial-sourced bayhead delta of the Leyre represents the progradational system deposited after maximum transgression since the rate of creation of accommodation is less than the rate of sediment supply (Fig. 8.3).

Assuming that significant changes occurred at mouths of the river during the second half of the eighteenth century in relation with pine tree plantation, decrease in the fluvial discharge and a lesser capacity of the Leyre River to transport (Bressolier-Bousquet 1991), the upper part of the Leyre estuary filling could have occurred in a time span shorter than 250 years, mostly during that recent period of extensive land-use changes. Thus, the cross-sectional areas of distributaries are considerably smaller than the antecedent valley particularly in the south part of the delta where the fluvial influence has been reduced as a consequence of land reclamation on both insular (Malprat) and continental (Le Teich) sides of the channel and confined to a narrow, steep-sided tidal channel.

Relative important velocities (up to 100 m^3/s) and high freshwater discharge (around 658 million m^3/year) still occur in the wider north distributary. As a consequence previous halophilic vegetation (Juncus maritimus) covering this part of the deltaic plain has been gradually replaced and relegated to downstream parts of the channel by extensive common reed beds (Phragmites australis) tending to form dense impenetrable monospecific stands (Fig. 8.6).

Because these communities represent an important habitat for water birds, they are widely converted into ponds ("lacs de tonne") and managed (cutting, land forming) for hunting purpose. Although not considered as a natural habitat of community interest, it belongs to a mosaic of different wetland habitats covering a total area of 5,175 ha which was designated in February 2012 by the Ramsar Convention as a wetland site of international importance for conservation.

The left side of the southern distributary located on the municipality of Teich and formerly reclaimed during the eighteenth century from the sea to create "fish reservoirs" became a protected nature reserve in 1972. Although the Teich Bird Reserve is relatively small (120 ha), its location on the flyway paths of numerous migratory birds and carefully designed experimental manipulation of habitat that control for complexity and surface areas allow a great richness of birds (260 species) and a high number of tourists (70,000 a year).

8.3.5 Certes and Graveyron Reclaimed Salt Marshes

Along the mainland sides of the lagoon both to north and the west of the bayhead delta, the distribution of fringe marshes was primarily related to the high drainage density of mainland inducing the formation of extensive interfluve (or headland) marshes between the valleys that drain into the lagoon.

Fig. 8.6 Funnel-shape mouth of Leyre delta north distributary (Eyre de Malprat). Discontinuous fringe of Spartina maritima communities backed by Juncus maritimus and followed by the expansion of the emergent Phragmites australis (Photo F. Bertrand 2010)

However, these mainland fringe marshes have been severely deteriorated under the impact of successive embankments on their landward side, harbor structure construction on their lateral side, and erosion processes on their seaward side (Le Nindre et al. 2006). As a result, the present shoreline of the lagoon appears very irregular, even crenulated in some places where fishponds (Certes, Graveyron) and earth platforms (Gujan) were created down to the distal edge of the march isolating shore-oblique relict marshes (Lanton, Gujan).

The preservation potential of the headland marshes is variable depending on the exposure to the open water of the lagoon. In front of the reclaimed marshes forming a prominent ridge, the distal edge of the marshes often has a small step onto the adjacent tidal flats experiencing a rapid landward recession (Bertrand and Goeldner-Gianella 2010) (Fig. 8.7).

Conversely the areas behind these low-lying reclaimed marshes are protected from bay waves and have maintained extensive marshes. If ambient suspended sediment concentrations are sufficient to trap substantial quantities of settling silts as in Lanton Bay, vertical accretion and salt-marsh progradation over the shallow adjacent tidal flats may occur forming dense patches and Spartina meadows. Considering that potential tidal marsh retreat along the mainland fringes of the Arcachon lagoon is presently limited by developed features and embankments, these remaining headland marshes represent a geomorphological landscape of high value which should provide protection measures for their preservation.

Meanwhile, the idea of using de-embankment to recreate salt marshes is under evaluation from evidence of spontaneous salt-marsh formation after recent accidental dike breaching of Graveyron and Malprat polders due to storm tides (Bertrand and Goeldner-Gianella 2010) (Fig. 8.8). This alternative considered here is to give up some of the created land to the sea by realigning the coast to an existing setback dike and to promote salt accretion in the former polder so that salt marshes can provide protection of the hinterland from waves and storms. This soft coastal defense technique aims to provide a long-term solution for reducing flood risk that has increased significantly along the mainland shore of Arcachon Bay owing to the acceleration of sea-level rise (0.35 mm/year over the period 1967–2010), the increase of storm intensity and extreme surges (within the lagoon), and, last but not least, the urbanization in low-lying areas.

8.3.6 Ile aux Oiseaux Mid-lagoon Marshes

Ile aux Oiseaux (2.7 km^2) is an isolated and yet most emblematic feature of the Arcachon lagoon landscapes (Fig. 8.9). Its origin is still not clearly elucidated. Assuming that the lagoon formed by the enclosure of an estuarine embayment by virtue of a migrating spit, the Ile aux Oiseaux could involve the stranding of either a former flood delta behind Cap Ferret migrating barrier spit or a beach ridge.

Fig. 8.7 Recession of low cliff at the transition between salt marsh dominated by Juncus maritimus communities and tidal flat in front of Graveyron polder. Erosion linked to changes in a larger-scale estuary process regime shaped by a combination of external sea-level rise, internal wind-wave forcing, and dredging of the Audenge channel (*right* background) (Photo F. Bertrand, 2010)

Fig. 8.8 Re-establishment of lower and middle marsh plant communities dominated by the perennial Puccinellia maritima and Halimione portulacoides within parts of the Graveyron abandoned reclamation. The headland of Certes reclaimed salt marsh (background) is still enclosed and "hardly" protected to permit fish farming development (Photo F. Bertrand, July 2010)

Fig. 8.9 Ile aux Oiseaux viewed from the east. Both south side (*left* foreground) and east side (*right* foreground) of the island are sheltered from the prevailing wind wave so that sediment movement is substantially lower than on the west side (background). It results in the accumulation of fine-grained sediment (muddy sand), the development of salt-marsh vegetation (*green*), and the relative stability of the island's shoreline. Nevertheless, the eastern part of the tidal flat or crassat assumes a strong tendency of retrogradation toward the west in connection with the widening of Mapouchet tidal channel where the highest velocities of the ebb tide occur (Photo courtesy of Philippe Dufour)

The identification around the island of the same underlying sediment facies (see above) than the Leyre valley fill (Féniès and Lericolais 2005) may confirm the abandon of such backbarrier features that have enhanced thereby the segregation of flow into ebb- and flood-dominated channels. Whatever the origin and precise nature of the Ile aux Oiseaux, the occurrence of marshes in the central section of Arcachon Bay must be determined by the changing character of a relict antecedent platform marsh into an accretional surface maintaining suitable elevations for marsh growth in relation with extensive tidal scour and fill processes near southwesterly highly dynamic part of the lagoon. Locally oyster reefs and oyster farm structures representing artificial obstacles may induce a decrease of the wave erosion and a deposition trend in the downstream part of some parks with regard to southwest wave propagation (Allard et al. 2009).

Although marsh surfaces and terrestrial habitats are relatively obvious subaerial geomorphological features, major parts of the island are subaqueous. Because the Ile aux Oiseaux has become a major destination for recreational boating and marine recreation, especially during the summer, subtidal seagrass communities are increasingly threatened by physical disturbance, e.g., sediment resuspension from propeller wash and anchor chain scour. This potential impact associated with the release of contaminants has led to the establishment of mooring regulations around the island in order to maintain some essential services provided by seagrass ecosystem including wave protection. Indeed, although considered as non-submersible, the upper part of the Ile aux Oiseaux has experienced severe flooding, i.e., far beyond the littoral zone, after the storm Xynthia hit the French coast on the 28th of February 2010. Thus, the preservation potential of these mid-lagoon marshes appears to be closely dependent on the degree to which sediment transfer and morphologic change will be able to offset submergence in this highly dynamic and heavily frequented part of the lagoon.

8.4 Conclusions

Considering the low-relief topography of Gascogny Lands which makes it difficult to have an overall view over the barrier island system and the increasing fragmentation of its landscapes in the last three centuries, the Arcachon Bay appears as one of the most unlikely "collage" of coastal landforms. However, the stratigraphic relationships of the various elements of the barrier system make it possible for a paleogeographic reconstruction of this outstanding and somehow unique combination of natural processes and land use along the French Atlantic coast. The description of these relations may be a valuable tool for popularization processes targeting the conservation of the lagoonal landscapes and the sustainable use of maritime public domain.

Acknowledgments The author thanks the University of Maryland (http://glcf.umiacs.umd.edu/data/glopem) for providing the Landsat ETM + data for analysis.

References

Allard J (2008) Enregistrements des changements environnementaux dans les sédiments littoraux: cas des Pertuis Charentais et du Bassin d'Arcachon. Ph.D. thesis, University of La Rochelle, 277 p

Allard J, Chaumillon E, Féniès H (2009) A synthesis of morphological evolutions and Holocene stratigraphy of a wave-dominated estuary: the Arcachon lagoon. SW France Cont Shelf Res 29(8):957–969

Auby I, Labourg PJ (1996) Seasonal dynamics of Zostera noltii Hornem in the Bay of Arcachon (France). J Sea Res 35(4):269–277

Bertrand F, Goeldner-Gianella L (2010) La vulnérabilité du littoral du bassin d'Arcachon à la submersion marine: le programme BARCASUB. In: Proceedings of the Journées « Impacts du changement climatique sur les risques côtiers », BRGM, Orléans, 15 et 16 Nov. 2010, pp 203–207

Bressolier-Bousquet C (1991) Geomorphological effects of land reclamation in the eighteen century at the mouth of the Leyre river, Arcachon Bay France. J Coast Res 7(1):113–126

Faugères JC, Cuignon R, Féniès H, Gayet J (1986) Caractères et facteurs d'un comblement littoral à l'Holocène Supérieur: passage d'un domaine estuarien à un domaine lagunaire (Bassin d' Arcachon, France). Bulletin de l'Institut de Géologie du Bassin d'Aquitaine 39:95–116

Féniès H, Lericolais G (2005) Internal architecture of an incised valley-fill on a wave- and tide-dominated coast (the Leyre incised valley, Bay of Biscay, France) Comptes Rendus. Geosciences 337(14):1257–1266

Florence C (2001) Long-term morphological modeling of a tidal inlet: the Arcachon Basin, France. Coast Eng 42(2):115–142

Froidefond JM, Legigan P (1985) La grande dune du Pilat et la progression des dunes sur le littoral aquitain. Bulletin de l'Institut de Géologie du Bassin d'Aquitaine 38:11

Ganthy F, Sottolochio A and Verney R (2010) Dynamique sédimentaire d'un estran lagunaire colonisé par un herbier de zostères (Bassin d'Arcachon, France) XIèmes Journées Nationales Génie Côtier – Génie Civil Les Sables d'Olonne, 22–25 juin 2010, pp 251–262

Lafon V, Froidefond JM, Lahet F, Castaing P (2002) SPOT shallow water bathymetry of a moderately turbid tidal inlet based on field measurements. Remote Sens Environ 81:136–148

Le Nindre YM, Bodéré G, Izac JL, Putot E, Levasseur J (2006) Etude pour le maintien de l'équilibre biosédimentaire des estrans de la côte sud du Bassin d'Arcachon. Rapport BRGM/RP 54814-FR. 129p + 4 annexes

Michel D, Howa HL (1997) Morphodynamic behaviour of a tidal inlet system in a mixed- energy enviroment. Phys Chem Earth 22:339–343

Plus M, Dumas F, Stanisière JY, Maurer D (2009) Hydrodynamic characterization of the Arcachon Bay, using model-derived descriptors. Cont Shelf Res 29(8):1008–1013

Sénéchal N, Sottolichio A, Bertrand F, Goeldner-Gianella L, Garlan T (2013) Observations of waves' impact on currents in a mixed-energy tidal inlet: Arcachon on the southern French Atlantic coast. J Coast Res SI 65:2053–2057

Tastet JP, Pontee N (1998) Morpho-chronology of coastal dunes in Medoc. A new interpretation of Holocene dunes in Southwestern France. Geomorphology 25:93–109

The Volcanic Chaîne des Puys: A Unique Collection of Simple and Compound Monogenetic Edifices

9

Pierre Boivin and Jean-Claude Thouret

Abstract

The volcanic range *Chaîne des Puys*, located in the French Massif central and last erupted 7,000 years ago, is potentially active. About 80 large monogenetic volcanoes form an alignment that parallels the Limagne fault, one of the major faults of the Western European Rift. The range shows a wide spectrum of volcano types: cinder cones, tuff rings, maars and domes, reflecting magmas from basalts to trachytes. Although some edifices and associated lava fields are known to be monogenetic, the simplicity of the features belies the complexity of compound cones often nested in tuff rings or maars, twin domes and maars overlapping cones. The volcanic range has been built up over 90,000 years, including two periods of heightened activity between 45,000 and 25,000 years and between 15,000 and 8,500 years. Volcanological research began as early as 1751 in the *Chaîne des Puys*, with major geological debates during the nineteenth century. A new impetus has been recently given to scientific research on the volcanic range. The *Chaîne des Puys* is renowned for its pedagogical interest and is visited by numerous students and teachers from France and abroad. Because the volcanic edifices have been used for grazing as far back as the Neolithic period and more intensively since the Roman Empire, they have been heavily deforested and eroded. The *Chaîne des Puys* became one of the French regional natural parks in 1977 and it has been protected ever since. Today a new incentive for protection is due to the candidacy of the volcanic range and the Limagne fault to the UNESCO World Heritage.

Keywords

Monogenetic volcanoes • Simple cone • Compound cone • Dome • Intrusion • Maar • Tuff ring • Lava field

9.1 Introduction

The monogenetic *Chaîne des Puys* shows an unrivalled diversity of volcanic landforms (Fig. 9.1). As young as 7,000 years and potentially active, this range is the most recent, albeit distant, aftermath of the formation of the Alps and Western European Rift. The central part of the range has an exceptional skyline encompassing 80 aligned edifices, which parallel the principal fault of the Limagne graben. After its scientific recognition during the eighteenth century, the monogenetic range has become a cultural symbol, transforming these 'green hills' into 'mountains that spit fire'. Since the Neolithic period, when it last erupted, the *Chaîne des Puys* has always been used by man, initially for crops and cattle as well as a source for building stones, sarcophagus and aggregate. The volcanic range also acts as an exceptional hydrological system, which supplies abundant drinking water to neighbouring towns and cities. Anthropogenic activities have deeply modified landscapes

P. Boivin (✉) • J.-C. Thouret
Clermont Université, Université Blaise Pascal,
CNRS UMR 6524, IRD, Laboratoire Magmas et Volcans,
BP 10448, F-63000, Clermont-Ferrand
e-mail: P.Boivin@opgc.univ-bpclermont.fr;
J.C.Thouret@opgc.univ-bpclermont.fr

M. Fort and M.-F. André (eds.), *Landscapes and Landforms of France*, World Geomorphological Landscapes,
DOI 10.1007/978-94-007-7022-5_9,

Fig. 9.1 Alignment of volcanic edifices forming the central part of the *Chaîne des Puys* (Credit: Jodie Way). From foreground to background, the aligned edifices are the Sarcoui dome (*1*) against the Strombolian cone of puy des Goules (*2*), the compound cone of puy Pariou (*3*) with a wide crater, the iconic Puy de Dôme (*4*) with its summit antenna, the nested Strombolian cone of puy de Côme (*5*) *on the right side* and in the distance the volcanic massif of Mont-Dore that was formed prior to the *Chaîne des Puys*

because historic episodes of enhanced erosion such as the early Middle Age have accelerated the geomorphic evolution of the range. In order to counteract the negative effects of human activities, the *Chaîne des Puys* has been protected as early as the eighteenth century through a process of property sharing, followed by reforestation and more recently by the creation of the regional natural park of the Auvergne volcanoes.

9.2 Geographical and Geological Setting

The *Chaîne des Puys* is the northernmost and youngest volcanic massif of the French Massif Central. The majority of monogenetic volcanoes are aligned on the north–south-trending granitic and metamorphic bedrock of the *plateau des Dômes*, which forms the western edge of the Limagne graben. The central part of the aligned *Chaîne des Puys* (Fig. 9.2) is about 30 km long and 3–4 km wide. Although the north–south range roughly parallels the Limagne fault on the western edge of the graben, volcanoes are located on NNE–SSW and sometimes NNW–SSE-trending segments. A few volcanoes located to the north (Gour de Tazenat, Puy de Chalard) and to the south (Puy de Monténard, Tartaret, Pavin group) extend the range beyond the central part.

The *Chaîne des Puys* has been formed by lavas that belong to an alkaline magmatic suite, which has been differentiated by fractionation and crystallisation (Villemant et al. 1981). The lavas, from basalts to trachytes, have provided a diversity of eruptive styles and positive and negative volcanic landforms (Boivin et al. 2009); as a result, scoria cones, domes, tuff rings and maars coexist in the range. If one considers neighbouring volcanoes around the *Chaîne des Puys*, the oldest edifice in the aligned maars of Enval–Volvic is 90,000 years old, while the maar of Clermont-Ferrand is even older (160,000 years).

Following these precursory volcanoes, the first paroxysmal activity apparently took place between 70,000 and 40,000 years. The corresponding tephra-fall deposits of basaltic and trachybasaltic compositions were identified by Vernet (1992) in the Clermont basin with 75 layers intercalated within the lacustrine filling of the Clermont maar.

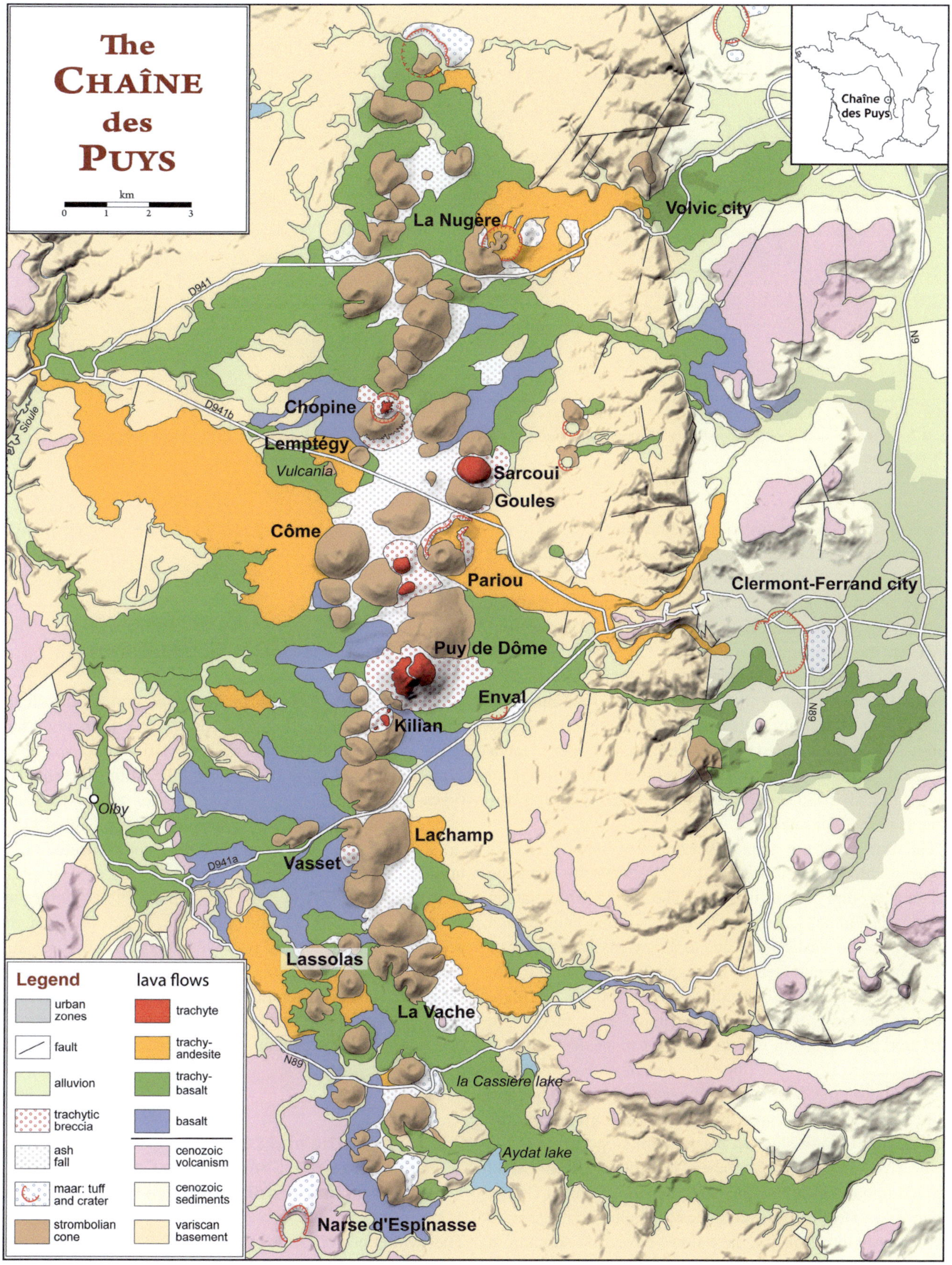

Fig. 9.2 Volcanological map of the *Chaîne des Puys* (Simplified after Boivin et al. 2009)

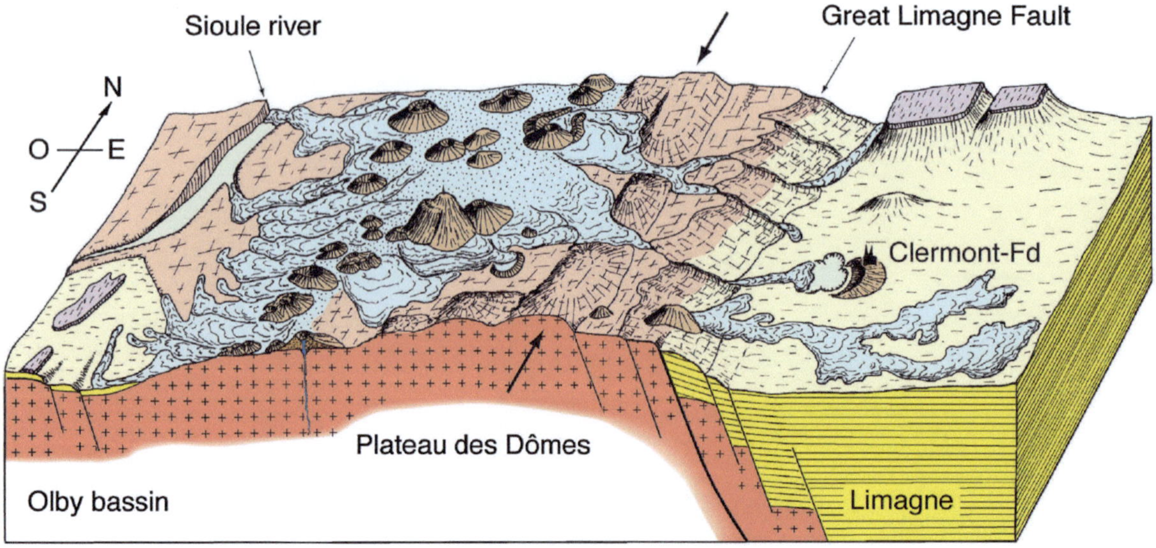

Fig. 9.3 Structural and geomorphic setting of the *Chaîne des Puys* (Modified after Boivin et al. (2009)). *Two arrows* indicate the axis of the plateau

This eruptive episode ends with the first output of trachyandesitic lavas (puys of Lemptégy 2 and Laschamp). Following a period of relative quiescence, eruptions became more explosive around 15 ka due to the production of evolved magmas of trachyandesitic and trachytic composition. However, the most recent eruption in the central part took place on the cones of la Nugère and Lassolas, with trachybasaltic lava flows, while the last eruption formed the Lake Pavin maar to the south with trachyandesitic tephras. A few tephras without identified sources suggest that more recent but small eruptions may have occurred between 6 and 4 ka (Juvigné et al. 1986; Fourmont et al. 2006), although this remains to be confirmed.

The altitude of the granitic basement of the *plateau des Dômes* is three times higher, at 1,023 m, than the Limagne graben at around 300 m (Fig. 9.3). The 15-km-wide plateau shows an asymmetric profile, the eastern slope being much steeper than the western side due to the extensive removal of sediment from the graben of Limagne. The axis of the plateau consists of north–south aligned hills 100–300 m high, about 4 km west of the Limagne plain. As a result, the narrower, high-gradient valleys have cut more deeply down into the eastern slope. The line of *Chaîne des Puys* volcanoes has been built up 2–3 km west of the plateau axis and 6–7 km of the Limagne fault. Lava flows have been guided by the pre-volcanic topography. On the western slope towards the Sioule river and the Olby graben, lava flows have formed large compound lava flow fields such as the *cheire* of Côme, which have extensively covered the basement. On the eastern side towards the Limagne, lava flows have first spread out between the cones and axis of the plateau before being channelled in narrow valleys as far as the Limagne graben, where larger valleys and widespread fans have enabled flows to spread out again.

9.3 Edifices: Diversity of Structures and Constructional Landforms

The *Chaîne des Puys* is made up of about 80 volcanic edifices readily identified on a topographic map, but a thorough geological analysis can add another 20 volcanoes. The interest and fame of this range lie in the fact that four types of so-called monogenetic edifices are concentrated in the same area, which is arguably unique worldwide: (a) simple scoria cones and (b) compound or nested cones, built up above tuff rings or maars (77 %); (c) simple or abutted domes, lava domes, low domes, Pelean domes and upheaved plugs (8 %); and (d) maars, which often encompass Strombolian cones or have been excavated on cone flanks (15 %). In fact, a majority of the volcanoes of the range show two or even three nested or superimposed structures: despite the monogenetic field type, most cones of the *Chaîne des Puys* are compound. Three iconic volcanoes are described below.

9.3.1 La Vache and Lassolas: Horseshoe-Shaped Twin Cones and Their Associated Lava Flow

The twin cones of La Vache and Lassolas are iconic volcanoes of the Chaîne des Puys (Fig. 9.4a). Their asymmetric, open-mouth shape makes the twin cones rare and famous.

Fig. 9.4 Three iconic volcanoes of the *Chaîne des Puys*. (**a**) The twin cones of puy de la Vache (2) and Lassolas (1) open to the SSE (Credit: Pierre Boivin). The c.8,500-year-old 'cheire of Aydat' lava flow has created the horseshoe-shape of the youngest cones of the central part of the range. (**b**) Nested and superimposed edifice of puy Pariou. *1* Early tuff ring, *2* Part of tuff ring (*arrow*) carried away by the lava-lake break-out producing a lava flow, *3* Lava flow of trachyandesitic composition, *4* Final, nested and superimposed Strombolian cone (Credit: Pierre Boivin). (**c**) The Pelean dome of Puy de Dôme, landmark of the Chaîne des Puys, as well as the temple devoted to Mercury on the SE summit (Credit: Pierre Boivin)

The asymmetry is due to a 16-km-long *cheire d'Aydat* lava flow that has carried almost half of the scoria cone away from the vent area. The twin cones, which are the youngest volcanoes of the central part of the *Chaîne des Puys*, encompass three units: two eruptive centres, tephra-fall deposits and one lava flow. Landforms are still well preserved today in the absence of run-off, but sheep grazing in recent times and visitors today became an erosional threat to the landscape.

The eruption of the twin cones has produced a trachybasaltic ash- and pumice-fall deposit, thicker and more widespread than any other Strombolian scoria fallout from the *Chaîne des Puys*. Shortly after the pumice-fall eruption, lava flows began erupting simultaneously at the puys de la Vache and Lassolas. They merged to form a single lava flow, which has been channelled towards the Limagne graben. The lava flow has blocked two tributaries, hence damming several lakes in the process. Lakes have been filled and dried up except for the Cassière and Aydat lakes. The eruption of puys de la Vache and Lassolas was the first in the *Chaîne des Puys* to be dated using the ^{14}C method. The charcoal, which has been dated, was burnt by the lava flow (Pelletier et al. 1959), and a large number of additional ^{14}C datings point to 8,300 ± 130 years.

9.3.2 The Compound and Nested Edifice of Puy Pariou

The apparent simplicity of the puy Pariou cone associated with a large lava flow (Fig. 9.4b) belies its complex structure and history. Its basis is a wide, sub-circular tuff ring structure of trachytic composition. Following the trachytic episode, a trachyandesitic magma filled the ring tuff. The northern part of the structure failed when a lava flow broke out from the lava lake and was channelled 8 km down in a narrow valley to the outskirts of Clermont-Ferrand. The imposing mass of the Strombolian cone, with a 300-m-wide and 93-m-deep crater, overlies the hidden structure. Thick ashfall deposits have mantled the area as far as the graben of Clermont-Ferrand. Although the Pariou's tephra deposits are always observed near the surface, they have been locally overlain by surge deposits of trachytic composition that stem from the Chopine extrusion followed by the Kilian vent. The well-dated Chopine and Kilian tephras suggest that Pariou eruptions occurred about 10,000 years ago.

9.3.3 Puy de Dôme: Two Abutted Domes with a Complex Internal Structure

The Puy de Dôme is a pivotal landmark of the *Chaîne des Puys*. Located in the centre and 500 m high, the characteristic bell-shaped landform overlooks the range without any rival (Fig. 9.4c). Several Strombolian cones have been buried under the dome mass. The Puy de Dôme that results from the accumulation of viscous trachytic lavas around the former vent shows two distinct geomorphic units: (1) looking east, the rocky mass with a flat summit and very steep flanks shows the typical landform of a large dome cluster and (2) looking west, the dome morphology mimics that of a smooth, regular cone with an average 33° slope and its highest top coinciding with the summit. The contact between 1 and 2 is underlined by a semi-circular depression on the summit plateau and by rocky cliffs that are cut to the north and south. Deep trenches excavated during the recent construction of the railway near the summit have displayed whitish or multicolor, loose, sandy deposit of ash and blocks, which has been weathered by fumarolic fluids. Ongoing research using muon radiography, electrical resistivity and microgravimetry suggests that this deposit is found in the bulk of the dome. The Puy de Dôme is 11,040 ± 230 years old. The edifice was later mantled by tephra from a late phreatic eruption (10,700 ± 100 years: Miallier et al. 2010) and by a several-metre-thick tephra deposit produced by an eruption of Kilian crater, at the south foot of Puy de Dôme (c.9,430 ± 120 years).

9.4 Space and Time Relationships in Growth and Degradation Episodes

The *Chaîne des Puys* may be considered as a single, temporal and spatial volcanic system. Figure 9.5a shows a symmetrical spatial distribution of lava compositions with respect to the range centre. The magma storage system has concentrated with more evolved magmas towards the central part of the range during the magma differentiation. The age of eruptions, as shown in Fig. 9.5b, suggests that lava compositions also follow spatial and temporal patterns. The eruptions younger than 15,000 years are found around the central range. The most recent eruptions with the most-evolved magmas have also been the most explosive events, which produced widespread pyroclastic-surge deposits (around Chopine, Kilian, Vasset and Pariou). The surge-producing eruptions have had repeated destructive impacts on landscapes and on food resources during the Epipalaeolithic and Mesolithic periods (Vivent and Vernet 2001).

The morphometry of the cones of the *Chaîne des Puys* edifices has been analysed using a DEM and the numerical 'MORVOLC' code (Shields 2010). The morphometric analysis displays six groups of cones: steep, smooth or breached cones; smooth, elongated or irregular, intermediate cones; and flat cones. This geomorphic variety also contributes to recognising the complex characteristics of the eruptions that formed these edifices. The discrepancy with classical monogenetic characteristic is due to vent migration, the number of eruptive stages and superimposed or nested edifice growth.

9.5 Posteruption and Present-Day Evolution

9.5.1 Erosional Processes and Evolution of the Slope System

The north–south alignment of the *Chaîne des Puys* volcanoes acts as a barrier facing oceanic air masses, which enhances the rainfall amount to as much as 1.2–1.3 m/year on the summit. This humid setting has paradoxically not given birth to any significant surficial river, owing to the extreme porosity of the volcanic material. However, the infiltration capacity can be exceeded when very intense rainfall occurs: run-off and gullying, which has been favoured by deforestation in historic times and has triggered spectacular erosion marks on a few volcanoes. The French meteorological office recorded rainfall intensity as high as 87 mm/m^2 during 1 h on the 26 June 2012 thunderstorm, which triggered mudflows and damaged the track of the new 'panoramic train of the domes'. Another documented example is provided by the puy Chopine, a trachytic, upheaved plug, which was carefully drawn by Laizer in 1808 (Fig. 9.6). The author has reported and dated the deep gullies that incised the puy during 1760 to the1800s. This observation confirms the effects of the intense deforestation in Auvergne during the eighteenth century (Legrand d'Aussy 1795).

The summit of the Puy de Dôme also displays the secondary effects of the human impact since the first century AD (Fig. 9.7). The tephra deposit that mantles the summit, which has been produced by the neighbouring Kilian crater, has been removed from nowadays bare flanks, while recent railway works carried out with heavy machinery have destabilised the road banks. Several platforms created below the summit have also deeply reworked the dome summit. During the twenty-first century, the forest is expanding again on all cones, which reduces the slope instability compared to that of previous deforestation periods.

9.5.2 Evolution of Anthropogenic Landscapes

Following its birth, the landscapes of the *Chaîne des Puys* initially experienced climatic changes that characterised the sharp transition from the glacial Würm period to the milder Holocene. Since 6,500 years, the human impact (Michelin 1996; cf. Fig. 9.8) has slowly become important and most significant about 2,700 years ago; large open meadows were created during the second Iron Age. The Roman conquest resulted in more dense housing and restructuration of landscape; the mid-altitude *Chaîne des Puys* yielded water, stone,

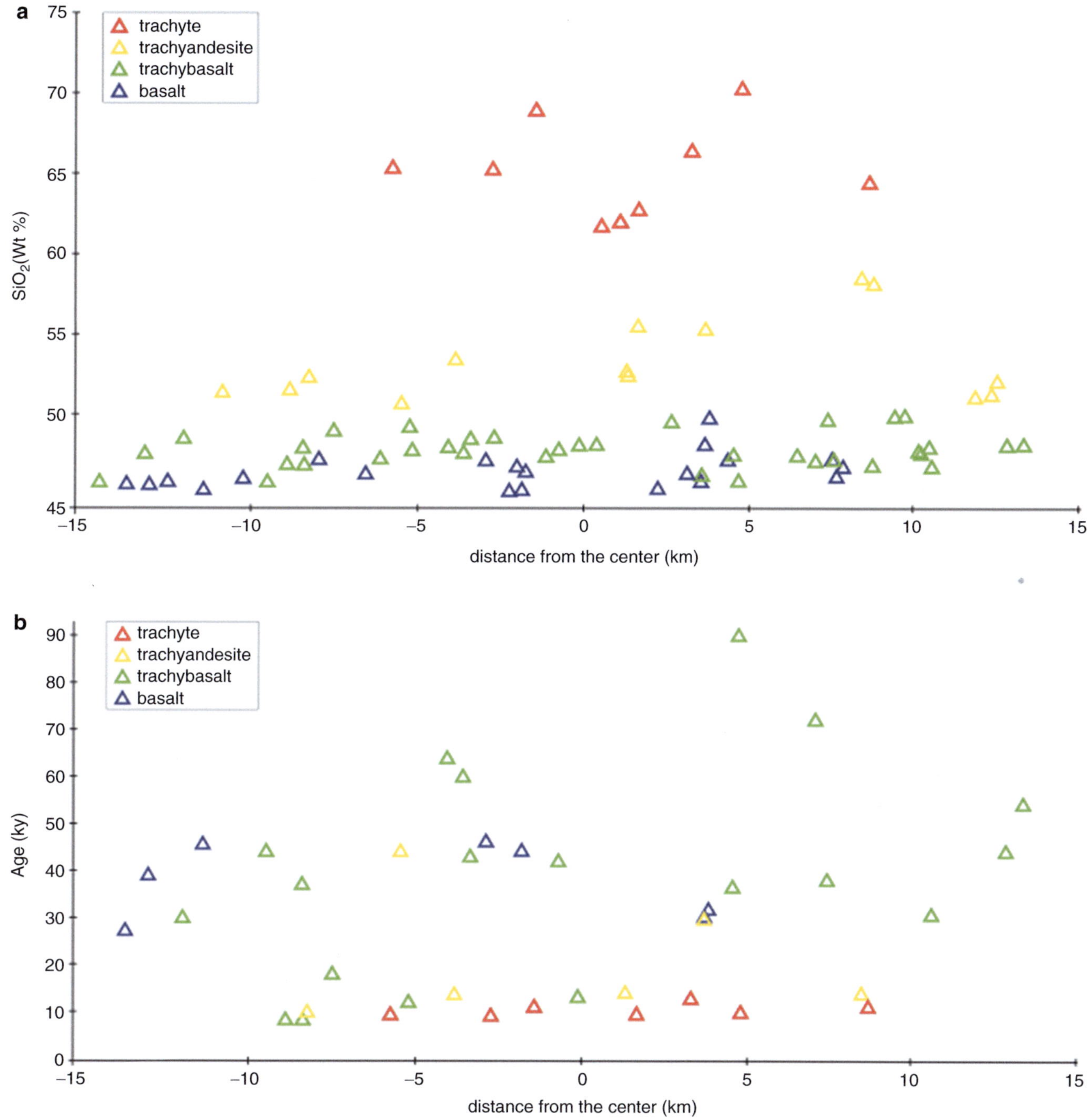

Fig. 9.5 Diagram showing the relationship between the distribution of volcanoes along the range and the SiO_2 content of their lavas (**a**) and the composition and age of their erupted lava (**b**). The centre of the range is about 1 km north of the Puy de Dôme

wood and grazing meadows for people living in the plain and cities. During the early Middle Age, the earlier organisation slowly disappeared and was replaced by the organisation led by churches and castles during the tenth century. During the twelfth century, land was collectively given to groups of villagers to form large properties, which become undividable, free from taxes and persist until today. During the fourteenth century, the forest, which had been previously overused due to the increasing population, was rapidly abandoned as a consequence of two large catastrophes: the Black Plague and the 100-year War, which would be later followed by the Little Ice Age. Starting with the seventeenth century, the population increased again, while the forest receded. Finally, around 1860–1870, the *Chaîne des Puys* was entirely cleared

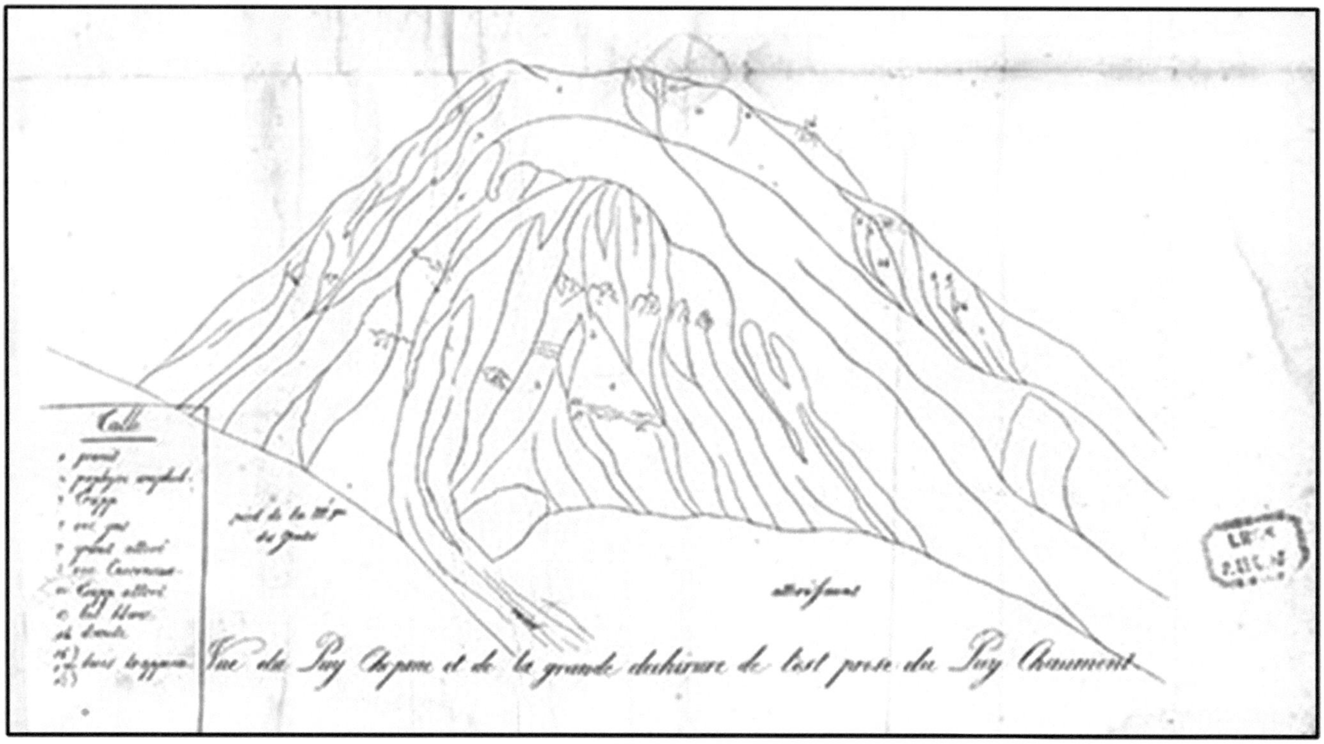

Fig. 9.6 Gully pattern created by erosion as observed by Laizer (1808) on puy Chopine during the early eighteenth century

Fig. 9.7 The Puy de Dôme. (**a**) Shaded relief map, based on high resolution (50 cm) LiDARverne (2011) showing the cumulo-dome of Puy de Dôme (Credit: CRAIG) (**a**). *1* Trail of possible Gallo-Roman age that provided access to the temple of Mercury, *2* Area that was eroded after the fourth century, *3* Small landslides on road slopes that were triggered when the access road was built towards the summit sometime around 1873, *4* Summit of Puy de Dôme, *5* Ruins of the temple of Mercury. (**b**) The 'treasure window' in the temple of Mercury (Credit: Pierre Boivin)

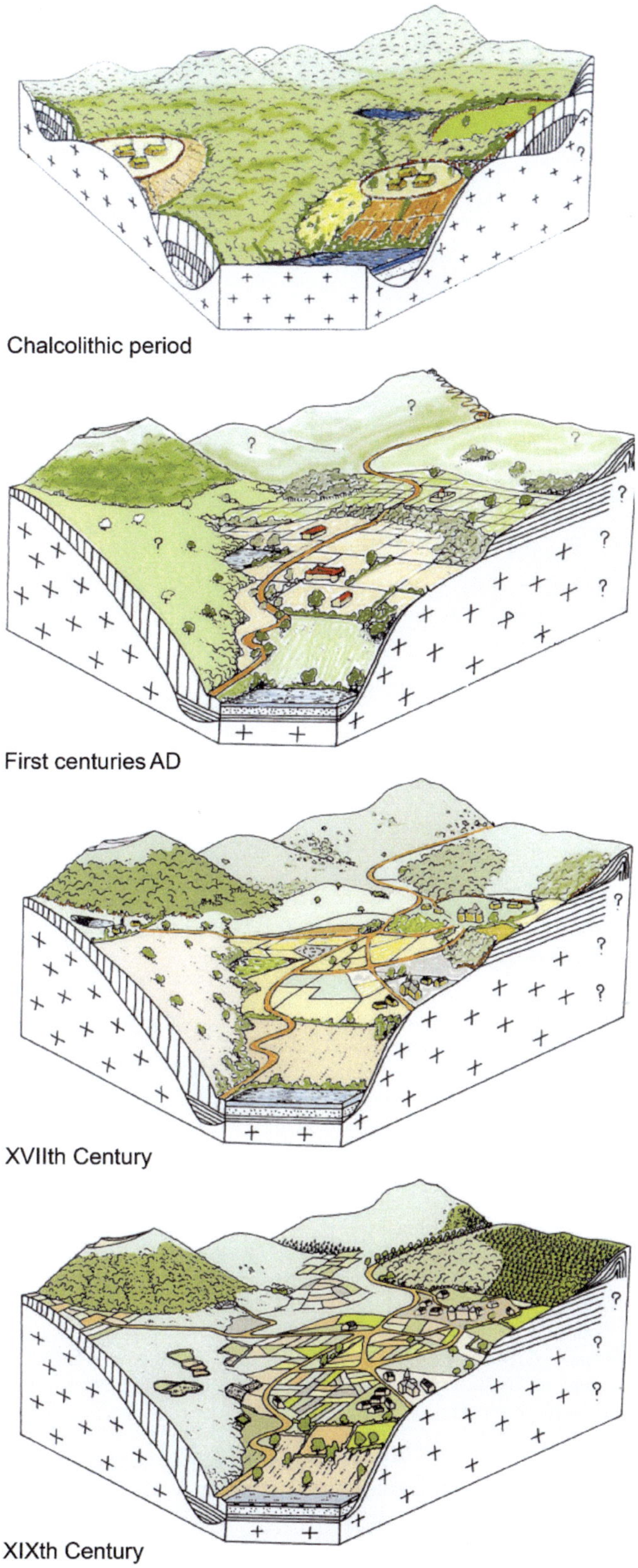

Fig. 9.8 Evolution of the land use in the *Chaîne des Puys* over the past 4,000 years (Modified after Michelin 1996)

and its meadows were overused. However, villages have maintained their population because the developing industry in nearby cities allowed active rural workers to work both at the factory and in the countryside. In the 1960s and nowadays, houses have expanded on the edges of the *Chaîne des Puys*. Construction is increasing, while agriculture is abandoned but tree plantations are implemented in many areas to yield new income.

9.6 Cultural Value and Tourism Attractiveness

The scientific significance of the *Chaîne des Puys* was discovered by J. É. Guettard in 1751. This historic recognition led to the development of European volcanology. Easy access and readily understandable landscapes explain why the range quickly became a visiting goal for all European geologists (Desmarest, von Buch, Poulett-Scrope, Dolomieu, de Saussure) and hosted numerous debates and controversies (neptunism–plutonism, 'uplifted' craters, volcanoes without craters, etc.). This fame persisted during the twentieth century through the detailed mapping of volcanoes and deposits carried out by Glangeaud (1913) and Camus et al. (1975). Today, scientific research is very active and considers the *Chaîne des Puys* like a laboratory volcano.

The position of Puy de Dôme in the centre of the range, and its size and isolation, has transformed it into a remarkable landmark, as well as a suitable geomorphic site for the building of the Mercury temple (Fig. 9.7b; cf. supra) owing to its visibility for 40 km around the edifice. This volcano has continuously illustrated the city of Clermont-Ferrand, in particular through advertisements, which have increased since the nineteenth century, while it has become the guardian spirit of the city. The construction of a 73-m-high tower for television broadcasting in 1956 reinforced the qualities of this landmark.

The *Chaîne des Puys* was a tool for developing volcanology as early as the eighteenth century and has a high educational value because its landforms are readily visible, from the scale of the edifices to that of small landforms on top of the lava flows. This visibility is endangered by the increasing growth of the forest. Operations are now undertaken by the regional natural park in order to stop the phenomenon: exploitation of forests and planned reforestation and increasing sheep grazing. Quarries that opened in the 1950s have been closed. In a few cases, quarries like Lemptégy and Volvic have remained open to public and schools, adding a very important pedagogical value. The Auvergne Region has also created Vulcania, an original recreational park that focuses on volcanism and Earth geodynamics in general.

Fig. 9.9 Tephra succession in the quarry of puy de Lemptégy. *Black layer*: scoria-rich, proximal tephra-fall deposit from Lemptégy 2 and *Whitish, coarse layer*: pumice-rich pyroclastic-flow deposit from Chopine protrusion (Credit: Pierre Boivin)

9.7 Conclusion

The *Chaîne des Puys* is a unique collection of volcanic landforms whose interest extends beyond the French and European setting. Its touristic and cultural appeal stems both from its aesthetic landscapes and scientific significance. The interest of the range has been enhanced by the fact that old quarries have been redesigned to welcome an increasing number of tourists. One can observe and understand, from the vivid colours of the quarry walls (Fig. 9.9), some elementary principles of geology, such as superposition and cross-cutting, and the diversity of volcanic products (lavas, bombs, scoriae and ash).

The *Chaîne des Puys* has been protected through a set of regulations in order to ensure a touristic development preserving the environment. The regional natural park of the Auvergne volcanoes has been established in 1977. The label 'Great Site of France' was bestowed to the Puy de Dôme in 2008. These protection regulations, which have applied to this outstanding site, contribute to support the candidacy of 'the Chaîne des Puys and the Limagne fault' to be inscribed on the UNESCO World Heritage list.

References

Boivin P, Besson J-C, Briot D, Camus G, Goër de Herve de A, Gourgaud A, Labazuy P, Langlois E, de Larouzière F-D, Livet M, Mergoil J, Miallier D, Morel J-M, Vernet G, Vincent PM (2009) Volcanology of the Chaîne des Puys. Parc Naturel Régional des Volcans d'Auvergne, Aydat, p 196 (with map)

Camus G, GoërdeHervede A, Kieffer G, Mergoil J, Vincent PM (1975) Volcanologie de la Chaîne des Puys. Parc Naturel Régional des Volcans d'Auvergne, Aydat, p 112 (avec carte)

de Laizer L (1808) Lettre à M. le professeur Jurine sur le puy Chopine, l'une des montagnes volcanisées qui forment la chaîne du Puy-de-Dôme. Imp. de Landriot, Clermont-Ferrand, 62 p. -2 plans h. t. -8°. Cote A 35568; fond Auvergne 2. Bibliothèque municipale et universitaire, Clermont-Ferrand

Fourmont A, Macaire JJ, Breheret JG, Argant J, Prat B, Vemet G (2006) Tephras in lacustrine sediments of the Sarliève marsh (French Massif Central): age and preservation. CR Geosci 338(16): 1141–1149

Glangeaud P (1913) Les régions volcaniques du Puy de-Dôme II – La Chaîne des Puys et la Petite Chaîne des Puys. Bull Serv Carte Géol France 135(XXII):241–496

Juvigné E, Bastin B, Gewelt M (1986) Découverte de retombées volcaniques d'âge holocène dans la Chaîne des Puys septentrionale (Massif Central, France). Revue des Sciences Naturelles d'Auvergne, Clermont-Ferrand 52:33–36

LiDARverne (2011) Modèle numérique de terrain au pas de 50 cm/pixel de la Chaîne des Puys. Collaboration LMV-OPGC, LMC, MSH, UBP, Augustonemetum. Coordination Centre Régional Auvergnat de l'Information Géographique (CRAIG). Scientific control P. Labazuy, LMV-OPGC

Miallier D, Boivin P, Deniel C, Gourgaud A, Lanos P, Sforna M, Pilleyre T (2010) The ultimate summit eruption of Puy de Dôme volcano (Chaîne des Puys, French Massif Central) about 10,700 years ago. CR Geosci 342:847–854

Michelin Y (1996) Les jardins de Vulcain, paysages d'hier, d'aujourd'hui et de demain dans la Chaîne des Puys du Massif Central français. Maison des sciences de l'Homme, Paris, p 155

Pelletier H, Delibrias G, Labeyrie J, Perquis M-T, Rudel A (1959) Mesure de l'âge de l'une des coulées volcaniques issues du puy de la Vache (Puy-de-Dôme) par la méthode du Carbone 14. CR Acad Sci Paris 214:2221

Legrand d'Aussy PJB (1795) Voyage en Auvergne. E. Onfroy, Paris, 6 vol, 265, 272, 210, 236, 243 and 263 p

Shields J (2010) The morphometry of the Chaîne des Puys. M2R Report, Laboratoire Magmas et Volcans, Université Blaise Pascal, Clermont-Ferrand, 42 p

Vernet G (1992) Message du volcanisme régional dans les formations quaternaires de Limagne occidentale (Massif Central Français). Minéraux denses et retombées. Doctorat d'Université. Université de Bordeaux I, Bordeaux, 342 p

Villemant B, Joron J-L, Jaffrezic H, Treuil M (1981) Distribution coefficients of major and trace elements: fractional crystallization in the alkali basalt series of Chaîne des Puys (Massif Central, France). Geochim Cosmochim Acta 45:1997–2016

Vivent D, Vernet G (2001) Apport de la palynologie à l'étude des retombées volcaniques tardiglaciaires et holocènes de la Chaîne des Puys enregistrées en Limagne (France). In: Juvigné E, Raynal J-P (eds) Tephras, chronologie archéologie, Coll 'Les dossiers de l'Archéo-Logis'. Edt Archéo-Logis/CDERAD, Goudet, pp 255–262

10 Velay and Mézenc Massif: Harsh Volcanic Highlands Around the Loire River Source

Emmanuelle Defive and Alexandre Poiraud

Abstract

The Velay and its Vivarais rim combine many historical aspects of the Massif Central, from its Hercynian basement to the Tertiary tectonic and volcanic reactivations following the Pyrenean and Alpine orogeneses, from the Mesozoic and Tertiary planations to the Plio-Quaternary excavation of the valleys. While some of the landforms are nearly intact, erosion has played a big part in their shaping at all scales, interacting with the entrenchment of the hydrographic network, volcanism and local tectonics. As a result, a great diversity of environments and landscapes emerged, which also comprises traces of human activity, man having settled early in these beautiful, harsh highlands, which are famous for their quality pasture lands.

Keywords

Volcanism • Le Puy Tertiary graben • Blockstream • Upper Loire catchment • Velay • Mézenc • High Ardèche • Devès plateau

10.1 Introduction

The Velay with its Vivarais promontories is a historical region located in the south-east of the Massif Central, at the latitude of Valence (45°N). During the *Ancien Régime*, i.e. before the French Revolution, it belonged to the former province of Languedoc, which was more oriented towards the Rhône valley than the present province of Auvergne. The geological and geomorphological coherence of the Velay and Vivarais complex originates from its shared basement and its tectonic and volcanic history, differentiated into the Tertiary basin of the Puy-en-Velay and the two volcanic horsts enclosing it: the Devès in the west and the Eastern Velay in the east. At the centre of the latter, which is dominated by Mont Mézenc (with 1,754 m the fourth highest summit of the Massif Central), the source of the Loire River is located at the foot of Mont Gerbier-de-Jonc, close to the drainage divide between the Loire side flowing to the Atlantic and the Rhône side flowing to the Mediterranean. Also, irrespective of the major topographic contrast separating the two sides, the geographic entities and landscapes (Fig. 10.1) are multifarious in their specific geological (Mergoil and Boivin 1993), geomorphological and climatic features but also in the modalities of their economic development and certain predominant cultural traits.

E. Defive (✉)
Département de géographie, Université Blaise Pascal, 29 Bd Gergovia, 63000 Clermont-Ferrand, France

GEOLAB, UMR 6042-CNRS/Université Blaise Pascal, Maison des Sciences de l'Homme, 4 rue Ledru, F-63057 Clermont-Ferrand, France
e-mail: emmanuelle.defive@orange.fr; emmanuelle.defive@univ-bpclermont.fr

A. Poiraud
GEOLAB, UMR 6042-CNRS/Université Blaise Pascal, Maison des Sciences de l'Homme, 4 rue Ledru, F-63057 Clermont-Ferrand, France

Bureau d'étude "Inselberg", 261 allée des Bauges, 73000 Chambéry, France
e-mail: apoiraud@yahoo.fr

M. Fort and M.-F. André (eds.), *Landscapes and Landforms of France*, World Geomorphological Landscapes,
DOI 10.1007/978-94-007-7022-5_10, © Springer Science+Business Media Dordrecht 2014

Fig. 10.1 Panoramic view of the Eastern Velay from the Puy basin. The reliefs dominating the volcanic plateau (*left:* Meygal, *centre:* Mézenc-Gerbier complex) present phonolitic domes. In the Puy basin, the *Rocher* of Polignac is a neck of volcanic breccia formed by the Surtseyan Plio-Pleistocene volcanism of the Devès. It was stripped of the Cenozoic sedimentary terrains due to the entrenchment of the Loire and its tributaries (Credit: E. Defive)

10.2 Geological and Geomorphological Setting and Evolution

10.2.1 Hercynian Crystalline Basement Dislocated in the Early Tertiary

The regional substratum consists of the granites and migmatites of the Velay (Fig. 10.2). This vast geological unit belongs to the roots of the Hercynian chain whose mountain reliefs covered all of Europe about 350 Ma, before erosion completely levelled it during the Palaeozoic-Mesozoic transition. The great axes of Hercynian fracturing (NW-SE and NE-SW) had a strong influence on the later tectonic history and on the orientation of the streams. In the Velay, as in the whole Western European plate, the Tertiary represents an important phase of tectonic reactivation, which was triggered by the Pyrenean and subsequently the Alpine orogenesis within the context of the Africa-Europe collision. The grabens of Puy-en-Velay and Emblavès thus developed by playback of the Hercynian faults. Their formation (Eocene and early Oligocene) was accompanied by the creation of a thick, essentially detrital sediment filling generated by partial erosion of the detritus covers on their rims. Hence these major tectonic structures were probably little expressed in the morphology before the tectonic reactivation of the late Tertiary and the Plio-Quaternary entrenchment of the hydrographic network. In the last 15 Ma, the Alpine thrust triggered the uplift of the eastern rim of the Massif Central by over a thousand metres and the reactivation of the local tectonic compartment. Only then did the horsts of the Eastern Velay in the east and the Devès in the west as well as the depression of the Puy basin develop. The excavation of the latter was limited by direct volcanic effects: damming of the valleys, lava intrusions and flows reinforcing the hillslopes and constraining their retreat.

10.2.2 An Extensive and Varied Neogene Volcanism, Coeval with the Alpine Thrust

Varied and extensive, the volcanism of the Velay and the Massif Central forms part of the system of Tertiary volcanic rifts at the rim of the Alpine arc.

Located east of the Loire River, the Eastern Velay, which overthrusts the Loire and Rhône sides, is the oldest volcanic region (13.5–5.7 Ma) of the discussed area. Its activity started with the emission of differentiated lava flows (trachytes and phonolites) in the north (Meygal, Emblavès, Yssingelais). It continued (11–8 Ma) with abundant lava flows of varied fluidities and progressive differentiation (from basaltic to trachytic), which form the basement of Mont Mézenc at the heart of the massif. These lava flows are followed by new trachytic and phonolitic occurrences, which mark the centre and the south of this volcanic region (8–5.7 Ma).

To the west, the crystalline horst separating the Loire and the Allier Rivers is covered by basaltic lava flows which poured out in abundance, essentially between 3 and 1 Ma, forming the Devès, the largest basaltic plateau in continental Europe. The principal eruptive crest, which holds the summit of the Devès, is aligned with a big Hercynian fault SSE-NNW (Villefort fault). Many Strombolian cones and other maars (explosion craters formed by clash of water and

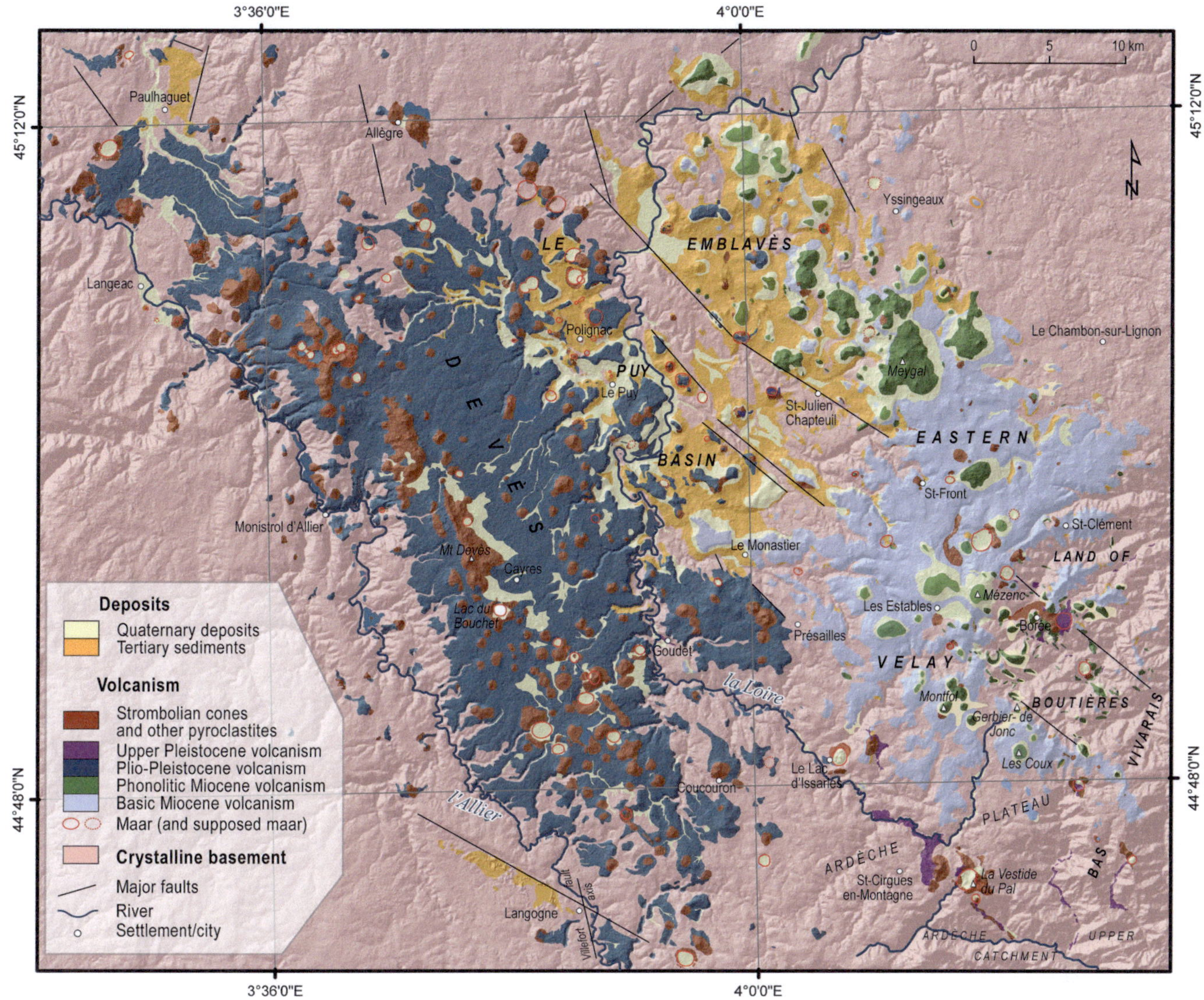

Fig. 10.2 The main geographic, geological and geomorphological entities of the Velay and its Vivarais promontories (Modified after Mergoil and Boivin 1993)

magma: phreatomagmatism) were progressively buried in the successive lava flows, and only the most recent ones are still visible, marking the plateau surface. In the Puy basin, which was being excavated at the time, the lacustrine complex generated by the volcanic dams caused a Surtseyan activity that formed the hyaloclastite breccias of the *Rochers* (rocks) of the Puy.

The Lower Vivarais volcanism formed a third complex with dispersed volcanoes which are superimposed on the basement or the older volcanic complexes at the SE rims of the Velay, on the Loire and the Rhône sides. Partly coeval with the volcanism of the Chain of Puys (approx. 330,000–40,000 years ago) but exclusively basaltic, this complex is aligned in three parallel, NNW/SSE axes (Rochette et al. 1993; Guérin and Gillot 2007). Nearly all eruptions started with a phreatomagmatic phase, often followed by a more classic Strombolian activity (Fig. 10.3).

10.2.3 Interactions with Climatic Evolution and Stream Entrenchment

The diachronic character of the Velay volcanism provides some features which facilitate an understanding of the stages and characteristics of its evolution, which it also complicates (Defive 1996; Defive et al. 2005). Thus, the little differentiated topography and the varied deposits beneath the lava flows of the Eastern Velay indicate that rather hot ambiences still dominated the late Miocene and that planation dynamics continued to prevail during the evolution of a relief that was

Fig. 10.3 Tuff ring at the southern flank of the maar of Borée-Echamps in the Boutières (151,000–181,000 Ma), presenting beautiful bomb sag or antidune structures. The cemented basaltic lapilli of the summit of the cross section mark the passage from phreatomagmatism to a classic Strombolian phase of activity (Credit: E. Defive)

doubtlessly not yet very uplifted. The entrenchment only emerged during the last 4 Ma, first as dale profiles, then, later than 2.5–3 Ma, as veritable gorges (Fig. 10.4). At the height of the Devès and the Puy basin, they are concealed by lava flows that have shifted the Loire further east, where the current gorge is less than a million years old. There are two combined reasons to this swift change during the Plio-Pleistocene transition from the former plains and dales to gorges morphology, with incisions dominating. It is above all a consequence of the maximal phase of regional elevation, caused by the uplift of the eastern rim of the Massif Central, thus creating a potential for entrenchment. This entrenchment was enhanced by the changeover from the former warm to the cold, Pleistocene climates during the Plio-Pleistocene transition, which modified the conditions of erosion, favouring incision. The sporadic occurrence of recent lava flows (late Devès and particularly Lower Vivarais volcanism) fitted onto the gorge flanks and sometimes onto the current thalweg, on both the Loire and the Rhône sides, shows the persistence of the volcanic disturbances that accompanied this evolution.

10.3 Landforms: Asymmetry of Complex and Diversity of Details

10.3.1 A Major Topographic Contrast Between the Rhône and the Loire Side

To the east and south of the Loire River, the Vivarais-Velay confines display a major topographic and scenic contrast between the Loire and the Rhône side, which is due to a markedly uneven dissection caused by the uplift of the eastern rim of the Massif Central.

On the Rhône side, the distinct altitude difference (1,000–1,500 m within 50 km linear distance) has favoured a vigorous entrenchment of the hydrographic network. Scarped massifs, scree, boulder fields and towerlike rock pinnacles mark the slopes, whereas the thalwegs are filled with rapids, cascades, potholes and other canyon-like features. The most impressive valleys are those where streams are cut into the counter-dips of the migmatites on the southern rim of the Ardèche plateau. The recent Lower Vivarais volcanism, the only one occurring here, has left but a discreet imprint: some sporadic structures

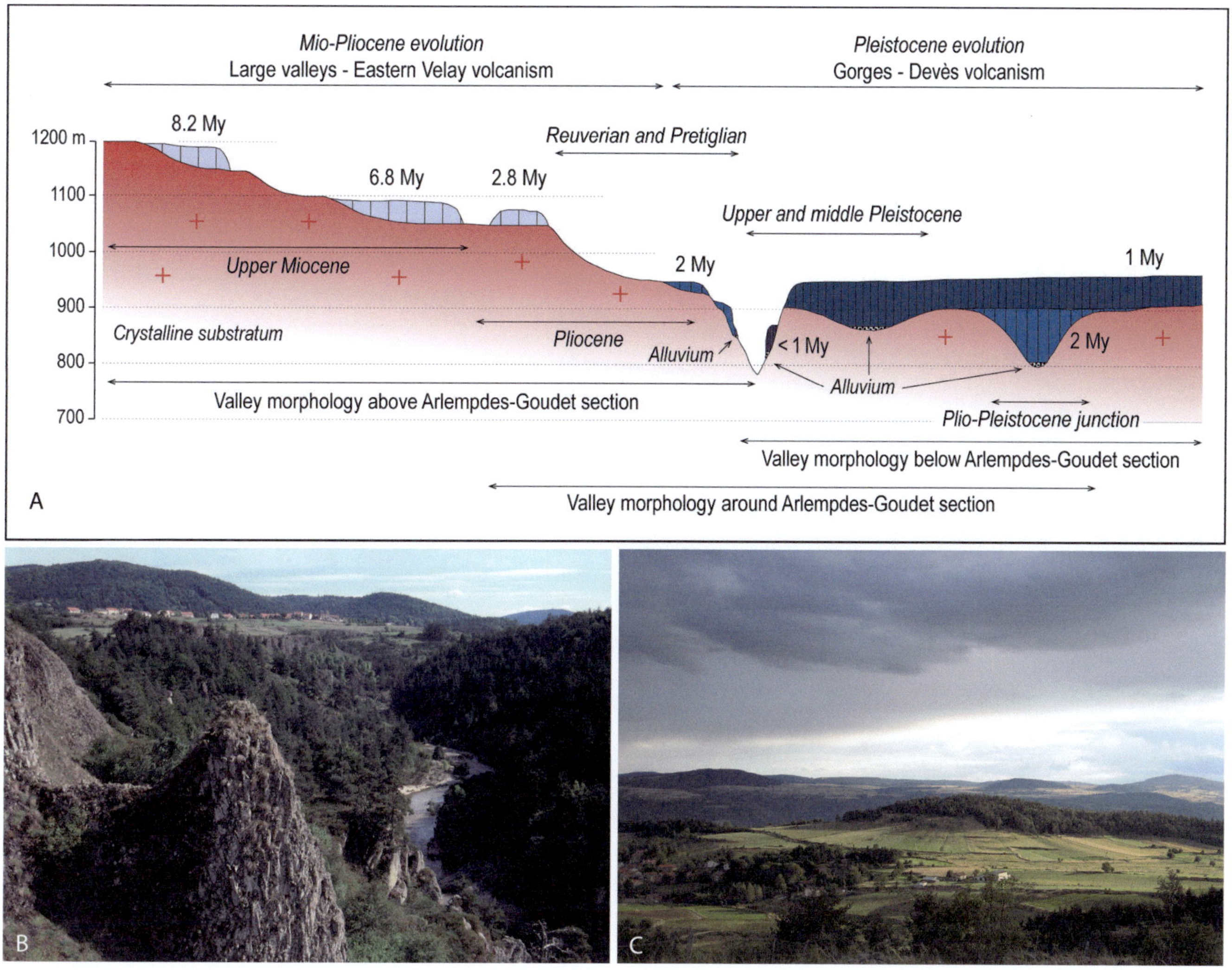

Fig. 10.4 Loire valley morphological history and volcanism. (**A**) Synthetic cross section. *Left:* Upstream from the area of maximum Devès extension, the Eastern Velay lava flows indicate the first stages of entrenchment; the lava flows coeval with the Devès mark the passage from the Middle Pliocene dales to the Plio-Pleistocene gorge. *Right:* Further downstream, in the maximum Devès extension area, the lava flows completely cover the Plio-Pleistocene gorge, shifting the Loire River further east where the current gorge, marked by late outpourings, was formed during the last million years. (**B**) Devès lava flows in upstream Loire gorge, at the level of the village Issarlès. (**C**) Well recognisable by their woody summit and their radial agricultural parcels, the *gardes* of the Devès are Strombolian cones worn by 1–2 Ma of erosion, resulting in their gentle slopes (Credit: E. Defive)

(cones and maars), which cause very local perturbations in the crest profile, and flats formed by lava flows on the valley floor. In the Boutières, where the entrenchment is a bit less distinct, the profile of crests and slopes is much more marked by the volcanic cover residues of the Eastern Velay plateaus, which originally stretched further towards the east on the basement. Retrogressive erosion, linked to the valley entrenchment, carved out its rim, which is sculptured in a series of cirques with steep slopes. This is an effect of the exploitation of the structural array (ancient cones or craters of maars) rather than glacial influence, quite feeble here.

To the north and west of this strong eastern rim, on the Loire side, the plateaus are much less dissected. Here we find the vastest and harshest but also the most typical areas of the highlands. In the south and NE, the transition between the volcanic core of the Eastern Velay (1,300–1,400 m) and its crystalline margins by progressive thinning of the volcanic cover is imperceptible. The resulting topographic regularity gives space to the classic multi-convex relief of weathered basement areas, separated from the main entrenchments and vales filled with redeposited granitic sand and varied occurrence of peatland. To the west, the plateaus decline progressively until they fade into those of the Devès or end in promontories (800–900 m) between the valleys cut into them. Narrow in their upstream section, these valleys widen downstream, where they

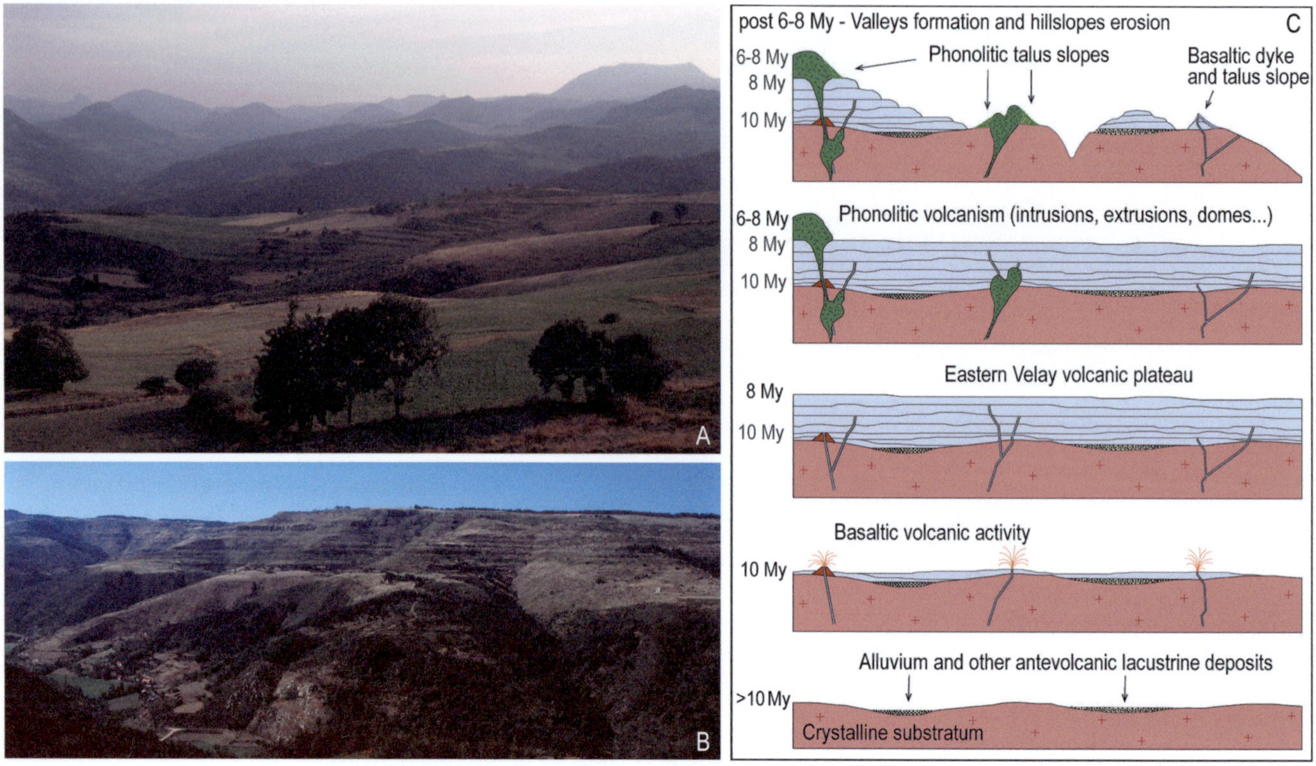

Fig. 10.5 Boutières morphology and evolution. (**A**) At the foot of Mont Mézenc (*top right* summit), the Boutières relief is characterised by numerous phonolitic *Sucs* stripped during the entrenchment of the valleys. (**B**) Saint-Clément plateau: the piles of lava flows form stepped slopes overlooking convexities of the crystalline basement. (**C**) Evolutionary stages of landforms since the Upper Miocene (Credit: E. Defive)

reach the sedimentary domain of the SE rim of the Puy basin, whose low points approximate 600 m.

10.3.2 A Great Variety of Volcanic Landforms

Volcanism is undeniably the main factor in the uniqueness of the Velay relief, whether the landforms were moulded directly by its varied activities or by differential erosion of buried structures.

The prismatic structure of lava flows favours the preservation of vertical profiles during lava flow retreat. This results in a stepped slope profile that is quite different from the convexities of the underlying basement, the covering lava flows generally creating a distinct flat (Fig. 10.5A–C). The lava piles thus flatten the irregularities of the substratum, moulding the vast plateaus of the Eastern Velay and the Devès. Subsequent erosion of various extents and forms has caused varying, age-related degrees of regularity.

The numerous phonolites of the Eastern Velay form the most original reliefs (Fig. 10.5A–C). They compose rock outcrops of different thicknesses surrounded by scree, named *Sucs*, *Roches* or *Monts*, in the local toponymy. On the side of Boutières, differential erosion during the excavation of the valleys has formed numerous dykes, ring-dykes and other protrusions, whereas domes and dome-flows dominate the plateaus on the Loire side, where the lava has reached the surface and accumulated directly beneath the emission points. To the north of the Massif (Meygal, Yssingelais, Emblaves), where the phonolites have more often penetrated the Tertiary sedimentary terrains, which were subsequently eroded, the plateau is replaced by a succession of massive mounds.

The Puy basin is another original relief, with its protruding rocks of hyaloclastite breccias (Fig. 10.1) representing volcanic necks, differential erosion having stripped their sedimentary gangue and the Surtseyan structures coeval with the Devès volcanism. The numerous, most recent Strombolian cones, called *gardes* (Fig. 10.4C), perturb the surface of this plateau and thus contribute to its formation. Their gentle slopes clearly distinguish them from the younger Strombolian cones of the Lower Vivarais volcanism (Figs. 10.4C and 10.6).

The maars, formed at the time of the Devès or later, are easily recognised as open, round depressions with steep ridges on the surface of the plateaus, which contain lakes or wetlands (Fig. 10.6). They are precious sediment traps, which together with the accumulated deposits upstream of the volcanic dams make the Velay a European reference area for palaeoenvironmental reconstruction of the Pleistocene (Reille et al. 2000).

Fig. 10.6 Upper Loire valley upstream of the zone of maximum extension of the lava flows of the Devès, showing two structures (73,000–84,000 years old) formed by the recent volcanism of the Lower Vivarais: at the centre, the water-filled maar of Le Lac d'Issarlès (138 m deep), in the *left* background, the breached Strombolian cone of the *Suc* of Bauzon, whose lava flow poured down to the Loire valley until Le Lac d'Issarlès (Credit: E. Defive)

10.3.3 An Omnipresent Periglacial Imprint

The cold Pleistocene periods have left their mark in the landscape erosion features of the predominant local periglacial environments (Valadas 1984). Influenced by cryoclasty and seasonal creeping, scree formed at the foot of *Sucs* and lava flows, gelifluction blocks formations truncating the former cambered and bedded granitic grus (*arènes fauchées litées*) and blockstreams (Fig. 10.7). These blockstreams rise in all directions, disconnected from the supplying upstream areas (predominantly phonolitic *Sucs* but also lava flow edges) and resting on locally very gentle slopes. They are the most original local periglacial landforms and little is known about their origin (hypothetically rock glaciers). They resemble the stone runs of the Falklands.

In the area of the Devès and the Puy basin and its southeastern margins, numerous plateaus or columns of basaltic lava flows have supplied scree at the foot of the cornices that is just as original, bearing in its midst big basalt slabs collapsed in all directions and progressively disintegrating while submerging into the scree (Fig. 10.8). At some points, these heterogeneous stocks were subsequently remobilised and removed from the source zones by various creeping or sliding processes (Poiraud 2012; Poiraud and Defive 2011). These processes were favoured by sediment formations and other layers of rock detritus located beneath the lava flows, which are mechanically fragile and prone to water logging and creeping, as well as by the periods of permafrost melting and some more humid Holocene phases (Subboreal). So the combined effect of this structural array, the entrenchment of the hydrographic network and the favourable climatic fluctuations has compromised the stability of the slopes, which show a chaotic relief due to huge ancient, mostly stabilised but also some sporadic contemporary landslides.

10.4 Landscapes: Between Nature and Culture

The harsh climates of the Pleistocene have not prevented man from frequenting the region since the Middle Palaeolithic (Neandertal). The shelters under lava flow rocks offered at least temporary accommodation; some sites have also been found on passes (Raynal et al. 2007). Today the Velay is a rural area with a low population density. After a maximum population in the eighteenth and nineteenth century and its corollary of nearly total deforestation, rural exodus and the decline of land use activity sustainably weakened these zones of harsh physical conditions with an aged population. It was against this background that at the end of the nineteenth century and in the 1950s, the RTM (Restauration des Terrains de Montagne, Restoration of Mountain Regions) services undertook large-scale operations of reafforestation (Reumaux 2003), which profoundly modified the landscape (Fig. 10.9). Agriculture remains one of the basic pillars of the local economy, complemented by quality production, family tourism and industrial and tertiary business in the nearest urban centres, where the population has concentrated. This economic trajectory developed in various ways according to the different sectors.

Fig. 10.7 Phonolitic blockstream, Les Estables (*ca.* 1,480 m). Note the blocks, tilted in all directions, and the open work structure. These structures stretch along the very gentle slopes and are disconnected from their supplying upstream areas. Their formation under periglacial conditions (possibly by rock glaciers) is still debated. Phonolitic domes appear further south (*left* background), dominating the plateau at the heart of the Eastern Velay (Credit: E. Defive)

Fig. 10.8 Cornice at the edge of a lava flow of the Eastern Velay, resting on a sandy to argilliferous Tertiary substratum (Présailles). The fracture network of the plateau facilitates the detachment of entire plates (megablocks), subsiding progressively, then creeping and submerging into the mass of sediment and frost-fractured small scree (Credit: E. Defive)

Fig. 10.9 Mont Mézenc (1,754 m, Eastern Velay) and the *teeth of Mézenc* in its north-western hillside, rhyolitic dykes stripped by erosion. Completely deforested and covered by heathland, the landscape of the late nineteenth century (*left*, RTM archives) differs from the current scenery (*right*), marked by the major reafforestation phases of the late nineteenth century and the 1950s (Credit: E. Defive (*right*) and archives RTM (*left*)).

10.4.1 The Rhône Side, Land of *Rayols*

With its tormented relief, the Rhône side still presents many constraints to access (slope, climate, soil poverty, flood risks due to the heavy Cevennes rains) and has thus always been a poor area. A valley area reclining on volcanic highlands, it is the land of the *Rayols*, the "men of the lowlands", reputed docile and obedient to kings and power in general. Traditional food production was based on sheep and goat farming, terrace cultures, chestnuts below 900 m and fruit trees in the low valleys. This agricultural system could only work with abundant workforce. The introduction of hydropower and the railway in the nineteenth century promoted the development of the spinning industry, but since the middle of the twentieth-century exodus, a declining agricultural activity and subsequently globalisation have defeated this economy, while the RTM afforestations have largely erased the traces of the agro-pastoral activities from the landscape. Against this background, it is remarkable that the Boutières region has developed a small, still active diversified industry based on local water bottling, mechanic parts fabrication and particularly plastic and costume jewellery manufacturing. The population of the Boutières thus consists of a small number of farmers and a majority of workers, who are among the best-paid ones in the Rhône-Alpes. The region thus seems to have managed relatively well to overcome the successive economic crises and deindustrialisation.

10.4.2 The Land of *Padgels*, "Men of the Highlands"

Above 1,150–1,200 m, the plateaus of the Eastern Velay are majestic pasture highlands. The savage *Burle* wind is typical here, carrying snow, causing snowdrifts and considerably hindering winter traffic (Serre 2001). It is its very harshness which makes this mountain climate a strong element of identity. Opposed by reputation to the *Rayols* of the valleys, the *Padgels*, "men of the highlands", consider themselves to be recalcitrant and disobedient. They are divided into two still hardly mixing plateau communities: the Protestants in the north (Meygal) and the Catholics in the south (Mézenc-Gerbier).

The only profitable agricultural activity in these very harsh climatic conditions today is haymaking, which benefits from favourable drying conditions in summer (Mediterranean influence). It has therefore been possible to establish a permanent high-altitude habitat that is able to endure the long hibernations (Fig. 10.10) and distinguishes the high plateaus of the Eastern Velay from the more humid Atlantic mountains of the western Massif Central, which are summer pastures. The communes in the heart of the Massif Central primarily live on cattle breeding, with a long-standing focus on speculation rather than subsistence (sale of cattle with specially bred traits, sale of meat and butter). The label for controlled farming "Fin Gras du Mézenc" (quality marbled beef, produced under that label from February to early June) has enabled the population to stay in business and maintain production even during the BSE crisis (Bonnefoy et al. 2011). This cattle breeding is also predominantly geared towards milk production, which is insufficiently valued in spite of the high pasture quality of the area.

Tourism is the other cornerstone of economic activity in the highlands. Apart from winter sports, which suffer from frequent thaws, the touristic attractiveness resides in the quality of its landscapes, its architectural heritage (*queyrats* with slated phonolite roofs, *pailhisses* with thatched, rye- or furze-covered roofs) and its academic heritage. The most visited site after the Puy (starting point of Camino de Santiago pilgrimage; Fig. 10.1) is Mont Gerbier-de-Jonc, where the Loire has its source (Fig. 10.11). Also the local lakes, mostly maars, are noteworthy sites. The geological,

Fig. 10.10 *Left*: Hayfield (foreground, with limousine cattle) and natural pastures (background) share a space marked by isolated farms, here at the foot of the phonolitic dome of Montfol at the centre of the Eastern Velay. *Right*: The *Roches* of Borée (Upper Boutières, 1,100 m), phonolite intrusions cleared by erosion and surrounded by scree. Situated at the foot of the *Sucs* and structured in cultivated terraces, the slopes were partly abandoned or reafforested (mark of the 1950s) in the twentieth century (Credit: E. Defive)

Fig. 10.11 Mont Gerbier-de-Jonc (*left* background) seen from the summit of Mont Mézenc. This phonolite extrusion overthrusts the Loire and the Rhône sides (foreground) at the level of the Boutières region. To the *right* of Mont Gerbier, a phonolitic table (Les Coux) dominates the plateau. The official source of the Loire, the longest river in France, is located at the foot of Mont Gerbier, a major site of regional tourism and an important academic and natural heritage (Credit: E. Defive)

geomorphological and volcanological heritage, nevertheless, is still largely underexploited.

10.4.3 Devès and Puy Basin: Milk, Lentils, Fallow Land and Urbanisation

At 800–1,200 m, where the climate is still harsh, the basaltic soils of the Devès plateau are fertile enough to form a bocage landscape and allow for an agriculture based on the production of milk and the green lentil of the Puy, which bears the labels AOC (Controlled Designation of Origin) and AOP (Protected Designation of Origin). The distinctive local architecture comprises houses made of black bricks (basalt) with Spanish tile roofs and bell-gable churches. The lower altitudes and continentalisation of the Puy and the Emblaves cause a milder climate, however, marked by frequent thermal inversions. Cereal, forage and milk production prevails here, but numerous agricultural parcels, particularly the ancient vineyards, have been turned into wasteland or suburban areas. The Puy-en-Velay is famous for its bobbin lace and has been one of its commercial centres for a long time. The lace was traditionally produced in the farms of the surrounding plateaus by the women, who gathered in *couvige* (women assemblies) to work at the *carreau* (frame for lace fabrication). The yarn and the *cartons* used as models were distributed by the *leveuse*, who also collected the lace and acted as intermediary between manufacturers and town merchants. This activity provided a considerable supplement to the revenues, particularly in Les Estables, a commune in the centre of the Eastern Velay, which is specialised in the fabrication of gold lace.

10.5 Conclusion

The rich natural and cultural history of the Velay and its Vivarais rim is undeniably marked by a volcanism that is noteworthy in its multifarious dynamics and products, its effects on erosion and stripped landforms, the resulting landscapes and soils as well as their exploitation and the local architecture. Except for some particularly noteworthy sites, such as the *Rochers* of Puy-en-Velay or Mont Gerbier-de-Jonc at the source of the Loire, this abundance is well known to experts but largely unknown to the wider public, who are rather attracted to the better publicised Chain of Puys, Mont Sancy or the Cantal. Aside from the influence of public relations and the relatively secluded position of the Velay, it is true that it has less volcanoes than the Chain of Puys, which is famous for its landforms (see Chap. 9). But the Velay offers some volcanoes which are just as young, and it comprises the region with the highest number of phonolites in Europe, which is most distinctive and unique with its domes, dome-flows and other structures stripped by erosion, as well as the hyaloclastite *Rochers* of Puy-en-Velay. This *other land of volcanoes* deserves to be better appreciated as heritage and discovered by the general public (Defive et al. 2004).

Acknowledgements We would like to thank Marie-Françoise André and Monique Fort for having proposed that we write a contribution to this book.

References

Bonnefoy B, Pochelon Y, Léogier JJ (2011) 1995–2011 Seize ans après, où en est le Fin Gras? Les Cahiers du Mézenc 23:9–12

Defive E (1996) L'encaissement du réseau hydrographique dans le bassin supérieur de la Loire. Contribution à l'étude des rythmes d'évolution géomorphologique en moyenne montagne volcanisée. Thèse, University of Paris I

Defive E, Mermet JC, Rieutort L (2004) L'autre pays des volcans: les paysages du Mézenc, entre oubli et résurgence. In: Bertrand D (ed) L'invention du paysage volcanique, collection Volcaniques issn 1633–3820. Presses Universitaires Blaise Pascal, Maison de la Recherche, pp 227–257

Defive E, Gauthier A, Pastre JF (2005) L'évolution plio-quaternaire du bassin du Puy (Massif central, France): rythmes morphosédimentaires et volcanisme. Quaternaire 16(3):177–190

Guérin G, Gillot PY (2007) Nouveaux éléments de chronologie du volcanisme Pléistocène du bas Vivarais (Ardèche, France) par thermoluminescence. C R Geosci 339:40–49

Mergoil J, Boivin P (1993) Le Velay, son volcanisme et les formations associées. Notice de la carte à 1/100 000. Géologie de la France 3:1–96

Poiraud A (2012) Les glissements de terrain du bassin du Puy-en-Velay: caractérisation, facteurs de contrôle et cartographie de l'aléa. Thèse de Doctorat, Université Blaise-Pascal, Clermont-Ferrand 2

Poiraud A, Defive E (2011) Morphology and geomorphological significance of relict landslides in the Tertiary basin of Puy-en-Velay (Massif Central, France). Géomorphologie: relief, processus, environnement 3:247–260

Raynal JP, Fernandes P, Le Corre-Le Beux M, Santagata C (2007) Le Paléolithique moyen de Haute-Loire (France): origines, diversité, affinités. In: Arts et cultures de la Préhistoire, CTHSth edn. Hommages à Henri Delporte, Paris

Reille M, de Beaulieu JL, Svobodova H, Andrieu-Ponel V, Goeury C (2000) Pollen analytical biostratigraphy of the last five climatic cycles from a long continental sequence from the Velay region (Massif Central, France). J Quaternary Sci 15(7):665–685

Reumaux E (2003) Une forêt et des hommes: l'entreprise de reboisement au Mézenc. Les Cahiers du Mézenc 15:31–46

Rochette P, Bertrand H, Braun C, Berger E (1993) La province volcanique pléistocène supérieure du Bas Vivarais (Ardèche, France): propagation de fentes crustales en échelons? C R Acad Sci 316:913–920

Serre F (2001) La neige dans le Massif central. Une contrainte pour la gestion des territoires ?. Presses Universitaires Blaise Pascal, Coll. "Nature et Sociétés", Clermont-Ferrand

Valadas B (1984) Les hautes terres du Massif central français. Contribution à l'étude des morphodynamiques récentes sur versants cristallins et volcaniques. Thèse de Doctorat d'État, University of Paris I Panthéon-Sorbonne, soutenue le 6 mai 1983, 2 tomes

The *Cirque de Navacelles*: A Major Geomorphosite in the *Grands Causses* Karstic Region

11

Martine Ambert

Abstract

The *Cirque de Navacelles* is a perfect, incised meander that sits at the base of the deep Vis river gorge; this gorge slices through the southernmost of the *Grands Causses* (limestone plateaus), the *Causse du Larzac*, south of the metamorphic and granitic Cévennes Mountains. Two panoramic viewpoints, opposite one another on either side of the basin, provide a magnificent view 300 m above this highly scenic location. Over time, this major *geomorphosite* has acquired a certain degree of scientific and touristic notoriety, but the understanding of its formation was until recently rather incomplete. The Cirque de Navacelles is in fact a unique example of an incised meander because it is much more complex than a simple bend in a watercourse. Its shape is typical of other fully formed meanders, but the way it formed is unique. Karstic contributions in the form of Holocene travertines lifted the thalweg and led to an eventual overflow. As time passed, changes in the landscape continued, with erosion creating a 10-m-high waterfall at the end of the meander. What is interesting about this site is that three different types of formations are interconnected at a single place, the gorge, the meander, and the waterfall, with each one corresponding to a different phase in the chronology of the site's history. Analysis of the creation of the gorges has also allowed researchers to include the gorges of the Tarn river in their studies and to show the impact on the karstic formations of the bordering plateaus.

Keywords

Gorge • Incised meander • Travertines • Grands Causses (*Great limestone plateaus*)

11.1 Introduction

The *Cirque de Navacelles* is an example of a perfect, very expressive incised meander, dug out of the base of the deep Vis river gorge, which slices through the southernmost of the *Grands Causses* (limestone plateaus), the *Causse du Larzac* (Fig. 11.1). Two panoramic viewpoints, opposite one another on either side of the basin, provide a magnificent view 300 m above this highly photogenic location. This major geomorphosite has acquired a certain degree of scientific notoriety, precisely because it is exceptionally representative in nature, to such an extent that it is often cited as an example in geomorphology and geology textbooks. Although the form of the site attracted considerable interest at a very early stage, an understanding of its formation would have not been achieved without the geomorphological investigations carried out by Professor Jean Nicod's team (Ambert et al. 1978). Since then, two benchmark studies have provided solid arguments to date the formation of the *caussenard* gorges and the karstic forms of the nearby plateaus (Ambert 1994; Bruxelles 2001). Using the *Cirque de Navacelles* as an example, this article will examine each of the interconnected formations that make up the landscape around this major site – incised meander and gorge – with the goal of integrating them within a wider perspective based

M. Ambert (✉)
Department of Physical Geography, UMR GRED – Université Paul-Valéry, Montpellier 3, Route de Mende, 34199, Montpellier Cedex, France
e-mail: m.p.ambert@wanadoo.fr

M. Fort and M.-F. André (eds.), *Landscapes and Landforms of France*, World Geomorphological Landscapes,
DOI 10.1007/978-94-007-7022-5_11,

Fig. 11.1 General overview from the perspective of Blandas (northern panoramic viewpoint) showing the entire *geomorphosite* (© Philippe Galant)

on the more recent data. Since this site also welcomes every year a great number of tourists, another objective will be to show how to enhance this tourist attractiveness by making the data more accessible to the general public in situ and by revealing the wealth of potential discoveries to be made there (Fig. 11.1).

11.2 Geographical and Geological Setting

11.2.1 Morphostructural Setting and Stratigraphy: Classic Foundations and Recent Developments

The Vis gorges slice through and subdivide the southern part of the *Grands Causses*. The term *Grands Causses* refers to the vast, high limestone plateaus of the Jurassic Period which extend to the south of the basement rocks of the Massif Central (Fig. 11.2). The gorge incision has exposed gigantic natural geological cross sections displaying limestone strata, deposited in an immense marine gulf formed during the Upper Jurassic. This must have been a shallow warmwater sea as evidenced by the presence of numerous coral reefs, with nearby limestone deposits having been locally transformed into dolomites due to the addition of magnesium from the seawater. A great number of bauxite and marine Cretaceous outcrops sit atop this main series, precisely in the upstream sector of the Vis river (Bruxelles 2001). The size and geometry of these deposits, the extent of which was unknown until recently, reveal that a long period of emersion occurred, when the first substantial karstifications formed; in the final phase, the sea returned to cover the limestone plateau during the Upper Cretaceous. Only residual remains of these Cretaceous rocks have been preserved in a synclinal trough, providing evidence of the significant degree of erosion that has taken place since then. Their discovery has nonetheless fundamentally renewed the understanding of the *caussenard* landscape and provided an additional marker in the chronology of its formation (Bruxelles 2001).

In terms of tectonics, the site has been influenced by the proximity of its Hercynian basement. The limestone cover has been fractured according to two orthogonal directions, but between the breaks, the limestone plateaus appear tabular, or at the most, just barely bent in the shape of vast synclines. The east-west faults are ancient and played a role during the Pyrenean compressive phase (mainly during the Eocene).

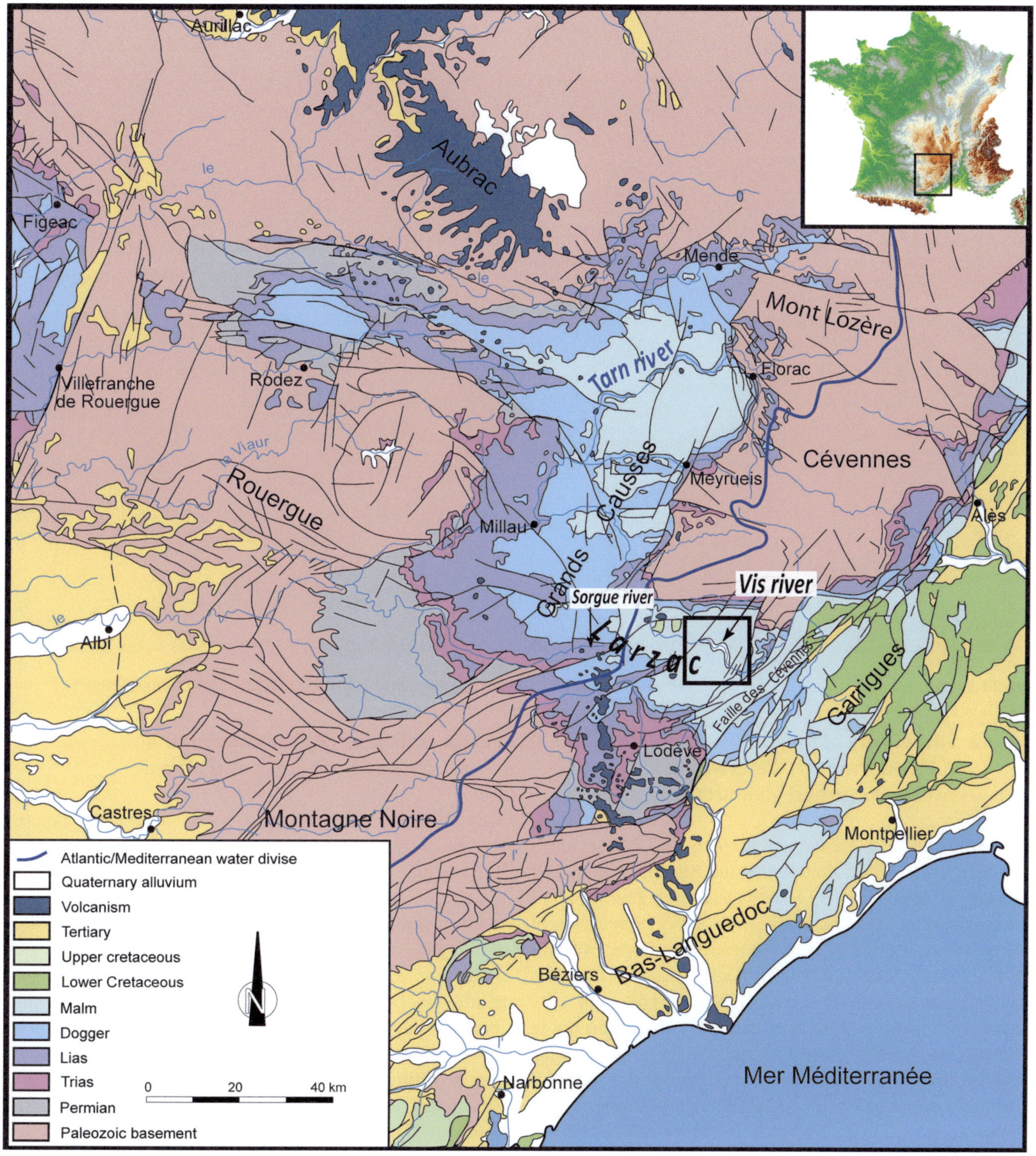

Fig. 11.2 Geological map (© Laurent Bruxelles after the geological map of France 1:1,000,000, BRGM 2003). The *black square* indicates the location of the site

The result, for this southern section of the *Causses*, was the creation of staircase made up of faults. These are barely visible on the plateau as a result of a subsequent leveling process; they do however tend to orient dissections of the borders. For instance, these faults account for the dissymmetry of the Sorgue and Cernon on the southwestern border of the Larzac. The fractures in the nearby Cévennes Mountains (NE-SW) date back to the distension phase of the mid-Oligocene and Miocene. This is why the Larzac and Blandas Causses are so closely huddled up against the Mount Séranne horst.

Fig. 11.3 Morphological contrast between the Bathonian dolomites in nested canyon and layered limestone above (© Philippe Galant)

11.2.2 Singularity of the Vis River Course: A Mediterranean Waterway with a Gorge and Meanders

In contrast with the more northern gorges of the Dourbie, Jonte, and Tarn rivers, which flow toward the Atlantic, the Vis is a tributary of the Hérault, a Mediterranean river. The Vis is for the most part of its course analogous waterway originating in the crystalline bedrock of the Cévennes, on the slopes of Saint-Guiral, 16 km southwest of Mont Aigoual. As soon as it reaches the limestone layer, the river disappears at the bottom of a dry valley. The Vissec sector is aptly named for *sec*, means dry in French! The river's course only becomes perennial in the downstream section of the powerful La Foux springs (France's sixth most powerful springs and the strongest-flowing springs of the *Causses*, with an average throughput rate of 2.5 m^3/s), 6 km away as the crow flies from the aforementioned losses.

The most majestic part of the gorges starts in Vissec, where the Virenque, another intermittently flowing river, adds its waters to those of the Vis and extends to Saint-Maurice de Navacelles-Madières. What makes this sector so unique is the presence of numerous meanders, some of which are still active as in the case of Vissec, and others that have been incised in the more distant past, on several levels above the thalweg.

11.2.3 Influence of Lithology on the Meandering Track and Slopes

This sinuous track is associated with the layered Oxfordian and Kimmeridgian limestones within which the Vis dug a deep furrow. While upstream and downstream of Vissec and Saint-Maurice de Navacelles-Madières, the Vis carves its path through more resistant carbonate rocks including Bathonian dolomites which hinder the erosion of the concave riverbanks and, as a result, the accentuation of any curve in its track. The minimal resistance of the layered limestone also explains a second morphological characteristic which makes the Vis gorges unique, that is to say its well-ordered banks (devoid of any structural irregularities) often covered with periglacial stratified scree deposits (*grèze litée*). Obviously the rock here is prone to frost shattering. The contrast with the Bathonian dolomites, visible in the nested canyon downstream from Vissec (Fig. 11.3) and extending to the springs at La Foux, is eloquent. The contrast between the layered limestone and the more resistant limestones and dolomites, found both in the gorges and on the plateaus, is the key to fully understanding the *caussenard* landscape (Ambert 2004b).

11.3 A Closer Look at the *Cirque du Navacelles* Meander

Although the *Cirque de Navacelles* is part of a series of meanders, some of which are quite evolved, it is nonetheless the most finished and representative of the series; it ends in a 10-m-high waterfall and as such is also the most photogenic among all meanders of the Vis gorges. Indeed, the other examples, located further upstream, were incised at an earlier time; as a result, they sit higher, are filled in with *grèzes litées*, and are harder to see.

This location is quite emblematic, and scientific investigations conducted to explain the *Cirque* formation and evolution have been used to strengthen its position as a unique geomorphosite.

These geomorphological studies coordinated by Professor Nicod, with the goal of mapping the Blandas Causse and the Vis gorges, on a 1:25,000 scale (Fig. 11.4), have provided interesting data to better understand the site (Ambert et al. 1978). A classic approach was adopted to map the site, based on a careful inventory of geomorphic features including sinkholes, poljes, as well as superficial formations. In the gorges, this systematic method revealed the existence of a terrace of Holocene travertines which continues well beyond the verdant ring of the *Cirque* itself. On this map, karstic forms appear in red as well as the karstic formations (mainly the polygenetic *terra rossa* complex). Fluviatile forms and formations, including Holocene tuffs, appear in green. Crosses in circles indicate abandoned meander loops.

11.3.1 An Abundant Holocene Travertinization at the Origin of the Overflow

Apart from its perfect shape, the *Cirque de Navacelles* is also unique in terms of location due to its karstic context which makes it a more complex incised meander than others due to a simple incision. Indeed, the massive accumulation of Holocene travertines (Fig. 11.5) raised the thalweg vertically by about 10 m and enabled it to overflow which makes the uniqueness of the site. The 10-m vertical travertinization development (carbonate precipitation process favored by slight temperature rise and by plant, moss, and algae respiration) results from the powerful La Foux springs, 5 km away from the Cirque. Malacological and palynological studies indicate that these travertines date back to the Holocene (Marcelin 1954). The presence of Mediterranean species such as *Laurus nobilis* and *Theodoxus fluviatilis* is particularly notable.

Geomorphological analyses have confirmed the age of this site: the anteriority of the periglacial shaping of the slopes is obvious, and two radiocarbon datings have also confirmed the Holocene age of this major travertinization phase (6,200 and 6,100 ± 100 B.P.). But most importantly, mapping the area revealed the extent of this deposit and its geometry, which had never been known before (Fig. 11.5b). From the village of Madières to the Navacelles waterfall, a clearly defined ledge is formed, 10 m above the current riverbed which tracks fully within the tuff embankment. This geometry is controlled both by the vertical development of the travertinization and the consecutive erosion. In other words, when the travertinization was at its maximum, the waterfall that can be currently observed did not exist. The entire bottom of the valley was also filled with travertines downstream from the *Cirque* to Madières.

11.3.2 Waterfall Formation: A Resumption of Erosion

The waterfall provides evidence for a resumption of erosion following the travertine accumulation, which must have only affected the downstream section and is also an indicator of trend reversal, or a sort of regressive erosion. This erosional process exhumed the Jurassic substratum from the travertine deposits (Figs. 11.5c and 11.6). When and how did this trend reversal occur? Few answers can be found in Navacelles but additional investigations of analogous phenomena in Languedoc and in Provence provided new insights. In Saint-Guilhem-le-Désert, one of the closest sites, the Verdus river, also dropped down by 20 m below a terrace of Holocene travertines. As indicated by the irrigation canal created by the monks upon their arrival at the site, this incision ceased around the year 1,000 AD (Ambert and Ambert 2001). The occurrence of interrupted travertinization and resumed vertical erosion of travertines is fairly common not only in the Languedoc region but also in Provence, where the human influence on the vegetal and geomorphological evolution can at least be discerned. This change is related to deforestation which is always accompanied by an increase in water runoff, at the expense of infiltration leading to an increase in terrigenous load. The results are a decrease in dissolved carbonates and the occurrence of more troubled waters that hinder the growth of algae and crust-forming moss, essential to the creation of travertine deposits. Hence, tuff formation is doubly hindered, and since the waters carry less matter, their flow becomes violent and is more apt to dissolve any accumulations. This violent water flow is also the explanation behind the waterfall creation as well as the main reason for the Navacelles meander formation, which allowed to the perennial nature of the river's short track.

The waterfall, just like the Navacelles incision, is unique, when one compares it to other incised meanders and waterfalls that can be found in similar formations. This rarity, inherent to the notion of heritage and geomorphosite, strengthens Navacelles's premier status among such locations.

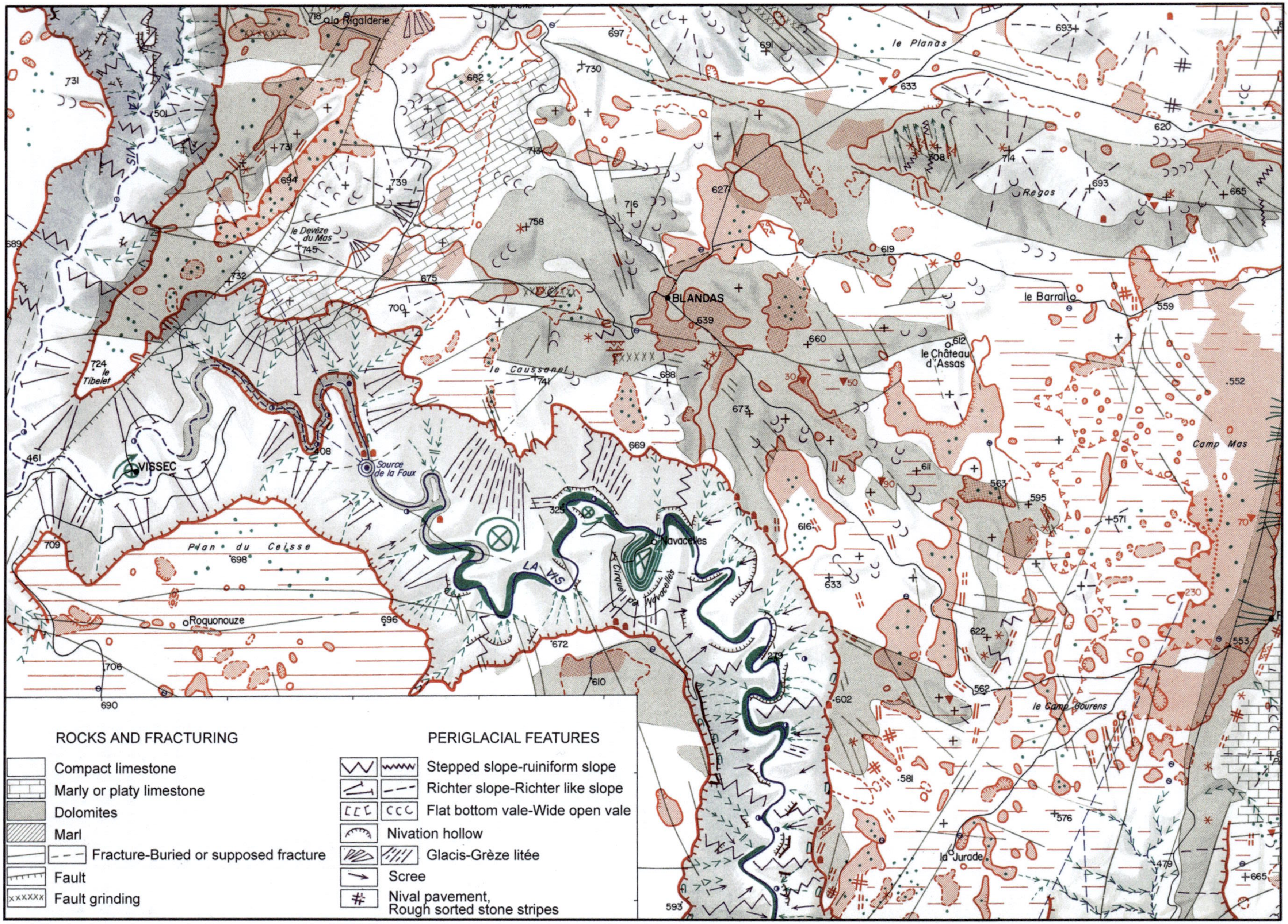

Fig. 11.4 Section of the geomorphological map of the Blandas Causse and the Vis gorges, 1:25,000 (© Guilhem Fabre after Ambert et al 1978)

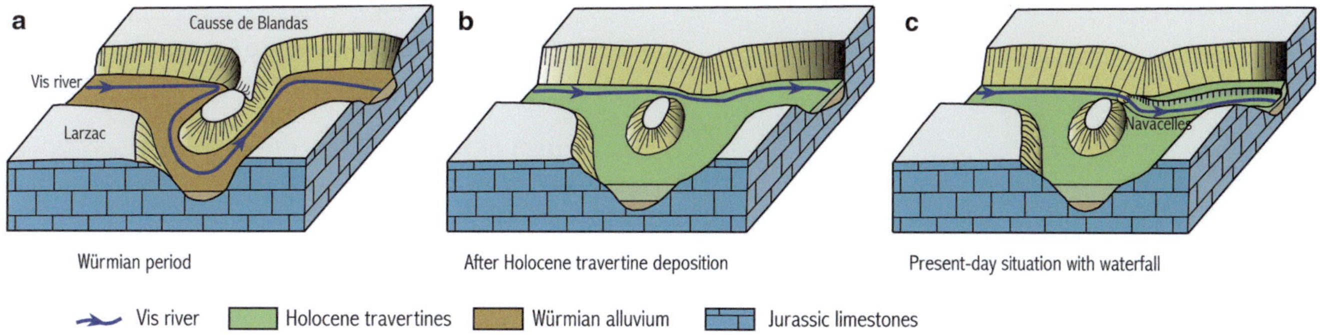

Fig. 11.5 Evolution of the meander. (**a**) Würmian period. (**b**) After Holocene travertine deposition. (**c**) Present-day situation with waterfall (© Paul and Martine Ambert/Claude Requirand)

Fig. 11.6 Close-up of the waterfall on Jurassic limestones and the travertine deposits (© Philippe Galant)

11.4 The Origin of the Vis Canyon: Debates About the Chronology

It had been considered for long that the formation of the gorges had only occurred during the Quaternary and was linked to a vigorous 1,000-m upheaval of the whole Causses-Cévennes area during the Early Pleistocene (Dubois 1970). Although the *Cirque de Navacelles* provides valuable information concerning the recent evolution of the Vis Canyon, studies of the older phases of the hollowing-out process that led to its formation are lacking. One thing is certain: the existence of the meanders clearly indicates that the gorge was formed above ground and not as a result of a ceiling collapse having affected underground galleries.

The nearby waterways of the Tarn, Hérault, and Orb rivers provide additional data and allow for a convergent and complementary vision of the chronology for the incisions in the southern *Causses*. A comparative analysis of neighboring valleys in which ancient river deposits are sealed up either by old travertines or by volcanism has allowed assessing the amount of the pre-Quaternary hollowing-out process (Ambert 1994).

The Tarn river valley is an exceptional example for studying the changes that one of the major canyons of the Grands Causses underwent (Fig. 11.7), for this valley provides three significant types of data. Millau's series of travertines indicates that the oldest deposit, the *Plateau de France*, which is attributed to the early Quaternary, is situated 250 m below the edge of the Larzac. Further upstream, the 5.75-million-year-old Azinières volcano seals off a topographical feature that lies 150 m below the edge of the gorges. The Miocene volcano Eglazines, incised by the Tarn, also indicates that the first phases of

Fig. 11.7 The Tarn canyon from the Point Sublime (© Martine Ambert)

the creation of these gorges have taken place during the mid-Miocene period (Ambert and Ambert 1995). Such data is precious for understanding the geodynamic context of this Atlantic Ocean tributary, barely affected by the Messinian regression. It indicates that the gorges were formed well before the crisis, whose impact has sometimes been exaggerated. Southward, the Escandorgue volcanoes, which cap the Southern Larzac, have fossilized at Bernasso, a narrow valley that was formed during the Pliocene. Recently, the timeline for the formation of these gorges was pushed back even further, based on an examination of land surveys of the area sealed up by the Azinières volcano (Giusti 2002).

Thus, scientists have a fairly good understanding of what occurred even if there is still some debate over when the formation of the gorges actually began, due to differences in methods used to fix the chronology. In any event, the recent upheaval that had been pointed as the key driver of their formation is no longer an essential factor to consider, at least not in terms of the recentness of the creation of the gorges, nor with respect to their impressive vertical development.

11.5 The Effect of the Gorge Formation on the Shape of the Surrounding Plateaus

The incision of the Vis and its tributary, the Virenque, had major and highly visible repercussions upon the evolution of the surrounding karstic forms.

11.5.1 Inactive Poljes

The two bordering poljes, that of Saint-Maurice de Navacelles, inserted in the Larzac Causse at the base of Mount Séranne, and that of Rogues, located north of the gorges, broke apart as a result of the progressive deepening of the river. Before the incision, this huge, 20-km-long plain used to flood on a regular basis as indicated by the presence of ridges with their western edges flattened out overlooking the bottom of the present-day plain. These benches (often >2 km) are essentially the ancient high-water mark for the floods and show the karstic truncation caused by dissolution. Furthermore, these rocky terraces are pitted here with sinkholes, circular depressions whose shape emphasized by crops or differences in vegetation. These depressions, quite numerous, reveal the extent of the draining off of the water at the expense of lateral corrosion, annihilated by the incision of the gorges and what this led to, the sinking of the underground networks. The last flood remembered at Rogues as at Coulet dates back to 1907 and only lasted several hours. In other words, these poljes are presently inactive because they sit too high above the underground aquifers. This is the primary characteristic of the network of *caussenard* poljes.

11.5.2 Remarkable Sinkholes

Throughout the valley of the Virenque, the right bank tributary of the Vis river, the large *sotchs* (local name of huge sinkhole) of Sorbs and Campestre are remarkable sinkholes, whose depth (up to 70 m) is unequaled in the *Causses* (Fig. 11.8). They are worth being visited for they are both captivating and somewhat mysterious. Indeed these abrupt, even breathtakingly steep forms commonly referred to as *swallow holes* are part of the Jurassic limestone sequence whose forms are often more subdued. In addition to this unique aspect, humans have transformed some of the *swallow holes*, such as the Campestre *sotch*, into caves designed for maturing a cheese that was labeled under the Roquefort *Appellation* until 1925.

In terms of the surrounding natural environment, this concentration of ten or so deep *sotchs* has been recently related to the presence of bauxite and marine Cretaceous outcrops, providing evidence for a long period of emersion, accompanied by the first substantial karstifications that were subsequently sealed

Fig. 11.8 The Saint-Férreol *sotch* (© Martine Ambert)

Fig. 11.9 La Baume Auriol exhibition (© Communauté de Communes Lodévois-Larzac)

by Upper Cretaceous marine sediments. Hence, these deep sinkholes are the result of the cumulative action of dissolution that took place over the last hundred million years. Here as elsewhere, the incision of the Virenque River reactivated the corrosive process under the limestone covering and triggered its subsequent draining away. Some *sotchs* of similar magnitude are still fossilized. This is certainly the case of the neighboring Luc Bas sinkhole, which superficially appears quite indistinct, but turned out to have very steep inner walls and to have undergone considerable drainage when surveys to explore its edges were conducted with a mechanical digger (Bruxelles 2001).

11.6 Cultural Value and Touristic Promotion

As early as 1912 Joanne, the local guide, praised the site of Navacelles, insisting on the "*contrast in the landscape between the monotonous limestone plateaus and the precipitous cliffs*." Descriptions like this one reveal the popular enthusiasm for such beautiful landscapes that led to part of the site being classified as a national park as early as 1943 and extended 40 years later.

Over the last century, this remarkable site has been one of the most visited. It is appreciated mainly for its natural beauty as demonstrated by a survey conducted in 1991 among 749 visitors whose 83 % that what struck them most was the breathtaking beauty of the place. This love of the site is also shared by the permanent residents of the area, as demonstrated by an ethnological study excluding heritage interpreters. For the interviewed local people, the *Cirque de Navacelles* appeared as the most emblematic of the southern Larzac region (Gomez and Ambert 2012).

The site is therefore universally recognized, and given this status, questions of who, how, when, and why people visit the site are the principal concern (Ambert and Guillot 2013). Every year some 250,000 visitors come to the *Cirque de Navacelles*. Since 1979, the high number of visitors has proven to be disproportionate with respect to the development of the facilities designed to welcome them. This led to one of the first *Opérations Grands Sites* in France whose goal was to improve the site management. Following 30 years of consultations and various actions, the recognition of Navacelles as a geomorphosite, with all that such a designation entails in terms of valorization, has taken strong hold. The Baume Auriol farmhouse located at the southernmost of the two scenic panoramic viewpoints has been renovated and transformed into a welcome center for the site. Here an exhibition that was opened to the public in 2007 (Fig. 11.9) reveals all the richness of the Vis gorges in their natural context and invites visitors to discover the area including the village and the waterfall. The hiking trail to La Foux allows visitors to discover another exhibition in harmoniously restored mills. Nature here is sublimated by the presence of man. The setting up of the Blandas viewpoint in 2013 with explanations on display, helps understanding the site formation and evolution. The recent inscription in 2011 of the area among the UNESCO's World Heritage sites as a *Living Cultural Landscape (agropastoralism)* should in the long term lead to further preservation, maintenance and valorization of these historically rich landscapes.

11.7 Conclusion

The *Cirque de Navacelles* adds to the perfection of its form the wealth of scientific information about its genesis and evolution. The karstic deposits explain the uniqueness of this incised meander, but the *Cirque* is also a deep gorge that captivates the visitor. What is interesting with this site is to show how three different formations interconnect – the gorge, the meander, and the waterfall – each one corresponding to a different type of geomorphological process. This site also represents an observatory from which it has been possible to witness the spread of geomorphological knowledge. After the fundamental phase that led to the aforementioned mapping of the site, an analysis of the type of information available to visitors compared to their wishes revealed that key geomorphological concepts concerning the primary elements of the landscape were only minimally transmitted through the different interpretative tools (Ambert 2004a). This led the author to actively participate in the spread of knowledge in situ. Such a contribution to the valorization of a site, a shift from basic research toward knowledge transfer and dissemination, was truly exhilarating.

Acknowledgments Laurent Bruxelles, Philippe Galant, and Claude Requirand are thanked for their contribution to illustration; Paul Rogers for the translation; and Marie-Françoise André who corrected the English text.

References

Ambert P (1994) L'évolution du Languedoc Central depuis le Néogène (Grands Causses méridionaux, piémont languedocien). Documents du BRGM, Orléans, n° 231, 210 p + 3 cartes géomorphologiques au 1/25 000. en couleur H.T.

Ambert M (2004a) Contribution à la connaissance du patrimoine naturel languedocien: enjeux, concepts et applications. Géomorphologie et Patrimoine. Thèse de doctorat, Université de Lyon 2, 325 p. + annexes 143 p

Ambert M (dir) (2004b) Hérault miroir de la Terre, Co-éditions BRGM-Presses du Languedoc, 160p

Ambert M, Ambert P (1995) Karstification des plateaux et encaissement des vallées au cours du Néogène et du Quaternaire dans les Grands Causses méridionaux. Géologie de la France 4:37–50

Ambert P, Ambert M (2001) Le Verdus et ses travertins (Saint-Guilhem-le-désert, Hérault), indicateurs de l'activité humaine et morphoclimatique au cours des dix derniers millénaires. In: "Eaux sauvages, eaux domestiquées" Hommage à L. Davy, Publications de l'Université de Provence, Aix, pp 271–284

Ambert M, Guillot M (2013) Enjeux de la conservation et de la valorisation du géomorphosite du cirque de Navacelles (Larzac méridional, France), Collection Edytem "Actes du colloque ISGM 2011"

Ambert M, Ambert P, Coulet E, Fabre G, Guendon JL, Nicod J, Orengo C (1978) Le Causse de Blandas (Gard). Présentation d'une carte géomorphologique à 1/25 000. Méditerranée 1–2:3–21

BRGM (2003) Geological map of France 1:1,000,000, 6th edn. Orléans

Bruxelles L (2001) Dépôts et altérites des plateaux du Larzac central: Causses de l'Hospitalet et de Campestre (Aveyron, Gard, Hérault). Evolution morphogénétique, conséquences géologiques et implications pour l'aménagement. Thèse de doctorat, université Aix-Marseille, 266 p. Document du BRGM n°304, Orléans, 2004

Dubois P (1970) Sur la morphologie et l'évolution souterraine de la région calcaire nord montpelliéraine. Actes de la réunion internationale de karstologie. Etudes et Travaux de Méditerranée, pp 37–50

Giusti C (2002) Le sud du Massif Central (France): implications morphogénétiques de l'activation d'une marge passive. Approche épistémologique et naturaliste. Thèse de doctorat, université de Perpignan, 550p

Gomez J, Ambert M (2012) Mise en évidence de l'existence du patrimoine géomorphologique dans l'imaginaire populaire en Larzac méridional. In: Giusti C (ed) Géomorphosites 2009: imagerie, inventaire, mise en valeur et vulgarisation du patrimoine géomorphologique. Paris Sorbonne Université, Paris, pp 156–161

Marcelin P (1954) Note sur quelques tufs du Département du Gard. Annales de l'Ecole nationale d'Agriculture de Montpellier, TXXIX, III-IV, 1–12

12 The Gavarnie Cirque: A Celebrated "Nature's Colossus"

Monique Fort

Abstract

Surrounded by peaks at height more than 3,000 m, the Gavarnie cirque forms, with two other cirques and the adjacent Spanish canyons, an exceptional landscape recognised in 1997 as a UNESCO World Heritage site for both its natural and cultural values. It is an outstanding geomorphosite straddling the Central Pyrenees, famous for its 1,500-m-high limestone amphitheatre, its waterfalls, and the *Brèche de Roland*. It results from a long evolution, including the formation of the Pyrenean range, the development of Quaternary glaciers, and a long history of human occupation and exchanges across the French/Spanish border. It has always been an inspiring place for many artists and is much appreciated by tourists and climbers.

Keywords

Calcareous cirque • Thrust nappes • Pyrenean glaciation • Waterfalls

12.1 Introduction

The Gavarnie cirque is a unique site along the Pyrenean range. Celebrated by many artists at the end of the nineteenth century, it is one of the most visited places in the Pyrenees. Surrounded by peaks at height more than 3,000 m on the French/Spanish border, the Gavarnie cirque forms, with the two cirques of Estaubé and Troumouse and the adjacent Aragonese canyons and the cirque of Barrosa, an exceptional landscape recognised in 1997 as a UNESCO World Heritage site for both its natural and cultural values (Fig. 12.1). It is sited in the central part of the collisional Pyrenean system, where a maximum shortening has occurred, as revealed by the thrust units exposed within the slopes surrounding the cirque. Both karstic processes and glacial erosion contributed to shape this outstanding, deep natural amphitheatre that "*resembles nothing that is found anywhere else*" (V. Hugo). Adjacent to the Mont Perdu (3,352 m), it is a place of legend and praise, where coexist a traditional pastoral landscape and a tourism-oriented economy sustained by both Catholic pilgrims from Lourdes and sportsmen attracted by this "*Nature's Colossus*" (V. Hugo).

12.2 Geographical and Geological Setting

When coming from the north of France, after the vast Landes plains or the Lannemezan plateau, the traveller abuts against compact, continuous relief towered by the Astronomic Observatory of the Pic du Midi de Bigorre. When getting closer, the Pyrenean range rises like higher and higher waves. It becomes difficult to clearly separate the mountain front from its foothills (administratively, it is the 500 m contour line which sets the limits); it is also difficult to discern those valleys that permit to penetrate deeper into the mountain and reach the top of this great barrier. The Gavarnie cirque stands in the very upper part of the Gave de Pau catchment, locally called the Lavedan country (Fig. 12.2). From upstream of Lourdes, this is a fairly deep valley, which also leads to other prestigious sites of the Central Pyrenees, i.e. Cauterêts valley,

M. Fort (✉)
Geography Department, UFR GHSS, CNRS UMR 8586 PRODIG,
University Paris Diderot-Sorbonne-Paris-Cité,
5 rue Thomas Mann, 75205 Paris Cedex 13, France
e-mail: fort@univ-paris-diderot.fr

M. Fort and M.-F. André (eds.), *Landscapes and Landforms of France*, World Geomorphological Landscapes,
DOI 10.1007/978-94-007-7022-5_12, © Springer Science+Business Media Dordrecht 2014

Fig. 12.1 The Gavarnie cirque seen from the northern trail up to the Refuge des Espuguettes (Credit: M. Fort). Note the stone circles, traces of earlier settlements of shepherds coming from both sides of the French–Spanish border

with the Pont d'Espagne and the Gaube lake at the foot of the steep Vignemale north face.

12.2.1 At the Crossing of Climatic Influences

The Gavarnie cirque and adjacent landforms correspond to the central part of the Pyrenees, where a number of >3,000 m high peaks are present: Mont Perdu, Astazou (3,012 m), Marboré (3,250 m), Taillon (3,144 m), and Vignemale (3,298 m), to name a few (Fig. 12.3). This dominating relief plays an important role in the climatic regime and determines a sharp contrast between the influences from the Atlantic to the North and the West and those from the Mediterranean to the Southeast. In the Lavedan valley, westerlies command humid fluxes and abundant precipitation, yet the topographic contrasts may locally create rain-shadow effects and temperature inversions. Mediterranean influences, locally known as the "*balaguere*", a foehn-like wind in Lavedan, bring sudden heat that generally precedes heavy rains or snowfalls. These contrasts are expressed in the rich, endemic flora and fauna specific of each altitudinal belt and are accentuated at high elevation by extreme climate, steep landforms, and bedrock nature (predominance of limestone).

12.2.2 Where Spain Tightens Against France

The east–west trending Pyrenean mountains represent the westernmost end of the Alpine–Himalayan collisional system (Fig. 12.4). They are the result of a long evolution during which the boundary of the Iberian and European plates evolved until the final compressional stage of the Pyrenean orogenic growth (Late Cretaceous to Tertiary times), responsible for the present-day asymmetry of the mountains. Indeed, the steep, narrower (≈45 km) northern flank on the French side contrasts sharply with the gently sloping southern Spanish side, where ridges elevation progressively declines down to the Ebro basin (≈100 km).

The general structure of the Pyrenees displays a fanlike, flowerlike arrangement: large thrust nappes lean towards the north on the northern side, while they lean towards the south on the southern side. They are flanked on both sides by two foreland basins, the Aquitanian and the Ebro basins

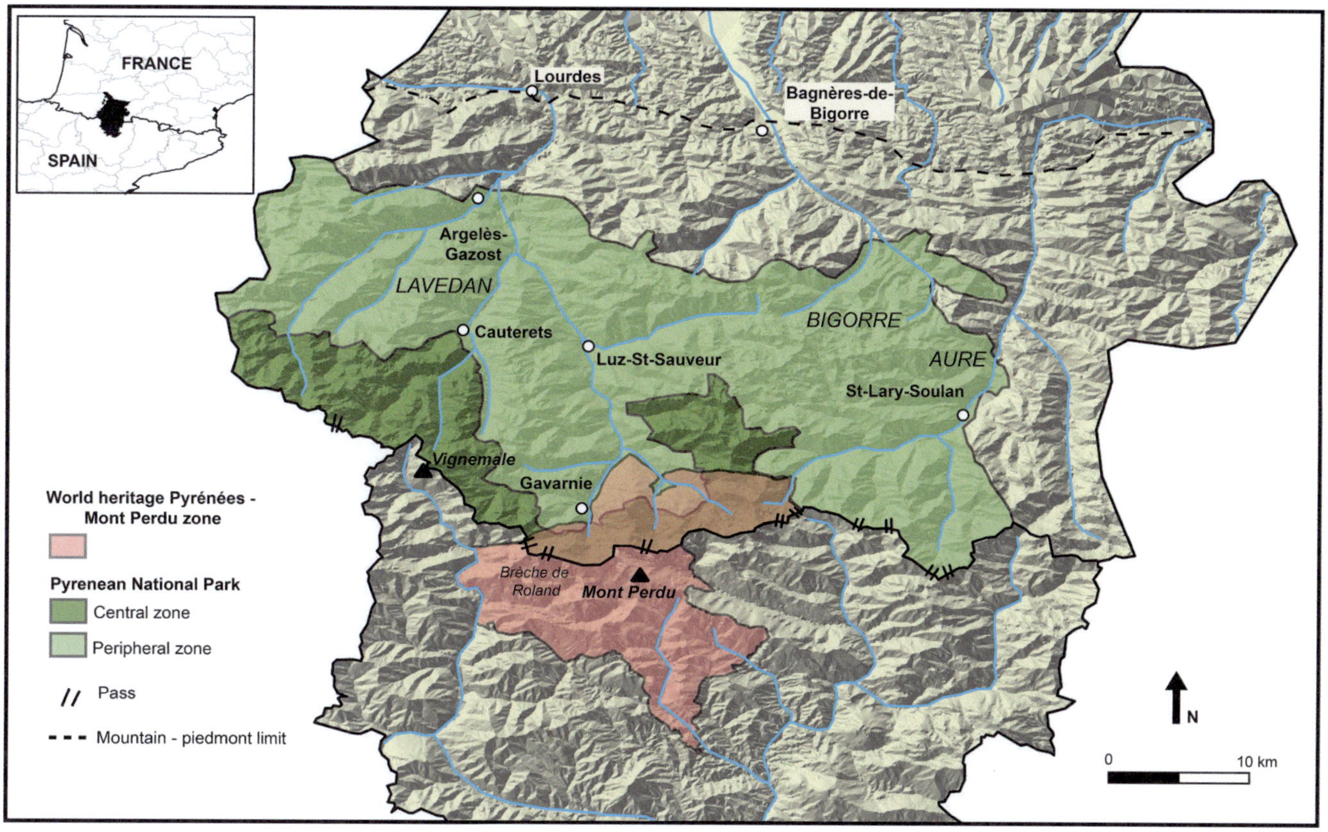

Fig. 12.2 Map of the UNESCO World Heritage zone: on the Spanish side, it includes the totality of Parque Nacional de Ordesa y Monte Perdido, whereas the French side extends over the core zone and the peripheral zone of the Pyrenees National Park. *Thick solid line* marks the international boundary

(Fig. 12.5a). In fact, as revealed by extensive seismic reflection profiling, the European wedge is overlain by the upper Iberian wedge, whereas at the lower crustal levels, the tectonic wedge of the European crust has indented the Iberian plate (Choukroune and ECORS team 1989; Teixell 1998). This tectonic wedging at different crust levels is one major thickening mechanism, at the origin of orogeny and uplift of the Pyrenees.

The Gavarnie cirque area is sited in a central position, the axial zone, where a maximum shortening occurred and where this geologic architecture is the most distinct and nearly horizontally emplaced. It is mostly composed of the Gavarnie–Mont Perdu nappe complex, a series of thrust units piled up on the autochthonous Palaeozoic crystalline rocks (migmatitic gneisses): this is particularly well expressed along the eastern side of the cirque (Fig. 12.5b). In detail, the Gavarnie–Mont Perdu nappe is composed of highly tectonised Palaeozoic material, with its tightly folded, Cretaceous and Eocene sedimentary cover that overthrusts to the south and culminates with Cylindre du Marboré and Mont Perdu. The extent of the Gavarnie nappe is very large, at least more than 50 km from north to south, and represents fairly well the tectonic forces that produced the Pyrenean range. Most of the uplift appears to have started in the Eocene period, but it continued during the Oligocene when most of the deformation occurred.

12.3 Landform Evolution

12.3.1 Long-Term Evolution

The Gavarnie and two adjacent cirques of Estaubé and Troumouse are the result of a long and complex history, which started well before the cold periods of the Quaternary era. It is generally considered that the drainage divide dates back to the Eocene: located in the axial zone – the area of maximum relief of the orogen – the divide appeared when the axial zone was uplifted by the Gavarnie thrust *s.l.* This led to active denudation, and eroded debris was transferred to the subsiding foreland, contributing to the very thick Mio-Pliocene molasses of the Ebro and Aquitaine forelands, with the Bigourdan piedmont close to the mountain (Teixell 1998; Vergés et al. 2002; Figs. 12.4 and 12.5).

In the Central Pyrenees, the original position of the divide in the mid-Eocene was located in the southern North

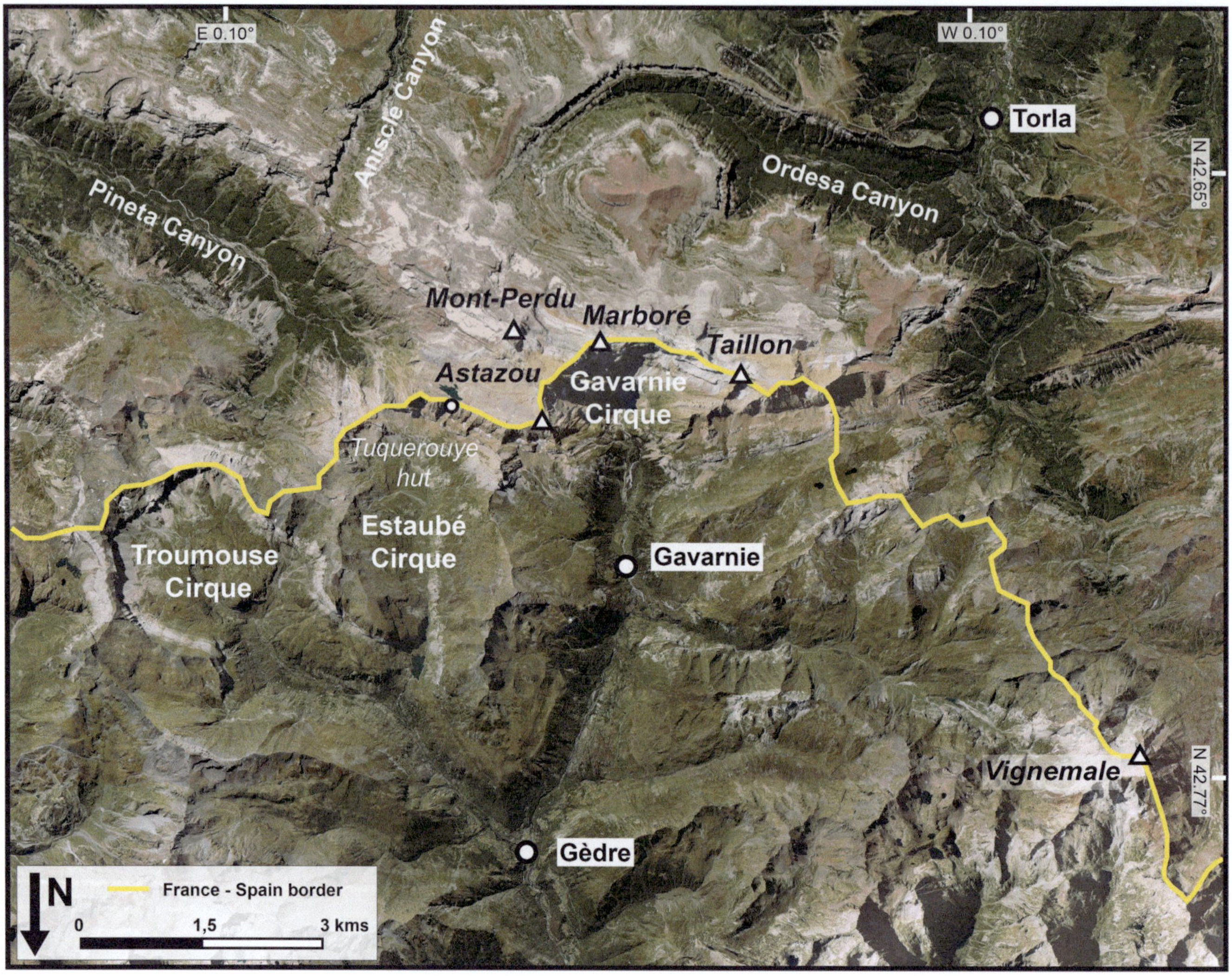

Fig. 12.3 The Gavarnie–Mont Perdu site includes spectacular cirque walls on the French side; two emblematic, still glaciated peaks (Mont Perdu and Vignemale); and three of the largest and deepest canyons on the Spanish side (Google map in the background, North oriented towards *bottom*)

Pyrenean Zone (retro-side of the belt). This is where exhumation was centred at that time, before erosion migrated southwards by tectonic control, divide migration being achieved by captures of the southern drainage network by the northern rivers, in particular with the source of the Garonne River in Val d'Aran (Babault et al. 2011). In the adjacent Aure valley (Fig. 12.2), the Neste River was also quite active: the Lannemezan megafan is the topmost relief feature where more than 2,000 m thick sediments are piled up in the subsiding foreland of the mountain, reflecting a significant incision of the rivers, probably favoured by both the steepness of the French side of the Pyrenees and the dominant moisture supply from the northwest, then by glaciation. In both the Val d'Aran and Aure valleys, granitic, joint substrate was deeply incised. In contrast, in the Gave de Pau valley, the presence of the thick and massive calcareous Gavarnie–Mont Perdu nappe did not allow further significant retrogressive erosion in its upstream part due to karstic, underground drainage network; indeed, the upper Gave de Pau valley evolved more like a calcareous "blind" valley, whose amphitheatre shape was then accentuated by Quaternary glaciations.

12.3.2 The Glacial History

During the Quaternary period, the Pyrenees have been extensively glaciated down to the piedmont (Calvet et al. 2011a; Delmas et al. 2011). During the maximum of the last glaciation (which occurred in the early part of the last glacial cycle), a large glacial tongue, more than 70 km long, fed by many tributaries from adjacent valleys, occupied the Gave de Pau valley and built up a complex, multilobate morainic system in the vicinity of Lourdes (Calvet et al. 2011b; Fig. 12.6). Around this area, 45-ka-BP-old morainic deposits are preserved and form a series of hills (≈600 m) perched above the present Gave de Pau valley floor (≈150 m).

With the retreat of glaciers that began before 34 ka BP under dry climatic conditions (Jalut et al. 1992), some

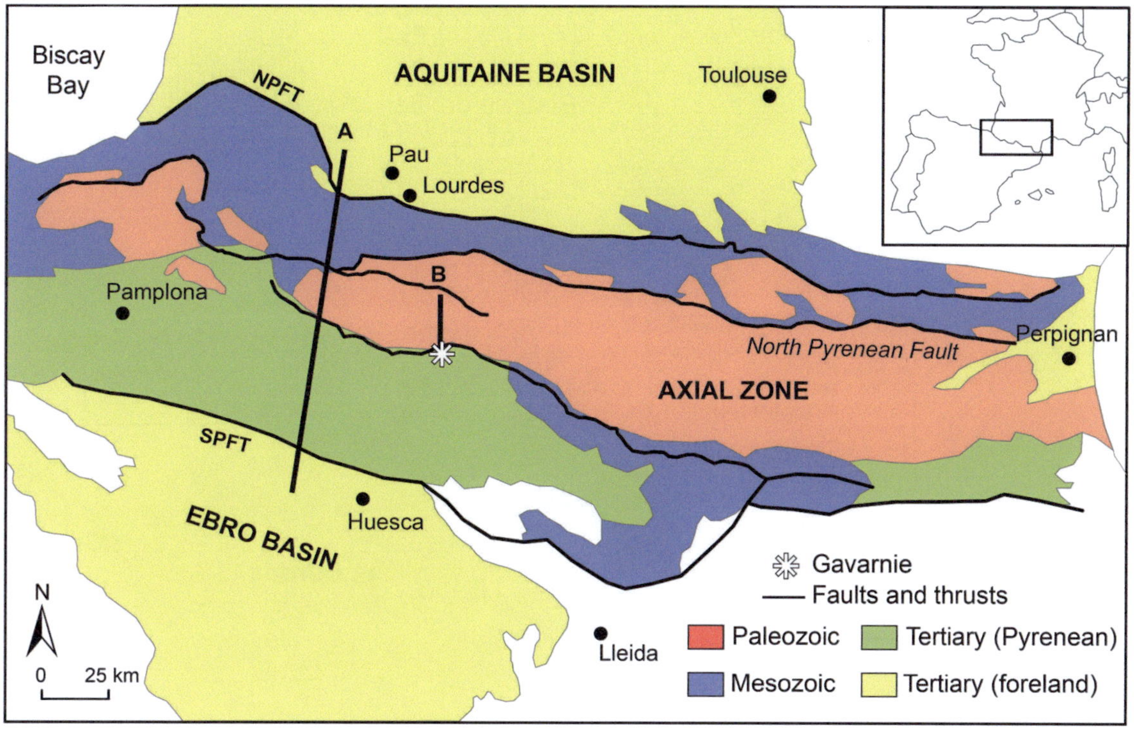

Fig. 12.4 Major geological entities of the Pyrenean range. A–A′ and B–B′ refer to the two cross sections of Fig. 12.5a, b

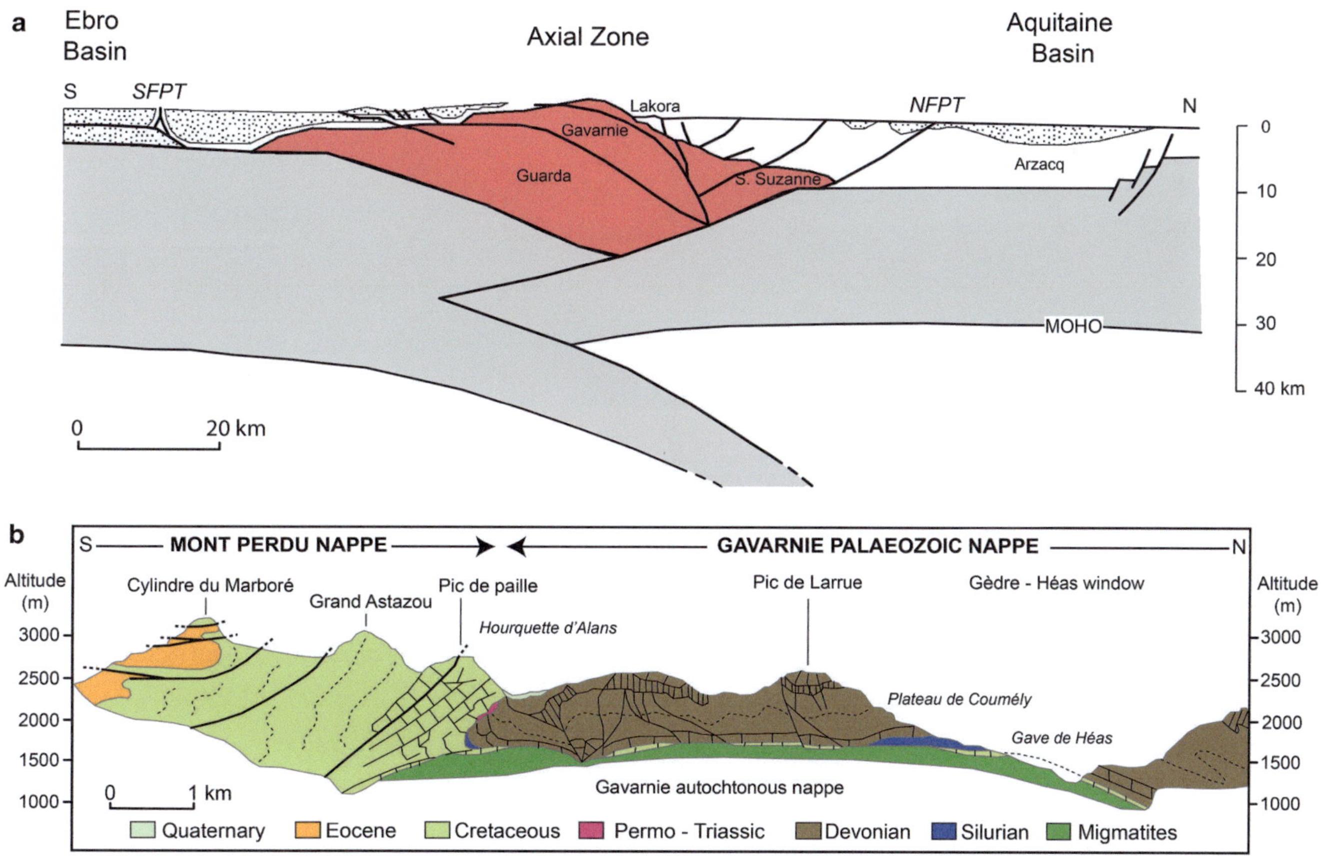

Fig. 12.5 Geological cross sections. (**a**) Across Western Pyrenean range (Adapted from Teixell 1998), with major structural units, i.e. from north to south: the Aquitaine retro-foreland basin, the North Pyrenean Frontal Thrust of the Pre-Pyrenees, the Axial zone with basement rock units, the South Pyrenean Frontal Thrust, and the Ebro foreland basin. (**b**) Detailed emplacement of the Gavarnie–Mont Perdu nappes (Redrawn and adapted from Gavarnie's Geological Map, 1:50,000, BRGM Edition 2002). Location of cross sections in Fig. 12.4

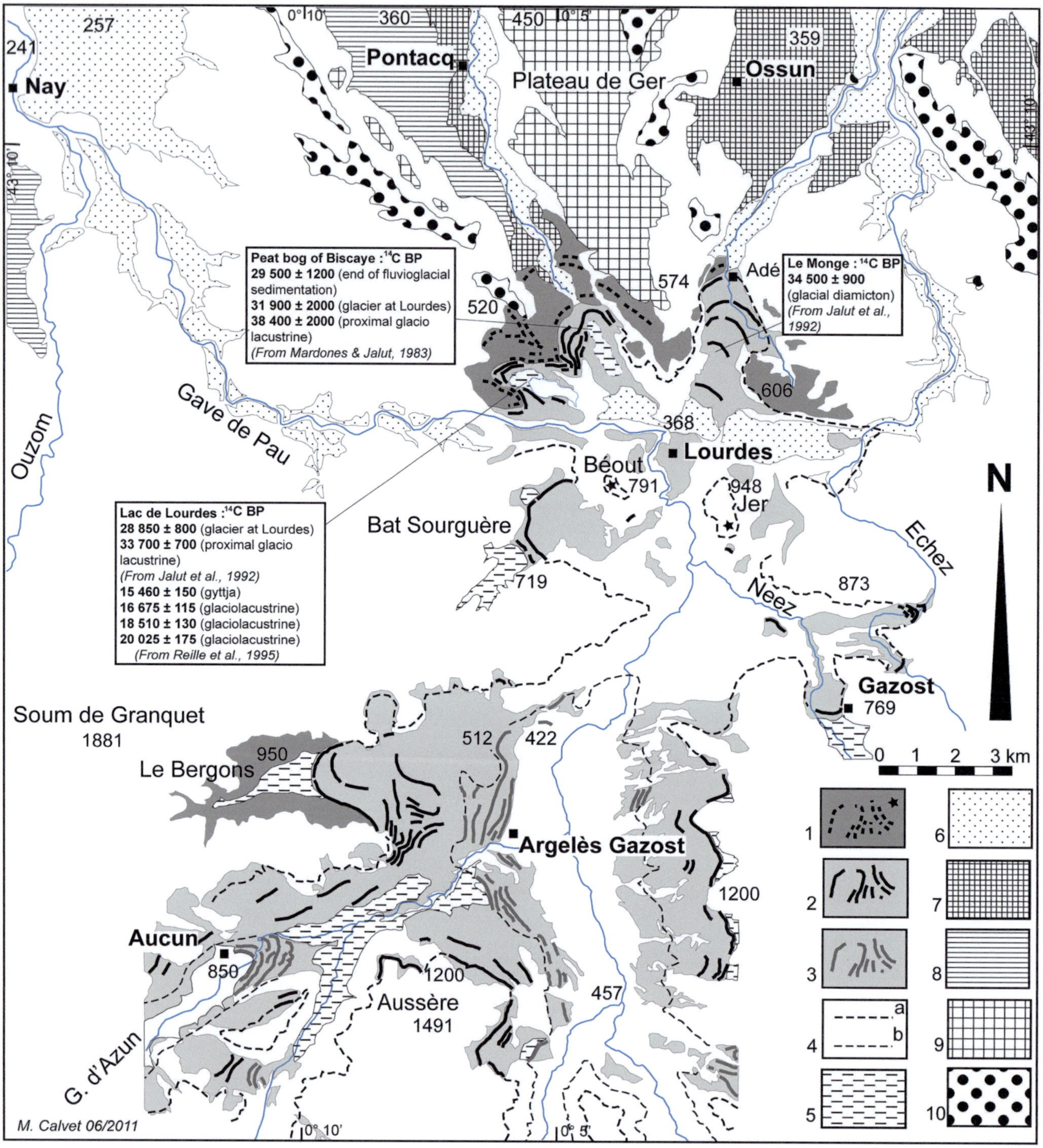

Fig. 12.6 Glacial extent in the Central Pyrenees, Gave de Pau lower valley around Lourdes (From M. Calvet, in Calvet et al. 2011b, with permission)

moraine-dammed lakes formed like the Lourdes Lake, while in the adjacent areas, glacial meltwaters deposited fluvioglacial sediments along the radiating valleys of the Plateau du Ger, north of Lourdes. The piedmont was probably ice-free at about 31 ka BP, but this phase was followed by a subsequent glacial readvance at 24 ka BP. This view was challenged by Reille and Andrieu (1995) who suggested that the ice cover lasted longer and vacated the piedmont zone between 16.67 and 14.46 ^{14}C ka BP only. It is most probable that glaciers advanced at different stages and reached similar positions repeatedly, as shown by clusters of very different ^{10}Be ages at about the same location (Rodés 2008). Then the

Fig. 12.7 Evolution of the Taillon Glacier (northern slope). (**a**) ca. 1900, old postcard (Association Moraine, with permission). (**b**) In 2010 (photo G. Nogué, Association Moraine, with permission). Note the readjustments (rockfall, scree, debris flow) of the rocky wall following deglaciation

glaciers receded rapidly, with a few traces of evidence left, such as lateral moraine of Argeles Gazost, because of their prompt removal by further erosion (Fig. 12.6). Yet, immediately below the crest zone, some traces of glacial fluctuations are preserved downvalley, suggesting the presence of Late Glacial and most probably Holocene moraines.

More recent Neoglacial events like in the Cirque de Troumouse have been recorded (Gellatly et al. 1992), between 4,955 ± 90 and 4,654 ± 60 BP, followed by more recent, yet still undated, events. Little Ice Age moraines are more frequently preserved in most massifs of the Central Zone. The modern glaciers disappear very quickly as observed in many places (René 2001). There are no more than 21 glaciers left in the Pyrenees, occupying a total area of 450 ha, whereas as recently as in 1870, the total glaciated surface amounted to 45 km^2. The gradual increase in temperature – a total of 0.9 °C from 1890 until now – implies that the Pyrenean glaciers will probably disappear by 2050 or even earlier. In the Gavarnie area, glaciers are essentially restricted to N to NE facing cirques at altitudes above 3,000 m a.s.l. (Mont Perdu or Taillon, 3,144 m; Fig. 12.7) or occur on high plateaus like in Ossoue–Vignemale. Nevertheless, glaciers are not what really have made the renown of Gavarnie site.

12.4 "Nature's Colossus"

The most striking part of the Cirque de Gavarnie is its 1,500 m high amphitheatre shape, widening from 800 m at the bottom up to 3,000 m at the top, in a series of three giant vertical steps (Fig. 12.1). Generally interpreted as a result of glacial erosion, it is a much more complex form where limestone and underground waters have played an important role in the overall architecture of this unique scenery. Two outstanding features are the main attractions of the Gavarnie cirque: the Grande Cascade and the Brèche de Roland.

12.4.1 Waterfalls

One remarkable feature of the Cirque de Gavarnie is its waterfalls in two or three drops, with an overall drop of 420 m, "taller than the Eiffel Tower" as locals say. The Grande Cascade, the lower one, is 281 m tall, and the breaking fall turns often into mist. As Victor Hugo has written (1855, posthumously published in 1891) in his poem "God": "*The twelve waterfalls that shadow hears complaining in this terrible and whirling cylinder, the mountain streams exhausted of foam and fighting, and flow, top torrent and smoke down*". The water at the origin of the waterfall is coming from the snowmelt and from a frozen pond located in Spain near the Mont Perdu, with the waters seeping and finding their way underground through the Cretaceous limestone strata before resurging right on the cirque rim. The annual discharge of the waterfall varies between 3 and 200 m^3/s when snowmelt is at its maximum. Indeed, the waterfall accentuates the vertical dimension of the relief, all the more as the adjacent limestone strata are also overturned and nearly vertical along the "*Arête Passet*" (Fig. 12.8). The waterfall is the source of the *Gave de Pau* stream, which collects other *gaves* (torrential streams) originated from the Barèges and Cauterêts valleys, hence being transformed in a fast-flowing stream in Lourdes that can be subject to severe flooding like in autumn 2012 and June 2013.

Fig. 12.8 Great Waterfall of Gavarnie. The water comes out of karstic network developed at the foot of the Grand Pic du Marboré (3,248 m). *Left* of the waterfall: note the folded, nearly vertical strata (Cretaceous limestones) of Arête Passet, a famous climbing road. Foreground: glacial rock-bar shaped into the Devonian sedimentary basement (Credit: M. Fort)

12.4.2 Brèche de Roland

The *Brèche de Roland* (2,807 m) is a natural gap, open in the upper layer of the cirque. The eye is attracted by its very spectacular geometry, 40 m across bordered by very slender walls over 100 m tall, cut into the Eocene limestone (Fig. 12.9). According to the legend, Roland, Charlemagne's nephew, threw his sword when he was mortally wounded in a battle against the Mores; the sword hit the cliff and opened the breach just at the border between France and Spain. In fact, this giant doorway is the result of a collapse of the rocky wall that was favoured by postglacial debuttressing. The pass is a classic way on its southern side to climb either Mont Perdu or the Taillon (3,144 m), which offers a vertiginous overlook down to the cirque bottom. It also marks the sharp contrast existing between the northern, moist Atlantic side of the mountain and the south, dry, and stony sloping plateaus of the Aragon, Mediterranean side, with their very deep canyons, that reveal natural geological cross sections cut into the multicoloured, Tertiary sedimentary strata.

12.5 Natural and Cultural Heritages

If the outstanding beauty of these landscapes has been celebrated for many centuries, the early occupation (40,000–10,000 BC) of these mountains is in fact testified by megalithic remnants, such as the Gavarnie stone circles (Fig. 12.1) and the Tella dolmen, or caves such as Anisclo or

BRÈCHE DE ROLAND
Dessin de Fr. Schrader, d'après nature.

Fig. 12.9 Schrader's drawing of the Brèche de Roland (2,807 m) (In Reclus 1877, p. 25). Franz Schrader was among the very first to promote "pyreneism" and to draw attention to endangered species

Escuain in Aragon, shaped by karstic waters (Fig. 12.2). The elevations of the crest lines were never a barrier, and high passes (called "*ports*") linking Bigorre (France) to Aragon (Spain) bear evidence of their frequentation by shepherds and extensive livestock commuting from one side to the other for grazing and sheep breeding (Galop et al. 2007). The Port de Boucharo (Puerto de Bujaruelo) west of Gavarnie was traditionally crossed by shepherds on their transhumance way from the grand canyon of Ordesa to the Gavarnie meadows, with halts known as "*hospices*" such as the *espitau/hospitales* of Gavarnie (Fig. 12.10), and summer shelters of pastoralists called "*cabanes*" or "*coueyla*" (hut), like the Cabane de Tuquerouye (Fig. 12.11), used by hunters and now used by recreational walkers, or the Aires *coueyla* in the Troumouse cirque, considered as cultural heritage and now rehabilitated and protected. Indeed, as in many valleys in the Pyrenees, French and Spanish living on both sides of the mountain share many commonalities and customary rights on a local and trans-Pyrenean basis, and landscapes still present cultural similarities like the use of terraces fields.

On the French, Lavedan side, the paucity of arable lands in the valley bottom favoured the development of a social life

Fig. 12.10 Gavarnie village. In the early days Gavarnie was a shepherd town and an important passage point between France and Spain (way to Boucharo port, Puerto de Bujaruelo). The hospitales is visible in the background (Early twentieth-century postcard, Public domain)

based on community of land, with exchanges and shares of farming production and grazing time in alpine pastures (Galop et al. 2007). This collective way of life persisted during the twentieth century when hydropower dams and tourist activity based on skiing development started providing a valuable, alternative income to villagers.

Actually, tourism started in the second half of nineteenth century, with mineral, sulphurous, often hot springs, mostly situated at the contact of the granitic basement and sedimentary rocks. Their exploitation as hydrothermal baths in Cauterêts, Barèges, Luz-St-Sauveur was largely popularised by Imperatrice Eugenie (1826–1920). Fairly differently, religious tourism is also a local specialty. The 18 apparitions of the Virgin Mary to Bernadette Soubirous young shepherdess, in 1858, changed the fate of the small town of Lourdes, which has developed into a major place of Roman Catholic pilgrimages. With good connection by rail and air, there are today millions of tourists who come each year to Lourdes. This attendance is beneficial to its hinterland of the Gave valleys, especially to the Cirque de Gavarnie, with two million tourists visiting the site each year. In wintertime, besides the Gèdre ski resort, Gavarnie has become famous for ice climbing on frozen waterfalls. In summertime visitors walk or ride a horse to discover the breathtaking walls and waterfalls issued from the bowels of the earth. In addition to the cirques, herding or hiking trails permit to discover the most striking geological and geomorphological features or to observe rare, protected species adapted to extreme environments, such as alpine rocky slopes near the Sarradets (plants *Androsace ciliata, Saxifraga iratiana, Artemisia genepi, Doroconum*) or caves (the endemic, cave-dwelling collembolan *Tricanthella frigida*). Edelweiss (*Leontopodium alpinum*) is the symbol of the Pyrenees National Park (PNP), whereas the local "isard" (*Rupicapra p. pyrenaica*), which had almost disappeared by dint of being hunted long, is again abundant since the PNP creation in 1967 and has become an integral part of the landscape (Fig. 12.2). On the Spanish side, in the Parque Nacional de Ordesa y Monte Perdido created in 1918 (2,100 ha, extended to 15,608 ha in 1982), the

Fig. 12.11 Tuquerouye refuge (2,655 m) at Gavarnie. This entirely stone-built hut is located at the *port de Tuquerouye,* along the trail to Spain across Monte Perdido to the Aragon canyons of Añisclo, Ordesa, Pineta and Escuain (Early twentieth-century postcard, Public domain)

VUE PRISE DE LA BRÈCHE DE ROLAND

Dessin de Fr. Schrader, d'après nature.

Fig. 12.12 View from the Brèche de Roland, drawn from life. Example of the delicate and precise drawings by Franz Schrader: Spanish side of the Brèche de Roland (2,807 m), with dry, southeast dipping limestone plateaus overlooked by the Casque du Marboré (3,006 m) on the *left* (In Reclus (1877, p. 77); Public domain) lammergeyer (*Gypaetus barbatus*) is in decline because of its dependence on pastoralism. The two adjacent national parks, together with the Site Classé and Classified Reserve of the Gavarnie, Troumouse and Estaubé cirques, with the Ordesa–Viñamala Biosphere Reserve and a WWF/IUCN Centre of Biodiversity, fully justify the UNESCO World Heritage status.

The attractiveness of the Gavarnie cirque also gave rise to summer entertainments such as the theatre festival which takes place at the entrance of the cirque for the last 27 years and attracts 20,000 people over 2 weeks. Since the inscription of the area as a World Heritage property in 1997, the committee of UNESCO has requested relocation of the theatre festival, incompatible with the outstanding universal value of the site. Despite efforts to minimise the impacts by festival organisers, their resistance to move the festival elsewhere creates a risk for the decommissioning of Gavarnie site as a part of the Mont Perdu World Heritage. In the adjacent valley, Gèdre villagers have accepted to restrict road access to Troumouse cirque in order to help the recovery of natural landscape and wilderness.

12.6 Conclusion

Considered the archetype of glacial cirques, the Gavarnie cirque results from a combination of both karstic and glacial influences, in the context of plate tectonics and complex convergence between Spain and France. The exceptional size of the cirque, taller than wider, generated many tales and legends, attracted writers and artists. If Victor Hugo was fascinated by natural forces, Franz Schrader (1844–1924) was probably the most emblematic Gavarnie's lover: he not only sustained the Gavarnie myth but also was the first to explore and explain the interactions between geology and climate at the origin of these dramatic landforms (Fig. 12.12) and to decipher the human imprints on the overall landscape. His exceptional skills in geography, cartography, drawing, and painting definitely made him the first "Pyreneist", whose graveyard rests forever on the lower slopes of Gavarnie Cirque.

Acknowledgements Warm thanks are due to Mélanie Pateau (Univ. Paris-Diderot), Marc Calvet (Perpignan University), and Pierre René (Association Moraine) for providing illustration.

References

Babault J, Van Den Driessche J, Teixell A (2011) Retro- to pro-side migration of the main drainage divide in the Pyrenees: geologic and geomorphological evidence. Geophys Res Abstr 13:EGU2011–EGU12567

Calvet M, Delmas M, Gunnell Y, Braucher R, Bourlès D et al (2011a) Recent advances in research on Quaternary glaciations in the Pyrenees. In: Gibbard PL (ed) Quaternary Glaciations, Extent and Chronology, a closer look Part IV, vol 15, Developments in Quaternary Science. Elsevier, Amsterdam, pp 127–139

Calvet M, Delmas M, Gunnell Y, Braucher R, Bourlès D (2011b) Unravelling the last glaciation in the Pyrenees: an overview of recent advances. Abstract XVIII INQUA Conference, Bern 21–27 July 2011. http://www.inqua2011.ch/?a=programme&subnavi=abstract&id=1002&sessionid=89

Choukroune P, ECORS team (1989) The ECORS Pyrenean deep seismic profile reflection data and the overall structure of an orogenic belt. Tectonics 8(1):23–39

Delmas M, Calvet M, Gunnell M, Braucher R, Bourlès D (2011) Palaeogeography and 10Be exposure-age chronology of Middle and Late Pleistocene glacier systems in the northern Pyrenees: implications for reconstructing regional palaeoclimates. Palaeogeogr Palaeoclimatol Palaeoecol 305:109–122

Galop D, Carozza L, Marembert F, Bal MC (2007) Activités pastorales et climat durant l'âge du Bronze dans les Pyrénées: l'état de la question à la lumière des données environnementales et archéologiques. In: Richard H et al (Dir) Environnements et cultures à l'âge du Bronze en Europe occidentale. Editions du CTHS, Paris, pp 107–119

Gellatly AF, Grove JM, Switsur VR (1992) Mid-Holocene glacial activity in the Pyrenees. Holocene 2(3):266–270

Hugo V (1891) Dieu, uncompleted Poem (Posthumous)

Jalut G, Montserrat Maerti J, Fontugne M, Delibrias G, Vilaplana JM, Julia R (1992) Glacial to interglacial vegetation changes in the northern and southern Pyrenees: Deglaciation, vegetation cover and chronology. Quat Sci Rev 11:449–480

Reclus E (1877) Nouvelle Géographie universelle, t.2: La France

Reille M, Andrieu V (1995) The late Pleistocene and Holocene in the Lourdes Basin, Western Pyrenees, France: new pollen analytical and chronology data. Veg Hist Archaeobot 4:1–21

René P (2001) Les glaciers des Pyrénées françaises: campagnes d'observation de septembre 2000. Réunion Société Hydrotechnique de France, sect. Glaciologie, Grenoble, 15–16 mars 2001, 16 p

Rodés A (2008) La Última deglaciación en los Pirineos: datación de super-ficies de exposición mediante 10Be, y modelado numérico de paleo-glaciares. Doctoral thesis, Universitat de Barcelona (unpublished), 238 pp

Teixell A (1998) Crustal structure and orogenic material budget in the West central Pyrenees. Tectonics 17(3):395–406

UNESCO World Heritage List, Pyrénées-Mont Perdu. http://www.mppm.org/ and http://whc.unesco.org/en/list/773/

Vergés J, Fernàndez M, Martínez A (2002) The Pyrenean orogen: pre-, syn-, and post-collisional evolution. In: Rosenbaum G, Lister GS (eds) Reconstruction of the evolution of the Alpine-Himalayan Orogen. J Virtual Explor 8:55–74

13 The Têt River Valley: A Condensed Record of Long-Term Landscape Evolution in the Pyrenees

Marc Calvet, Yanni Gunnell, and Magali Delmas

Abstract

While coinciding with a unique biogeographical hotspot where the Mediterranean and Atlantic biomes meet in the Eastern Pyrenees, the Têt river catchment displays a spectacular suite of tectonic, glacial, periglacial, fluvial and hillslope processes. Prominent fragments of smooth, pre-Quaternary Paleic relief survive where glacial erosion has not serrated the interfluves, and well-preserved exposures of both continental and marine basin fill sequences provide a compelling record of post-orogenic landscape evolution and base-level changes since the Early Neogene. The heritage value of the physical landscape is enriched by a palimpsest of agrarian, pastoral, industrial and cultural vestiges from the Neolithic to recent history.

Keywords

Eastern Pyrenees • Ancient landforms • Paleic erosion surfaces • Neotectonics • Faceted fault scarps • Glacial and fluvial landscapes • Knickpoints • Physical and cultural landscapes

13.1 Introduction

The natural landscapes of the Pyrenees became the focus of scientific investigation by geologists, geographers and other scholars (e.g. Palassou 1781; Ramond de Carbonnières 1792) before gaining appeal as a cultural, aesthetic and touristic attraction in the nineteenth century (Besson 2000). Unique signatures of this orogen's history occur in its eastern segment, where the mountains are breached by a sequence of intermontane tectonic basins and strike-parallel fluvial valleys, such as the Têt, which have ushered in Mediterranean climatic regimes deep into the orogen's crest zone (Figs. 13.1 and 13.2; cf. also Calvet 1996, and Calvet and Gunnell 2008). Accordingly, the landscape of the Eastern Pyrenees has been attracting geomorphologists since the late nineteenth century, with scholars from Germany (A. Penck; H. Ashauer; W. Panzer), the Netherlands (A. J. Pannekoek; H. Boissevain; L. U. de Sitter), Switzerland (F. Nussbaum), Spain (Ll. Solé Sabaris) and France (R. Blanchard; M. Sorre; P. Birot; F. Taillefer; G. Viers). Detailed references can be found in Birot (1937) and Calvet (1996).

The Têt is a Mediterranean river with the dubious reputation of having recorded the most devastating floods in French history. The river rises in the Carlit massif, which is surrounded on two sides by the Neogene half-grabens of Cerdagne and Capcir, and rapidly descends towards the Mediterranean Sea through a gorge that follows the E–W and NE–SW strike of the Têt fault (Fig. 13.2a, b). This recently active fault controls the shape of the Conflent and Roussillon Basins. An itinerary down the Têt valley provides excellent opportunities for appreciating some key landscape features that have helped to disentangle the tectonic and climatic contributions to landscape evolution in the Pyrenees.

M. Calvet (✉) • M. Delmas
Department of Geography, Université de Perpignan-Via Domitia, EA 4605 Médi-Terra, 52 avenue Paul Alduy, 66860 Perpignan cedex, France
e-mail: calvet@univ-perp.fr; magali.delmas@univ-perp.fr

Y. Gunnell
Department of Geography, Université Lumière-Lyon 2, UMR 5600 CNRS, 5 avenue Pierre Mendès-France, 69676 Bron cedex, France
e-mail: yanni.gunnell@univ-lyon2.fr

M. Fort and M.-F. André (eds.), *Landscapes and Landforms of France*, World Geomorphological Landscapes,
DOI 10.1007/978-94-007-7022-5_13,

Fig. 13.1 Mt. Canigou, seen from the Roussillon Basin in spring after heavy snowfall. Note the glacial cirques in the crest zone. Behind the peach groves, note the rounded Aspres hills covered by Mediterranean scrub, or maquis (Credits: M. Calvet)

13.2 The Pyrenees: Old Orogen, Young Mountain Range

13.2.1 Three Distinctive Features

The *topographic identity* of the Eastern Pyrenees arises from the presence of low-relief erosional surfaces capping the highest massifs of the crest zone. This has been a major enigma in relation to mountain-building processes.

Their *morphostructural identity* is puzzling because the boundaries of the mountain topography do not coincide either geographically or historically with the distribution of convergent crustal deformation. For example, the orogen extends eastwards as far as Provence, but the orogenic crust later underwent rapid extensional thinning in the Western Mediterranean. To the west, the Cantabrian Mountains occur as a topographic continuation of the Pyrenees, but they correspond to a south- instead of north-dipping crustal underthrust. In the Aquitaine basin, the outer fold and thrust belts of the orogenic wedge are buried beneath weakly deformed Paleogene and younger conglomeratic sequences at substantial distances from the range front. In the Ebro foreland, the mountain topography includes intensely folded and overthrust sequences of Paleogene conglomerate. Such discrepancies between the topographic relief of the Pyrenees as a mountain range and its synorogenic structural fabric require the incidence of significant overprints by post-orogenic tectonic processes, as discussed in Sects. 13.3 and 13.4.

The *morphoclimatic identity* of the mountain range hinges on the convergence between Atlantic and Mediterranean meteorological influences. Precipitation contrasts between the north- and south-facing range fronts are sharp because of

Fig. 13.2 (continued) (b) Structure and geomorphology of the Têt watershed. Key: 1 Paleogene thrust faults and sheets, 2 Other faults, major Neogene normal faults associated with faceted scarps, 3 Lower Miocene clastic deposits of the Conflent and Roussillon basins, 4 Upper Miocene clastic deposits of the Cerdagne Basin, including Col de Fins formation in the Conflent Basin, 5 Pliocene (marine and continental) deposits of the Roussillon Basin, 6 Quaternary fluvial and glacifluvial gravel terraces, 7 major frontal moraines and associated till deposits. ML Mont-Louis, V Villefranche-de-Conflent, Vt Vernet-les-Bains, PG Pla Guillem, P Prades, R Rodès, I Ille-sur-Têt, A Arles-sur-Tech. Numbered geomorphosites: 1 Carlit plateau, 2 Col de La Perche knick-point, 3 Conflent Basin and Mt. Canigou, 4 Orgues d'Ille-sur-Têt

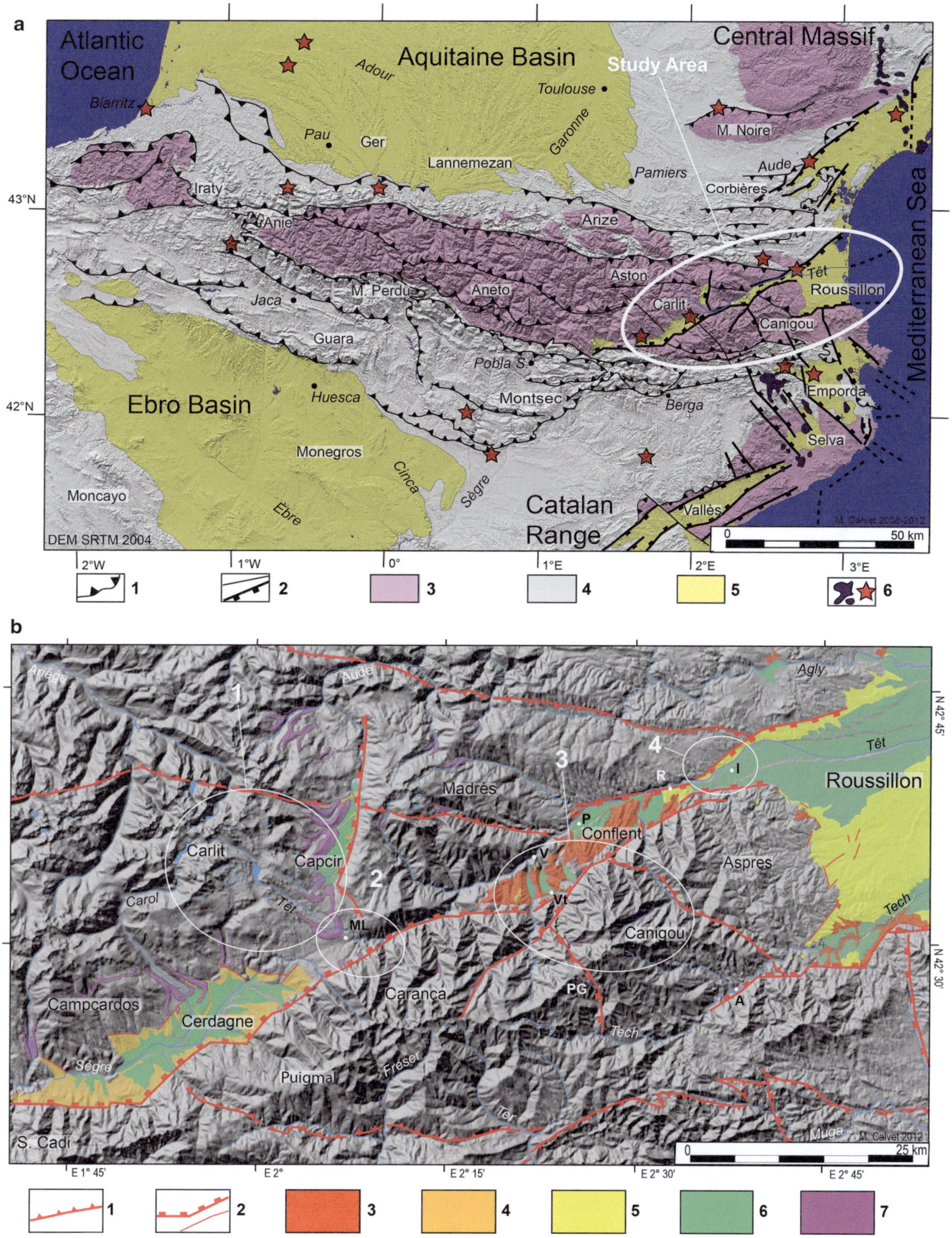

Fig. 13.2 The Pyrenees, with location maps of the Têt watershed area. (**a**). Structure of the Pyrenean range. Key: *1* Paleogene thrust faults and sheets, *2* Other faults, major Neogene normal faults, *3* Hercynian basement, *4* Sedimentary marine cover (Trias to Eocene) and Paleogene continental molasse, *5* Neogene continental and marine sediments, *6* Neogene and Quaternary volcanism, Quaternary neotectonic markers.

Fig. 13.3 The Paleic surface at 2,800 m (0.25 km^2), here at the northeastern base of Pic Carlit (2,921 m). Note the patterned ground with upturned rock slabs. In the background: distant crest zones of Campcardos, Andorra and other massifs in Ariège (Credits: M. Calvet)

the ~1,000 km long Cantabro-Pyrenean topographic barrier to Atlantic weather systems. These conditions also prevailed throughout the Pleistocene, with an imprint left in today's scenery by the glacial and fluvial incision of bedrock.

13.2.2 Geological Setting

Convergence, collision and underthrusting of the Iberian Plate beneath the European followed in stages between the Late Cretaceous and the Oligocene, continuing until ~20 Ma in the west (Vergés et al. 2002). Estimated crustal shortening is 165 km in the Central Pyrenees, 125 km to the east, and ~80 km to the west. The resulting orogenic wedge is an asymmetrical, south-vergent stack of Hercynian basement thrust sheets. These form large outcrops in the Axial Zone (Fig. 13.2a), which in the east consists mainly of granite, schist and augengneiss and includes tightly folded Paleozoic sedimentary rocks including Devonian limestone. Triassic to Eocene marine shale and limestone form the belts of external sierras, which are much narrower in the north than in the south. These basement cover rocks converge westwards and merge at the Pic d'Anie (Fig. 13.2a), where they still form an envelope of cap rock over the Axial Zone. South of the Axial Zone, their outcrops locally form the highest summits of the mountain range, e.g. at the Pic d'Anie, Mont Perdu and Serra de Cadí.

The Pyrenean orogeny has produced a lithospheric root >100 km thick, and the crustal root itself exceeds 50 km beneath the Central Pyrenees. These attributes change rapidly eastwards due to relief-transforming processes during the Neogene. In the Middle Oligocene, rifting and southeastward drift of the Corsica–Sardinia block marked the opening of the Western Mediterranean back-arc basin, forming the Gulf of Lions. The crust thins rapidly just 20 km east of the current Roussillon coastline, and much of the lithospheric root has disappeared beneath the Eastern Pyrenees (Gunnell et al. 2008). In the Central Pyrenees, geophysical evidence suggests that the root is hot and partially melting. A dozen extensional basins were generated in the Eastern Pyrenees, fragmenting the topography and progressively zipping the orogen open from east to west down its middle between the Late Oligocene–Early Miocene (Roussillon, Conflent, Vallès-Penedès and several other smaller basins in the Corbières) and the Late Miocene (Cerdagne, Capcir, Seu d'Urgell, Emporda, Selva). The older basins (Vallès, Roussillon) were reactivated during the Late Miocene and Pliocene (Fig. 13.2a, b).

13.3 Landscapes and Geomorphosites Along the Têt Valley

13.3.1 Low-Relief Plateaus of the Carlit Massif

The summit of the Carlit massif (2,921 m) provides the best panoramic vistas of the area. It allows direct appreciation of two generations of Cenozoic erosion surfaces, both of which have been lightly retouched by the cold-climate surface processes of the Pleistocene. The most elevated plateau consists of a series of flat-topped summits, the most spectacular occurring ~2,800 m a.s.l. immediately to the NE of Pic Carlit (Fig. 13.3). These vestigial plateaus, which bevel outcrops of schist, hornfels and granitic rocks, define a Paleic (or skyline) surface. Further west, these Paleic remains are even more widely preserved (3 km^2 in the Campcardos massif), likewise at elevations between 2,700 and 2,900 m. These summit surfaces were not covered by glaciers. Large patterned ground features, currently inactive and partly covered by soil supporting alpine grass- or heathland, are widespread.

Fig. 13.4 The Miocene Pediment in the Carlit, looking south. Here (2,000–2,200 m) it is weakly degraded by glacial erosion, forming fjell scenery. On the skyline, note the Puigmal massif and another vestige of the Pediment (*P*), distinguishable on the southern edge of the Cerdagne basin (itself not visible) (Credits: M. Delmas)

The steep edges of the Paleic surface in the Carlit are scalloped by numerous embayments, and the entire unit stands above a second shelf of low relief between 2,400 and 2,100 m known as the Pediment (Calvet and Gunnell 2008). This younger surface functioned as a fjell during the Pleistocene, currently with a cover of heathland, peat and lakes over glacial deposits (Fig. 13.4).

This double tier of erosional plateaus occurs in most massifs of the Eastern Pyrenean crest zone, but it occurs also at increasingly lower mean elevations in the external sierras north, east and south. These surfaces have been age bracketed by cross-correlating stratigraphic, palaeontological and thermochronological criteria (Calvet 1996; Calvet and Gunnell 2008; Gunnell et al. 2009). In the Carlit and all other sites sampled in the Pyrenees, apatite fission-track (AFT) and apatite helium (AHe) dating together indicate that the entire population of erosion surfaces shares the same regional geomorphic history. Presumably because of declining energy in the topographic system, denudation decreased from syn-orogenic peak rates during the Paleogene to minimum rates ca. 30–25 Ma. Near the shores of the Mediterranean, the Pediment grades topographically to extant outcrops of littoral sediments of Middle Miocene age. On limestone outcrops, e.g. in the Corbières, surface karst cavities affecting the Pediment have preserved rich concentrations of micromammalian fossil remains with palaeontological ages consistently spanning the entire Miocene. Such constraints provide a Middle Miocene terminal age for the Pediment. By deduction, the Paleic surface was formed earlier, between 30 and 20 Ma in agreement with the thermochronology.

The plateaus are remarkable examples of transient landscapes, not only where these low-relief landforms have been preserved for millions of years in a dynamic, steepland environment but also where substantial depths of in situ grus occur and reach several tens of metres in some elevated massifs of the crest zone. Quaternary glaciers have selectively scoured away the weathered rock, with glacial till covering residual saprolite profiles at sites scattered across the Carlit fjell (Fig. 13.4). Around the Carlit, the Pediment and the Cerdagne and Capcir basin floors have provided local base-level conditions well suited to the preservation of terminal moraines (Figs. 13.2b and 13.5). The remains of several Middle and Late Pleistocene glacial cycles have been identified and mapped in the area (Calvet et al. 2011). The youngest

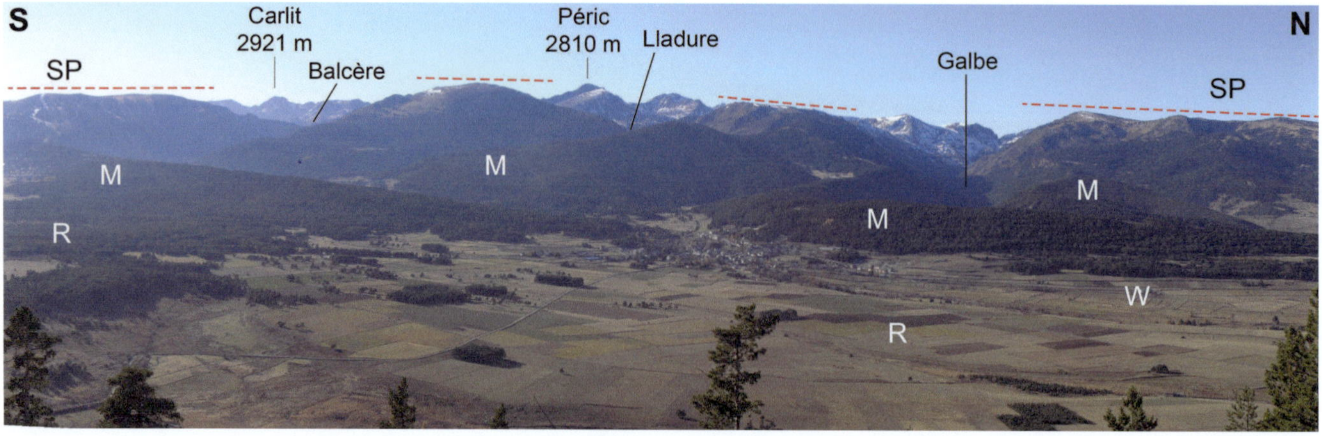

Fig. 13.5 Terminal moraines in the Capcir basin, here viewed towards the west. In the background, the Carlit massif (*SP* Paleic surface) and, from south to north, the glacial troughs of Balcère, Lladure and Galbe. The frontal and lateral moraines (*M*) of Würmian and Rissian (MOIS 6) ages are covered by pine forest; they meet with cultivated and partly forested glacial outwash fans in the foreground. *W* Würmian fan; *R* weathered Rissian (MOIS 6) or Middle Quaternary fans (Credits: M. Calvet)

frontal moraines at Mont-Louis (Têt glacier) and in western Cerdagne (Querol glacier) have yielded ages of 21–24 ^{10}Be ka (Delmas et al. 2008).

13.3.2 The Conflent Basin: The Imprint of Fluvial Knickpoints and Active Tectonics

Before it reaches Mont-Louis, the Têt valley is lightly incised, and its frontal moraines were deposited on the edge of the Col de La Perche plateau (1,577 m, cf. Fig. 13.2b). The latter consists of a population of coalescing rock pediments below the Carlit Pediment. Their Pliocene age is inferred from the fact that they bevel both the Hercynian bedrock and uptilted Upper Miocene sedimentary rocks of the Cerdagne fill sequence (Calvet 1996; Calvet and Gunnell 2008). Beyond Mont-Louis, the Têt cuts a deep fluvial gorge, and its bedrock channel descends in just 4 km from 1,600 to 1,100 m a.s.l. Highway N116 follows the gorge and offers spectacular views of this large knickpoint (Fig. 13.6).

The Conflent Basin (Figs. 13.2b and 13.7) is an asymmetric Neogene half-graben filled by coarse continental clastic deposits of Aquitanian to Early Burdigalian age. It also contains suspected Upper Miocene sediments (Col de Fins sequence) and, in its distal eastern part, Lower Pliocene cut-and-fill deltaic and debris-flow sediments confined to a Messinian ria. These younger deposits spread out eastwards into the Roussillon Basin. The Têt river has cut two short epigenetic gorges into underlying Devonian limestone (at Villefranche) and granite (at Rodès). Incision occurred by superimposition of the river on the clastic fill and is thus a post-Pliocene feature.

The most outstanding geomorphological feature of the area, however, is Mt. Canigou: its northern flank forms a 2 km high fault scarp with remarkably fresh-looking triangular facets at its base. These can be inspected by following the D27 from Escaro to Vernet and Taurinya (Fig. 13.8). This segment of the Têt fault has not been substantially displaced since the Middle Pleistocene, its fresh appearance being instead ascribable to the strike and steep dip of the faulted augengneiss. The stages of recent uplift can be reconstructed by comparing clast provenance in the Conflent Basin stratigraphy with the Hercynian outcrop stratigraphy of Mt. Canigou (Fig. 13.7, cf. supra): rifting ca. 23–18 Ma promoted rapid denudation, which explains why Mt. Canigou, unlike the Carlit and other massifs of the flat-topped crest zone, exhibits no vestiges of the Paleic surface. Even remains of the Middle Miocene Pediment are scarce, but a notable vestige occurs along the crest (Fig. 13.9). Overall, tectonic relief on the north face of Mt. Canigou was formed during the last 10 Ma because none of the Lower Miocene clastic deposits in the Conflent Basin correlates with the currently outcropping bedrock exposures of Mt. Canigou. Such intense and rapid unroofing of the Canigou massif explains the present-day dominance of serrated alpine relief (Fig. 13.1), sharpened by Pleistocene glaciers and out of character with the flat-topped scenery elsewhere. Steep slopes and the position of Mt. Canigou as an elevated outpost exposed to autumnal Mediterranean storms also explain the widespread scars of intense torrential activity in the area. Boulder fans generated by the catastrophic floods of 1940, 1763, 1421, 1264 and 878 can be appreciated around Vernet-les-Bains.

13.3.3 The Orgues d'Ille Geomorphosite: An Epitome of Late Cenozoic Landscape Evolution

Lodged against the northern fault of the Roussillon, the site known as *les orgues* d'Ille-sur-Têt (Fig. 13.10) records a

Fig. 13.6 The Têt knickpoint of Mont-Louis as seen from Highway N116. Note the La Perche Pediment in the background and the beginning of the Têt gorge below the town of Mont-Louis (Credits: M. Calvet)

condensed geomorphological history of the Eastern Pyrenees. It can be reached by routes D2 and D21. The 'organ pipes' are earth pillars in deeply dissected badland terrain, providing a vivid reminder of the intensity and flashiness of Mediterranean rainfall regimes. The natural vertical exposures also reveal the stratigraphy of the Pliocene, which based on mammalian fossil deposits elsewhere in the Roussillon Basin (type locality of the continental chronostratigraphic stage known as the Ruscinian) only represents the lower portion of the series (5.3–3.5 Ma). This clastic sedimentation corresponds to the Zanclean stage, which occurred during an abrupt marine transgression itself following the abrupt Late Messinian regression. The sediments thus clearly fill the Têt palaeovalley, which became a ria during the Early Pliocene (Clauzon et al. 1987). Transition to the Lower Pliocene marine facies, distinctive by its shell-rich content, occurs a little to the east of Ille, in the badlands area of Néfiach. At Ille, the blue-to-grey facies is deltaic and the texture sandy clay with pebble beds (best inspected in the northern exposures of the site). Pebble roundedness indicates intense rolling in a shoreface environment. Locally, collapse breccia containing large, angular boulders of granite represents lateral, short-range input from the granitic footwall. Ochre-coloured, high-energy topset beds >100 m thick overlie the deltaic sequence in the southern exposures of the site. The facies and its preserved bedforms are diagnostic of a braided river system in which sand and gravel sequences alternate with thicker channel bars containing cobbles up to 1 m in diameter. Such a thick and coarse-grained clastic sequence clearly indicates coeval mountain-building processes in the Canigou area, which based on the relative abundance of augengneiss pebbles was the main source of debris for the Lower Pliocene formation.

Neotectonic features affecting the Pliocene sequence are another highlight at Ille. Part of the deformation was synsedimentary, but the entire Pliocene sequence is tilted northwards towards the master fault, and the oldest Quaternary alluvial terrace deposit of the Têt caps the entire sequence with a marked angular unconformity (Fig. 13.10). The Pliocene beds are internally displaced by a series of N 40°E step faults which have lowered the continental topset beds and placed them on a level with the marine deltaic sequence. Sheared cobblestones and local fault polish features reveal left-lateral displacements, sometimes with a reverse-fault component. The tectonic style of crustal deformation has, therefore, been complex, probably involving several episodes. The topmost Pleistocene terrace deposit itself appears vertically offset by ~10 m over the length of the outcrop exposure.

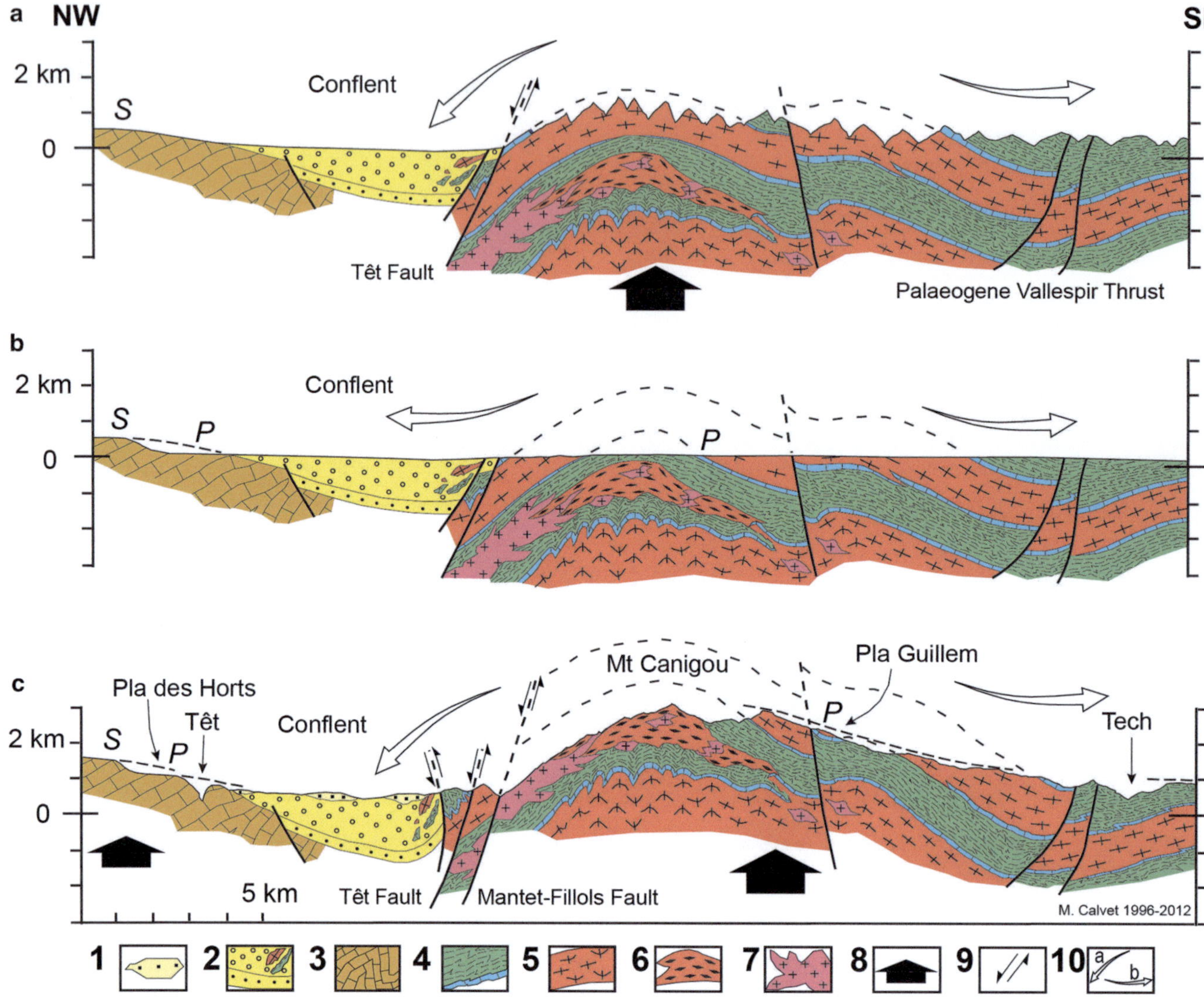

Fig. 13.7 Post-orogenic evolution of Mount Canigou in the Neogene. (**a**) Kilometre-scale uplift was accompanied by coarse clastic infilling of the Conflent Basin between 25 and 18 Ma. Sediment input from the north (Madres massif) was subdued after Aquitanian time due to limited uplift of the northern flank (Marquixanes arkoses). Meanwhile, the southern flank recorded an influx of schist clasts and olistoliths followed by an exclusive input of olistoliths and boulders of augengneiss. (**b**) Relative tectonic stability and opportunities for pedimentation (18–10 Ma). Erosion only supplied fine-grained sediment to the Roussillon and Mediterranean basins (sandy and argillaceous molasse). (**c**) Latest stage of kilometre-scale uplift (10–0 Ma) responsible for the present-day relief. Denudation has reached the core of the Canigou metamorphic dome structure. Micaschist, granite, migmatite and Casemi leptynite, which together form the entire exposure of the current scarp, all occur in clastic material at the Col de Fins (Upper Miocene) and in the Conflent Basin (Pliocene). Key: *1* Col de Fins Formation, *2* Aquitanian and Lower Burdigalian, *3* Devonian marble and limestone, *4* Lower Paleozoic micaschist and basal marble, *5* Augengneiss and migmatite among the lowest outcrops, *6* Casemi leptynites, *7* Two-mica granite core, *8* Neogene vertical uplift, *9* Active fault, *10* Clastic input to the Conflent Basin (a), and clastic outflux towards the Roussillon Basin and the Mediterranean rifted margin (b). *S* Paleic surface, *P* Middle Miocene Pediment

13.4 History of the Landforms

The geomorphological history of the Eastern Pyrenees begins with the peak of Paleogene continental collision, which based on apatite fission-track and apatite helium data, involved the denudation of at least 10 km of crust. Denudation rates subsequently declined, and by the Oligocene–Miocene transition, a regional land surface of low relief prevailed. The short geological time to completion of this Paleic surface can be explained by the narrowness of the Pyrenean range in the Paleogene, by its fragmentation into discrete tectonic blocks due to extensional tectonics, and possibly also by a dense lithospheric root pulling downwards and delaying the full potential of isostatic rebound. Rifting during the Early Miocene generated substantial local relief in certain areas such as Mt. Canigou (Fig. 13.7, cf. supra), but this was an

Fig. 13.8 The Canigou fault scarp near Taurinya. The best preserved faceted spurs (centre) are 500 m high. The Têt master fault occurs exactly at the base of the facet. This north-facing slope of Mt. Canigou was weakly glaciated and is deeply incised by fluvial gorges. Note the Fillols avalanche corridor (1.4 km of local relief), known to produce infrequent, high magnitude avalanches. One occurred in January 1986 (Credits: M. Calvet)

exception. Other morphotectonic units were not affected with comparable magnitude. Rifting initiated a cycle of topographic rejuvenation in response to newly forming marine (Mediterranean) base levels, with the younger generation of pediments (e.g. the Carlit fjell) forming 200–500 m below the Paleic surface at initial elevations that would not have exceeded 800 m even in the most internal parts of the orogen's catchments (Gunnell et al. 2008).

It follows that at least two thirds of the topographic relief we see today formed almost entirely during the Late Cenozoic, i.e. after 10–12 Ma, in several stages. The driving mechanism of rapid crustal uplift in the Eastern Pyrenees is ascribable to melting of the dense lithospheric root (Gunnell et al. 2009), accompanied by sustained crustal extension along the Mediterranean back-arc margin. During the latest Miocene, N–S convergence resumed but was too short-lived to contribute measurably to mountain building. Regional uplift of the Pyrenees, elevating gently undulating land surfaces such as La Perche to altitudes exceeding 1,500 m, caused rapid incision of the valleys during the latest Pliocene and Quaternary. Stages of fluvial incision can be reconstructed from five generations of alluvial terraces in most catchments, with three of these directly correlating to glacial moraines. Fluvially driven landscape evolution in the Têt catchment dominated largely over glacial, the latter remaining slight and localised, with an equilibrium line altitude between 2,200 and 2,000 m involving mostly cirque glaciers and glacial denudation rates (around 5 m for the last 100 ky glacial cycle) plotting in the lower bracket among known global rates (Delmas et al. 2009). The largest Würmian outlet glaciers did not exceed 20 km in length and never extended below the 1,200 m contour. These conditions are best appreciated by comparison with the north-facing side of the Pyrenees, where the Ariège glacier was ~70 km long and descended to elevations of 370 m near Foix.

Fig. 13.9 Vestiges of the Miocene Pediment (Pla Guillem, 2,300 m), looking west from Mt. Canigou. In the background: crests of the Carança massif (2,881 m), and deeply dissected Tech catchment (Credits: M. Calvet)

13.5 Conclusions

The geomorphosites of the Têt catchment provide exceptional insights into deep geomorphological time and hence into the evolution of large landscape features, but the impact of mankind provides additional cultural value to finer details of the scenery. Being located at a bioclimatic crossroads between the Mediterranean and Atlantic biomes, the area is a local biodiversity hotspot. It has attracted botanists since the Renaissance and was the birthplace of phytosociology under pioneers such as C. Flahault, J. Braun-Blanquet and H. Gaussen. The draw to local natural resources is highlighted in herds of ibex depicted by Magdalenian hunters on a large schist boulder near Campôme (Conflent Basin) and in the widespread legacy of stone engravings by migrant herdsmen since the Neolithic – with co-occurrences of Iberian, Medieval and Modern inscriptions on single rock slabs (Abélanet 1990). Shepherd's stone huts in the Carlit have been continuously occupied for millennia (Rendu 2003). Catalan bloomeries (Medieval blast furnaces), antique slagheaps, charcoal combustion sites, drift-mining adits and wrought iron strap hinges on eleventh- and twelfth-century Romanesque chapel doors all further testify to prosperous iron ore smelting activity around Mt. Canigou.

Being situated in the marches of the Carolingian empire, the area also performed as political barrier, with high population densities of Iberian refugees already operating in the ninth century as an effective deterrent against Muslim invaders (Bonnassie 1990). The villages subsequently functioned as a rearguard to the Reconquista. As a dynastic homeland of Barcelonian nobility, the Têt valley became the cradle of national identity and nurtured Catalonia's accession to statehood. The ribbon of large abbeys surrounding Mt. Canigou, with cloisters built from a variety of locally quarried marbles, testifies to this glorious past: Saint-Martin-du-Canigou, Saint-Michel-de-Cuxa, Serrabonne, Sainte-Marie d'Arles and Ripoll. The myth of a Catalan nation rooted in these mountains is still alive today and met its strongest resurgence in the nineteenth century. This cultural renaissance was immortalised in *Canigò* (1886), an ode by Jacint Verdaguer to holy Mt. Canigou and to the legends of the Reconquista.

Fig. 13.10 The Orgues d'Ille badlands in Pliocene deposits. To the *right*: *grey* sand- and pebble-rich deltaic deposits (*sd*), with local collapse breccia (*cb*). To the *left*: ochre channel bars of the topset unit (*fb*). In the centre, faults mark the contact between the two facies. Note crowning outcrops of the oldest Quaternary terrace (*T*), possibly displaced by the faults. The topsets are tilted to the north. In the background: Aspres hills and Mt. Canigou, with glacial cirques and Würmian moraines below the crest. On the slopes of Mt. Canigou, note gullies formed during the torrential 1940 flood (Credits: M. Calvet)

References

Abélanet J (1990) Les roches gravées nord-catalanes (Thèse doctorat 3e cycle, 1977). Revista Terra Nostra, Prades

Besson F (2000) Le paysage pyrénéen dans la littérature de voyage et l'iconographie britannique du dix-neuvième siècle. L'Harmattan, Paris

Birot P (1937) Recherches sur la morphologie des Pyrénées orientales franco-espagnoles. Baillière, Paris

Bonnassie P (1990) La Catalogne au tournant de l'an mil, croissance et mutation d'une société. Thèse doctorat état 1973. Albin Michel, Paris

Calvet M (1996) Morphogenèse d'une montagne méditerranéenne: les Pyrénées orientales. Thèse doctorat état 1994. Documents du BRGM n°225, Orléans

Calvet M, Gunnell Y (2008) Planar landforms as markers of denudation chronology: an inversion of East Pyrenean tectonics based on landscape and sedimentary basin analysis. In: Gallagher K, Jones SJ, Wainwright J (eds) Landscape evolution: denudation, climate and tectonics over different time and space scales. Geological Society, London, Special Publications, 296, pp 147–166

Calvet M, Delmas M, Gunnell Y, Braucher R, Bourlès D (2011) Recent advances in research on Quaternary glaciations in the Pyrenees. In: Ehlers J, Gibbard PL (eds) Quaternary glaciations, extent and chronology, a closer look, Part IV, vol 15, Developments in Quaternary Science. Elsevier, Amsterdam, pp 127–139

Clauzon G, Aguilar JP, Michaux J (1987) Le bassin pliocène du Roussillon (Pyrénées orientales, France) exemple d'évolution géodynamique d'une ria méditerranéenne consécutive à la crise de salinité messinienne. C R Acad Sci Paris II 304(11):585–590

Ramond de Carbonnières (1792) Voyages et observations faites dans les Pyrénées; pour servir de suite aux Lettres de W. Coxe, sur la Suisse. Dumoulin, Liège

Delmas M, Gunnell Y, Braucher R, Calvet M, Bourlès D (2008) Exposure age chronology of the last glaciation in the eastern Pyrenees. Quat Res 69:231–241

Delmas M, Calvet M, Gunnell Y (2009) Variability of Quaternary glacial erosion rates – a global perspective with special reference to the Eastern Pyrenees. Quat Sci Rev 28:484–498

Gunnell Y, Zeyen H, Calvet M (2008) Geophysical evidence of a missing lithospheric root beneath the Eastern Pyrenees: consequences for post-orogenic uplift and associated geomorphic signatures. Earth Planet Sci Lett 276:302–313

Gunnell Y, Calvet M, Brichau S, Carter A, Aguilar JP, Zeyen H (2009) Low long-term erosion rates in high-energy mountain belts: insights from thermo- and biochronology in the Eastern Pyrenees. Earth Planet Sci Lett 278:208–218

Palassou B (1781) Essai sur la minéralogie des Monts-Pyrénées. Didot-A. Jombert-Esprit, Paris, reprint: 2007, La Découvrance, La Rochelle

Rendu C (2003) La montagne d'Enveitg, une estive pyrénéenne dans la longue durée. Thèse doctorat, 2000. Trabucaire, Canet

Vergés J, Fernàndez M, Martínez A (2002) The Pyrenean orogen: pre-, syn-, and post-collisional evolution. In: Rosenbaum G, Lister GS (eds) Reconstruction of the evolution of the Alpine-Himalayan Orogen. J Virt Explor 8:55–74

The Vosgian-Alsatian Side of the Rhine Graben: A Unique, Tectonically Controlled and Manmade Landscape

14

Jean-Michel Carozza

Abstract

Large-scale Alsatian landscapes are dominated by structural organisation inherited from the Cenozoic rifting of the Upper Rhine Graben (URG). This results in a subdivision into three main units, from west to east: the Vosges Mountains, the Vosges foothills and the Ill-Rhine plain. In each of these three units, the weight of structural control, i.e. tectonic or lithological, remains strong. Transition from the northern Vosges to the southern Vosges appears dominated by differential uplift and transition between crystalline rocks to Triassic sedimentary cover. In the Vosges foothills, tectonics factors are lesser than lithological one. Despite a common hilly shape, frequent lithological changes give rise to a mosaic of 'terroirs' of complementary interest, but chiefly dedicated to vineyard. The Ill-Rhine plain is organised along a downstream gradient under regional tectonic control. There, despite its young age and strong anthropogenic impact, the stigmata of tectonic remains marked in the landscape. These strong structural constraints and their consequences have been exploited through historical times as one of the most recognisable cultural landscape of France.

14.1 Introduction

The Rhine Graben and surrounding Vosges Mountains form one among the most prominent and well-known example of continental rift structure in Europe. Together, they constitute the western half part of the Upper Rhine Graben (URG). URG extends over 300 km long in a south-north direction between the cities of Basel in the south to Frankfurt in the north and has an average width of 50 km (Fig. 14.1). The German Black Mountains on the east flank complete the topography and give to the rift system an almost symmetrical structure. If the geological framework of the URG is well known, the origin and chronology of the Vosges Mountains/Rhine Graben relief formation are still debated (Fig. 14.2).

The geomorpho-diversity of the Vosges Mountains/foothill and Rhine plain is exceptionally rich and largely related to its Upper Neogene-Pleistocene to present geomorphological and tectonic evolution. This landscape shaped by tectonics shows a mosaic of both natural and cultural landscapes, from the wild peaks of the high sandstone Vosges covered by dark forests to the luminous landscapes of the Vosges foothills and its incomparable vineyard. Down to the Rhine valley, millennia of human actions have shaped the Ill-Rhine plain; yet it is where is still preserved one of the largest and wilder alluvial forest in Europe.

14.2 Geological Structure

Since the works of Elie de Beaumont in the early nineteenth century, the Vosges and adjacent Vosges hills – Rhine Graben – are described as a continental extensional feature. Geological works in the 1970s and 1980s recognised the half-graben asymmetric structure of the Vosges Mountains and its relation with the Middle to Upper Cenozoic extensional phase that largely affected the European continent in relation with Alpine collision (Sittler 1974; Illies 1975) thereafter named European Cenozoic Rift System (ECRIS).

The initiation of the URG, as recorded in the Middle Eocene (Lutetian) deposits, was followed by a paroxysmic

J.-M. Carozza (✉)
Department of Geography, University of Strasbourg,
3 rue de l Argonne, 67083 Strasbourg Cedex, Strasbourg, France
e-mail: carozza@unistra.fr; jmcarozza@yahoo.fr

M. Fort and M.-F. André (eds.), *Landscapes and Landforms of France*, World Geomorphological Landscapes,
DOI 10.1007/978-94-007-7022-5_14, © Springer Science+Business Media Dordrecht 2014

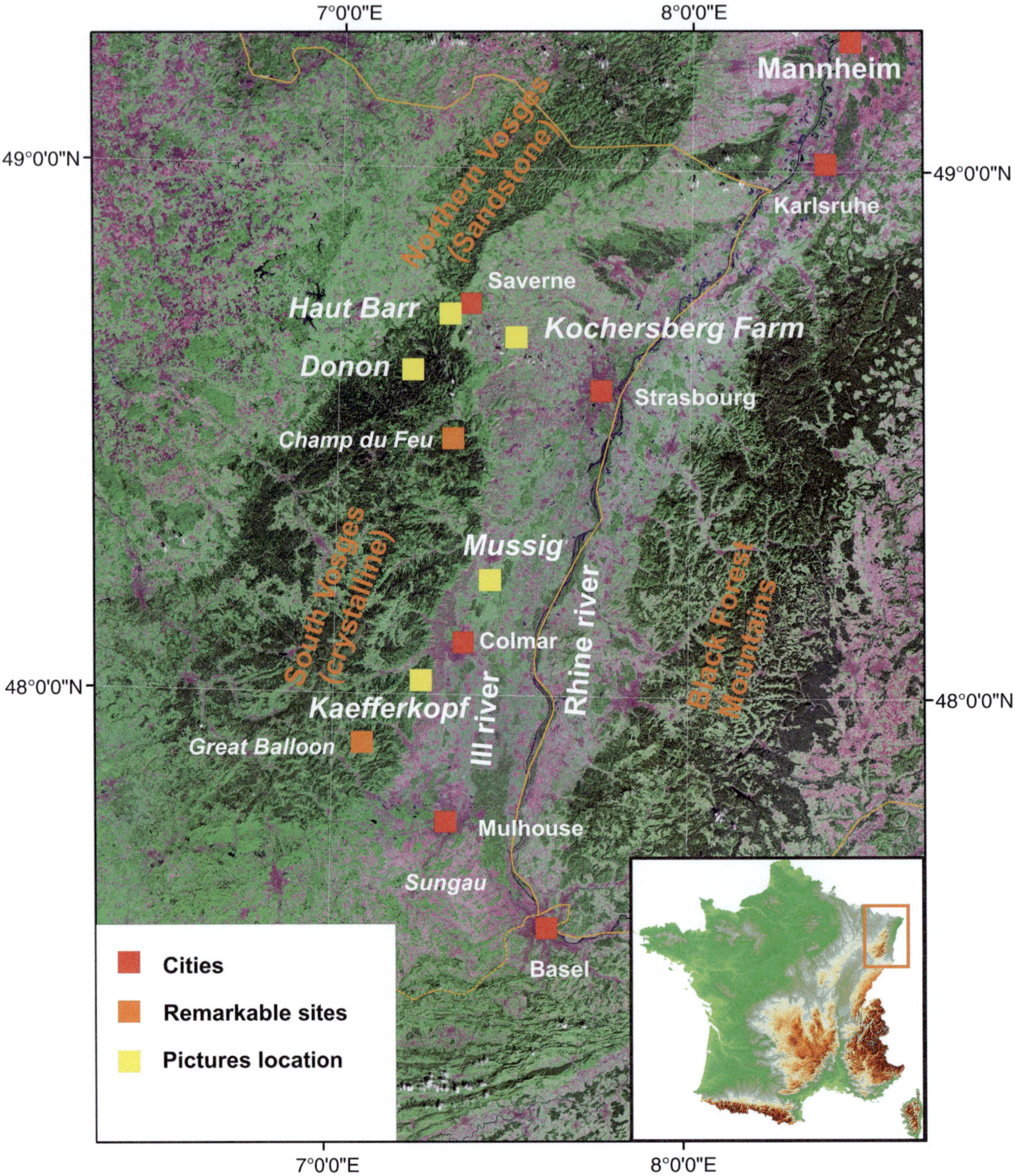

Fig. 14.1 Landsat satellite picture of the Upper Rhine Graben and main locations (J.-M. Carozza)

tectonic event from Upper Eocene to Early Oligocene (Priabonian). Subsidence affected subbasins and developed from the south to the north of the basin, whereas the rift shoulders individualised as evidenced by the petrography of the edge conglomerate deposits. During Upper Oligocene to Lower Miocene, the northern part of the URG was mainly affected by a new phase of active subsidence. Middle Miocene evolution is characterised in the southern part of the

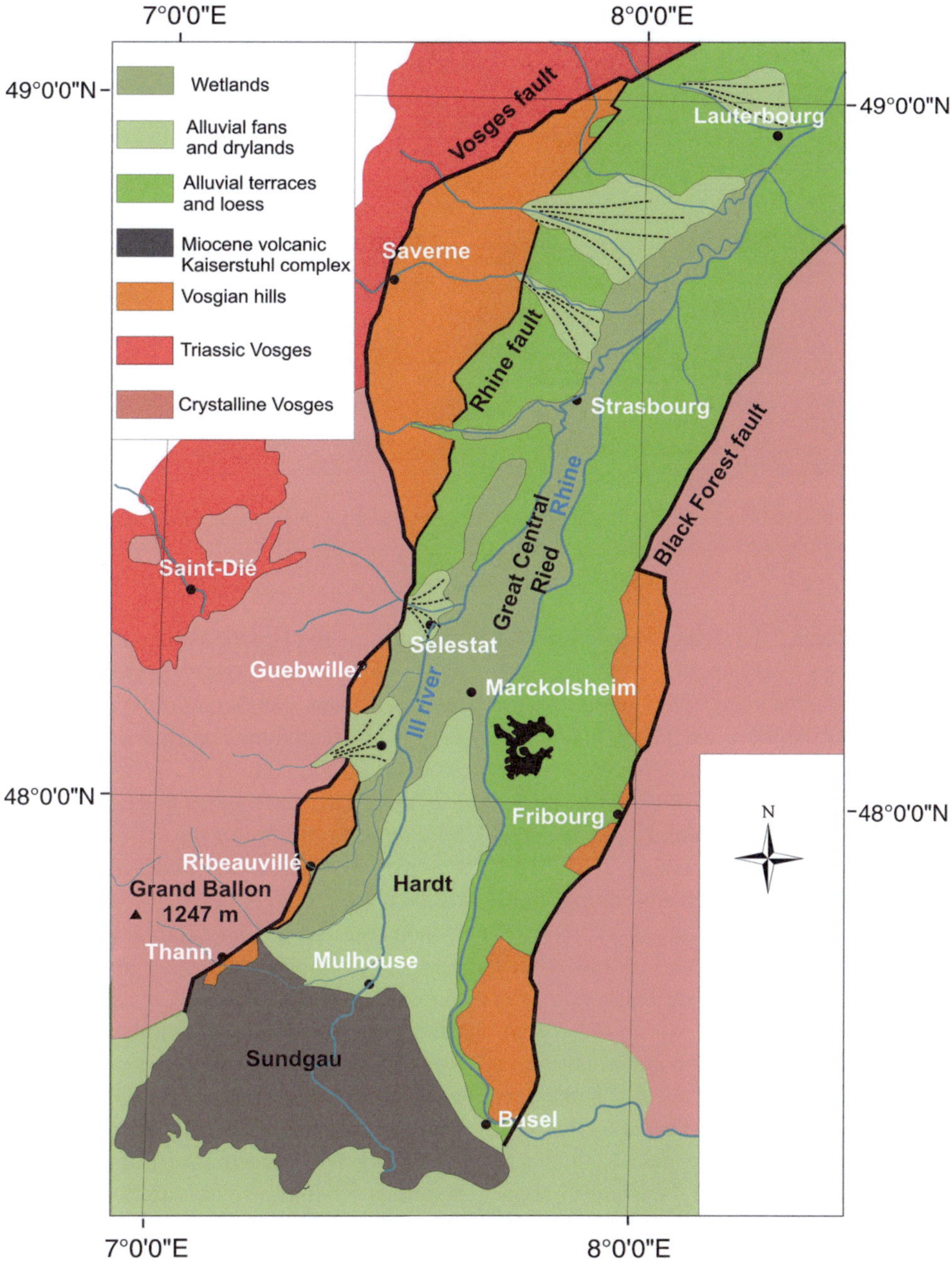

Fig. 14.2 Simplified geological map of the Rhine Graben (J.-M. Carozza, adapted from various sources)

URG by erosion, which affected rift shoulders (the Vosges and Black Forest Mountains). From Upper Miocene to present, change in tectonic regime to NW-SE compression led to a transpressive tectonics and sinistral fault reactivation. Upper Pliocene to Pleistocene tectonics resulted in differential uplift at the origin of present-day relief in the Vosges Mountains and the Rhine graben.

Total Cenozoic infill of the graben is more than 3.5 km thick especially in the northern URG (Karlsruhe basin); this implies a tectonic upthrow more than 4.5–5 km between the graben floor and the rift shoulders. Upper Pliocene to Pleistocene deposits are especially thick in the southern URG, south of Colmar (Sundgau conglomerate). This syntectonic formation was affected by faulting and folding associated with Jura deformation (Giamboni et al. 2004). Major change in regional drainage occurred during Messinian-Upper Pliocene with the piracy of the Upper Rhine, which was successively a tributary of the Aar-Danube (>4.2 Ma) and Aar-Doubs (>2.9 Ma) rivers before flowing into the Rhine Graben (see synthesis *in* Preusser 2008). The role of tectonics and erosion processes is yet unclear in this evolution.

Fig. 14.3 The Donon isolated hill (1,009 m a.s.l.). The Higher Donon is flanked by the Lower Donon on its east side. Both are Triassic residual hills lying on the pre-Triassic erosional surface (J.-M. Carozza)

Based on these considerations, the western French side of the URG, which corresponds to the region of Alsace, was divided into three main units – the Vosges Mountains *s.s.*, the Vosges foothills and the Ill-Rhine plain (Théobald et al. 1977). Each of them is separated from the other by a major tectonic feature, the Vosges fault and the Rhine fault, respectively. On the west, the Vosges Mountains has an almond shape elongated in an almost meridian direction. The Vosges Mountains can be divided into two relatively simple units based on lithological and geomorphological criteria. The southern part corresponds to an exhumed and eroded pre-Hercynian basement relieved from its Palaeozoic and Mesozoic sedimentary cover, whereas the northern part, from Molsheim and Champ du Feu to Mannheim, is a north tilted dissected hilly plateau of Palaeozoic sandstone (i.e. Triassic). The Vosgian foothills are a complex structural unit, which forms a transition from the Vosges Mountains to the Ill-Rhine plain. In the southern part of the Rhine Graben, the foothills zone is merely absent. Elsewhere, its width varies between a few Kilometers to more than 13 km and looks like a hilly area and offsets more or less horizontal levels. This unit is bounded by the Rhine fault, often a blind fault, with the Ill-Rhine alluvial plain extending eastwards. Each of these units exhibits remarkable landscapes.

14.3 The Northern Vosges Mountains

If the south crystalline Vosges are a remarkable landscape as a result of the importance of inherited pre-Triassic landscape and glacial imprint (see Chap. 16, this volume), the northern sandstone Vosges are also a typical and fascinating forest landscape, at the junction of the provinces of Alsace and Lorraine.

North of the Bruche valley, the Vosges still have part of their sedimentary cover (Middle Triassic Buntsandstein sandstone and conglomerate) preserved above crystalline basement, due to lower uplift and higher resistance to erosion processes of the sandstones. In the northern part of the crystalline Vosges, the sandstones first appear in the form of isolated hills directly resting on the Hercynian- and Permian-planed basements, and lithological contrast and shaping gave rise to residual relief, like in the Schneeberg, the Mutzig rock or the Donon (Fig. 14.3). This last one is a remarkable site on both geological and archaeological point of view, which explains the attractiveness of these landscapes. The site looks populated from the Neolithic period and Bronze Age, but during Roman times, between the second and third century AD, a place of worship of great importance has developed there as evidenced by the numerous votive offerings found around.

Further to the north, the Buntsandstein sandstone cover becomes continuous, and landscapes of the Vosges appear as a vast plateau tilted towards the northwest and incised by broad valleys with straight slope, such as the Zorn or the Moder valleys. Due to the poor quality of acid soils developed on sandstone, slopes and plateaus remain under forest and gave birth to one of the largest and beautiful Alsatian forest (beach groves and fir forests, *Fagus sylvatica* and *Pinus sylvestris* or *Picea abies'* single-crop farming). The wilderness of the landscape is reinforced by the presence of scraggly sandstone outcrops and conglomerate cliffs, like at the Dabo rock.

These valleys contribute to the openness of the landscape and have formed long-standing major communication axis between the Germanic and the Gallic world, as evidenced by the Gallic oppidum of Saverne and its remarkable defence system, a *murus gallicus* (Celtic wall or rampart) described by Julius Caesar in his *De bello gallico*.

Fig. 14.4 Built during the eleventh century AD, the Haut-Barr castle rests upon the Saint-Odile conglomerate (Upper Middle Buntsandstein). Towering the hillslope, it controls access to the Lorraine plateau and Paris Basin and to the Zorn valley (*right*)

Upper hillslopes are often reinforced by highly resistant Upper Middle Buntsandstein – Saint-Odile conglomerate – shaped into tower-like typical landscapes. This is particularly dramatic at the Haut-Barr castle (Fig. 14.4). There the Vosges fault towers by nearly 200 m the Saverne foothills landscape.

14.4 The Saverne Fault 'Field' and Its Cultural Heritage

14.4.1 A Faulted Zone

The Vosges foothills, at an altitude of about 400–200 m a.s.l., form a transition zone between the Vosges and the Rhine Graben. This unit is bounded on its west side by the Vosges fault and on its eastern flank by the Rhine fault. The width of this unit is variable and four subunits are individualised: the Thann, Guebwiller, Ribeauvillé and Saverne fault compartments (Fig. 14.2), this latter being the largest and the most complex one (Figs. 14.1 and 14.2). The geomorphology of these compartments depends both on the relative position of the two main faults (i.e. the Vosges and Rhine faults) that determines the width of the faulted zone and on the sedimentary rocks cropping out. Nearly 30 km long and 20 km wide with an almond shape, the morphology of the Saverne fault 'field' consists of a hilly relief corresponding to a dissected limestone plateau in the west, such as the Bastberg hills, that gives way to a series of little dissected plateaus in the east (Kochersberg and the Outre-Forêt region). Sittler (1974) showed the structural complexity of the Ribeauvillé fault zone and the resulting high variety of '*terroirs*' (Fig. 14.5).

14.4.2 A Land Prone to Vineyard

Forming a discontinuous band from south to north of the Rhine Graben, the Vosges foothills are the field of election of the Alsatian quality vineyards in particular around Ribeauvillé and Guebwiller where their more southern situation and climate are most appropriate to quality grapevine cultivation (Fig. 14.6). This is on west facing hillslopes that the most famous Alsace vintage was created. Since the Middle Ages, the culture of the vine is organised around villages that have preserved their typical architecture, both in the plains and on the hillslopes. Among the 51 vintages, '*Grand Cru*' of Alsace, limestone and marly limestone are the dominant substrates (respectively in Schoenenbourg, Eichberg and Brudertal, for example). Sandstones (Kessler), granite (Sommerberg) or volcanic rocks (in the rare Rangen of Thann) also give rise to unique wines. There, the question is for the winemaker to adapt the better local grape variety (riesling, gewurtztraminer, sylvaner, muscat and white, grey and black pinot) to the soil's properties to give rise to balanced wines. Wealth resulting from wine production together with the limited effects of the Thirty Years' War in these regions favoured both the preservation of the traditional architecture inherited from the Middle Ages. Many of these villages still have their mediaeval fortifications wall and a densely populated city centre inherited from Late Mediaeval to Early Modern periods (Fig. 14.7).

Despite a less favourable situation, vineyards are not absent from Saverne fault zone, and the vineyards of the 'Couronne d'Or' and its great vines like the limestone Steinklotz '*Grand Cru*', or high-quality *terroir* like Bruderbach – on dolomitic sandstone – have long supplied the city of Strasbourg.

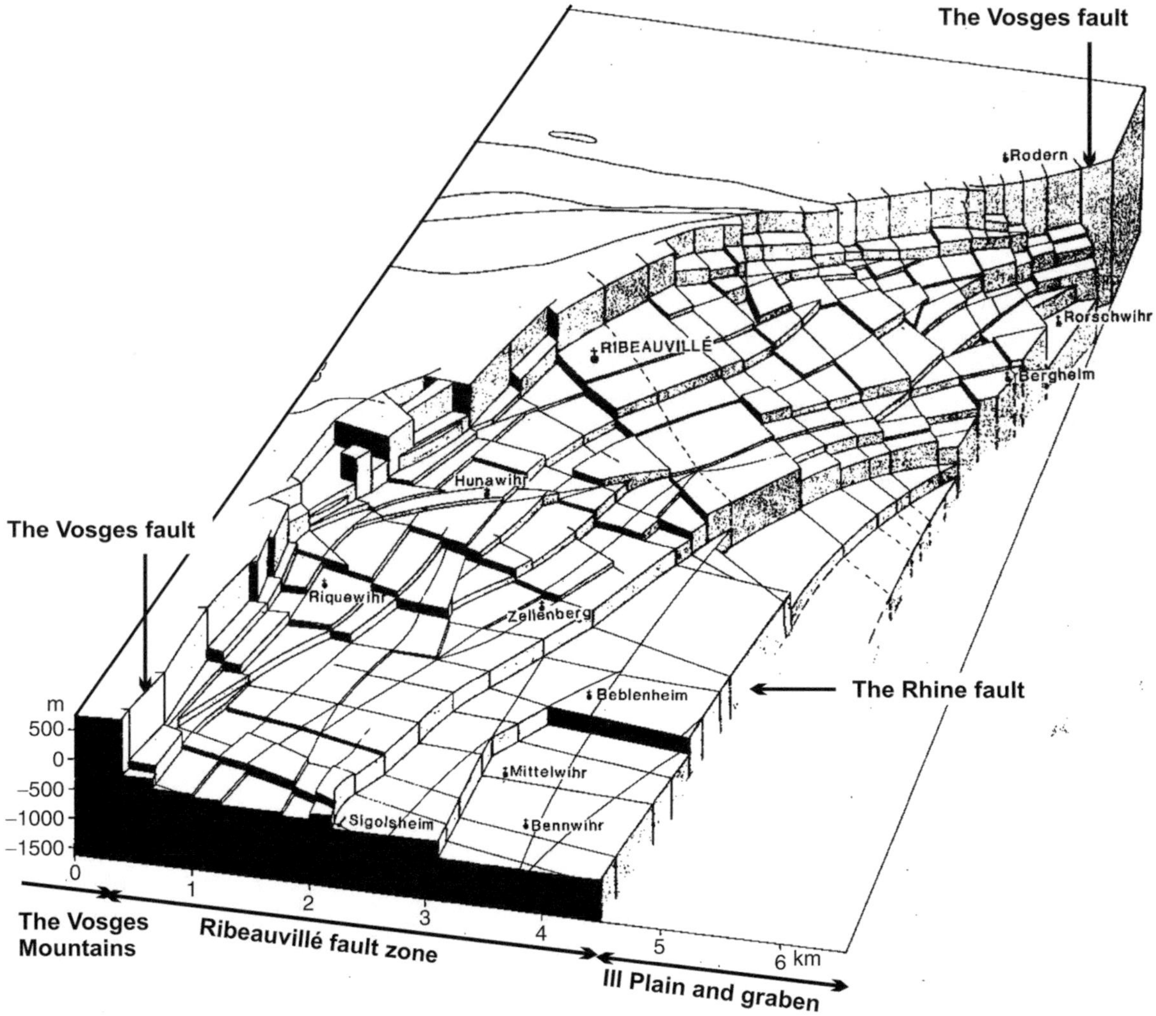

Fig. 14.5 Block diagram of the Ribeauvillé fault zone (Modified after Sittler 1974). Geological investigations have shown the structural complexity of the Vosges foothills zone, which explains the extreme variability of vineyard 'terroirs'

14.4.3 Lower Foothills Terraces: An Active Tectonic Area with Evidence of Early Man Occupation

East alluvial terraces present a thick loess cover that hides the structural framework. Thanks to loess quarrying like in Achenheim, loess sequences of major palaeoclimatological, paleontological and archaeological interest were discovered (Wernert 1957). There, four major climatic cycles covering Middle to Upper Pleistocene (last 600,000 years) are recorded, separated by palaeosoils; these are a European reference site for climate evolution during the Pleistocene (Lautridou et al. 1985). Stratified lithic industries allowed the description of the transition from Lower Palaeolithic (300,000 years BP) to Early Middle Palaeolithic (Mousterian, 200,000 years BP, Junkmanns, 1995).

Archaeological soils are affected by normal faulting that attests Middle to Upper Pleistocene tectonic activity of the URG (Baize et al. 2002) and its control on the local topography of the Vosges foothill and Ill-Rhine alluvial plain. Indeed the region is still of high seismic activity (Ritter et al. 2009), especially in the southern part of the URG as attested by the 1356 AD Basel earthquake and its consequences in Alsace (Fig. 14.8). Some people believe that the half-timbered architecture, which is present in Alsace, is a consequence to earthquake hazards because of its paraseismic properties.

Very fertile black soils developed on loess terraces were the first to be occupied and cropped during Early Neolithic time (~5400 BC) by the 'Danubians', a Central European cultural affinity group. Alsace was one of the first regions in France to experience Mesolithic to Neolithic transition as a consequence of the influence of the southern regions on autochthonous people. From there arises a potential Initial Neolithic as early as ~5800 BC (Jeunesse and Schnitzler 1993) documented by both ceramics (La Hoguette and Limbourg) and early palynological evidences of cereal cropping. The

Fig. 14.6 The Vosges foothills around Sigolsheim and the Weiss valley. Above the village of Ammerschwihr, locates the "Grand Cru" on Kaefferkopf granitic soils. The forest approximately outlines the Vosges fault positioned along the hillslope (Picture courtesy of P. Paul)

Fig. 14.7 Typical farming house of the Kochersberg (see Fig. 14.1 for location): house with multiple roof or home-court (Mehrdachhüs)

first Early Neolithic settlements are well known for their characteristic habitats (large rectangular houses) and vast necropolis like in Lingolsheim, Entzheim or Rosheim villages. It seems that the first human occupation within the alluvial Rhine plain did not appeared prior to end of the Neolithic times.

Fig. 14.8 Sebastian Münster (1488–1552). The consequences of the Basel 1356 earthquake (From *Cosmographia,* wood carving, around 1544)

14.5 The Ill-Rhine Plains: Between Hardt and Rieds

The Alsace plain, 20 km wide, corresponds to the Upper Pleistocene and Holocene alluvial plain of the Ill River on the west part and the Rhine River on the east. In the south of the URG, the two alluvial plains are parallel, separated by a wide gravel terrace. This broad and flat interfluve is named the Hardt. Further north, the two rivers form a vast wetland (the Central Ried) until they joined north of Strasbourg.

14.5.1 The Hardt

The name 'Hardt' means in Germanic languages the dry forest by opposition to the alluvial one. The geomorphology of the Ill-Rhine valley is strongly influenced by Upper Pleistocene to Holocene alluvial response to climate and by further subsidence/uplift within the graben. From Basel to Rhinau, the Weichselian (around 20,000–12,000 years BP according to Théobald et al. 1978) terrace slope exceeds present-day valley gradient. As a result, the upstream part of the valley in the Hardt area is 10 m higher above the plains of the Rhine and Ill rivers, which channels are then incised. This pattern may result from the relative uplift of the area between Basel and Marckolsheim highlighted by levelling comparisons and geomorphological data. In addition, a possible contribution of salt tectonics is likely to explain the uplift of the central terrace and the isolation of the Ill River from the Rhine River over 100 km length (Liaghat 1999).

The Hardt appears as a large and flat stony terrace, associated with the Weichselian alluvial fan of the Rhine River (Théobald et al. 1977) partially recovered by loess deposits and reworked by torrential flows of the Rhine during Early Holocene, especially in its northern part. During the Holocene, paedogenetic processes are dominant on this surface and led to the differentiation between sectors underlain by pebble and/or gravel, which often have red soils, and areas blanketed by more or less reworked loess that support grey soils. This alluvial formation bears today remarkable dry forest (*Carpinus* sp. and *Quercus sessiliflora*) and unique xeric, steppe-like grasslands, related to both climatic and soil dry conditions.

14.5.2 The Rieds

Further north, the Strasbourg-Marckolsheim area is, in contrast, an area of local subsidence. Alluvial deposits vertically stack and the fluvial pattern of the Rhine channel changes from braided to anastomosed. North of Strasbourg, continued subsidence is recorded by the Zorn, the Moder and Laur alluvial fans (Vogt 1992).

The term 'ried' derived from the Alsatian word '*Rieth*' meaning reed refers to an alluvial wetland forest area, which is now widely cultivated or converted into grassland. These

Fig. 14.9 Palaeochannel and protohistoric tumuli in the area of Mussig during flooding. Tumuli (small round hills) are dated from Middle Bronze Age to Iron Age. Some of them were destroyed by tillage (foreground), whilst others were protected by grassland (Picture courtesy of X. Boes)

zones correspond to the Holocene floodplain of the Ill River and tributaries on the one hand and Rhine River on the other. Rivers of less importance like the Thur or the Bruche issued from the Vosges Mountains also develop ried-like wetlands. Carbiener (1983) distinguished different entities by combining sediments sources (i.e. Rhine or Ill deposits) and grades of vegetation and soil evolution. Five major units are well distinguished in a west-east gradient (Fig. 14.1) and named according to the dominated colour of their soils (black, grey, brown, blond and brown-grey ried). A strong correlation exists between these units, their geomorphology, age and present-day hydrology. Among the most important features of the Central Ried are the 'phreatic rivers' – rivers without watershed and exclusively fed by groundwater. The most significant ones are the Blint and the Zembs rivers. The present-day Blint River channel could correspond to a former Ill channel before its westward migration to present-day position during the end of Early Holocene in response to continuing subsidence. Channel migration history is still poorly documented, but minor relief features (i.e. palaeochannels and natural levees) inherited from Holocene evolution control the present plain landforms and current hydrological processes (Fig. 14.9).

Recent geoarchaeological investigations, both in the large Central Ried and in the Zorn-Moder Ried, have provided a better understanding of the complex evolution of these units. In particular, the role of channel migration together with hydrological variations especially during the Late Bronze Age–Early Iron Age – begins to be perceived. Despite the intensification of human occupation of the plain during Middle Bronze Age as evidenced by the development of tumulus culture like in Ensisheim or Mussig (Fig. 14.7), anthropogenic influence seems of little importance on riparian and river dynamics until the second part of the first millennium. Major hydrological changes are documented during the second century AD and also during the fourteenth century, but its incidence on human occupation is poorly understood. The traditional houses and farms of the '*Grand Ried*', mostly built during the eighteenth and nineteenth centuries, are adapted to repeated flooding of the Ill and the Rhine rivers. They are usually in a high topographic position and basements are rare because of the position of the water table. To protect the walls from moisture, they were built on a bottom plate and walls were reinforced by dense half-timbering.

Major transformation of the plain landscape results from the nineteenth-century damming of the Rhine by the engineer G. Von Tulla. Von Tulla's project was to merge all the channels into a single channel and a straight course to favour vertical incision hence to reduce flooding and wetlands. This project has succeeded beyond expectations of its designer and deeply simplified the flood plain pattern. Despite this significant anthropogenic intervention, the *rieds* still remain a characteristic Alsatian landscape of major ecological interest at a European scale.

14.6 Conclusion

The landscapes of URG remain largely marked by the role of recent tectonics in shaping both megaforms and mesoforms. From a geomorphological point of view, the Vosges and their associated units – Vosges foothills and plains Ill-Rhine – are young landforms, and most of their morphology is related to recent, i.e. Plio-Pleistocene tectonics. The thickness of alluvial Pleistocene deposits within the URG attests the Pliocene to Lower Pleistocene major Rhine Graben floor subsidence and Vosges shoulders uplift phase. Despite these elements, direct evidence of tectonic faulting during the Pleistocene remains rare in the URG, and recent tectonic activity is mainly deduced from scarce geomorphological and seismological data, difficult to unravel due to superficial shaping by both natural and mostly man-induced erosion processes.

More than 7,000 years anthropogenic impact, from Early Neolithic to present, have reshaped this structurally controlled landscape into a cultural landscape, in balance between natural constraints and opportunities on the one hand and history on the other. From these arises the uniqueness of the Alsatian landscapes.

Acknowledgements This contribution is a tribute to Henri Vogt (1930–2012), who introduced me to the geomorphology of Alsace. I am grateful to Monique Fort, who offered me to contribute to this book and for her careful and patient proofreading.

References

Baize S, Cusching M, Lemeille F, Granier T, Grellet B, Carbon D, Combes P, Hibsch C (2002) Inventaire des indices de rupture affectant le Quaternaire, en relation avec les grandes structures connues, en France métropolitaine et dans les régions limitrophes. Mém Soc Géol Fr 175, 142 p

Carbiener R (1983) Le grand Ried Central d'Alsace: écologie et évolution d'une zone humide d'origine fluviale rhénane. Bull Ecol 14:249–277

Giamboni M, Wetzel A, Nivière B, Schumacher M (2004) Plio-Pleistocene folding in the southern Rhine graben recorded by the evolution of the drainage network (Sundgau area; northwestern Switzerland and France). Eclogae Geol Helv 97:17–31

Illies JH (1975) Recent and paleo-intraplate tectonics in stable Europe and the Rhine graben rift system. Tectonophysics 29:251–264

Jeunesse C, Schnitzler B (1993) Les premiers agriculteurs. Le Néolithique en Alsace. Les collections du Musée Archéologique, tome 2, Musées de la Ville de Strasbourg, 125 p

Junkmanns J (1995) Les ensembles lithiques d'Achenheim d'après la collection de Paul Wernert. Bull Soc Préhist Fr 92:26–36

Lautridou JP, Sommé J, Heim J, Puisségur JJ, Rousseau DD (1985) La stratigraphie des loess et formations fluviatiles d'Achenheim (Alsace): nouvelles données bioclimatiques et corrélations avec les séquences pléistocènes de la France du Nord-Ouest. Bull Assoc Fr Quat 22:125–132

Liaghat C (1999) Etude de la déformation récente et actuelle à l'apex des diapirs salifères. Thèse, Université de Savoie, p 235

Preusser F (2008) Characterisation and evolution of the River Rhine system. Neth J Geosci 87:7–19

Ritter JRR, Wagner M, Bonjer K, Schmidt B (2009) The 2005 Heidelberg and Speyer earthquakes and their relationships to active tectonics in the central Up-per Rhine Graben. Int J Earth Sci 98:697–705

Sittler C (1974) Les Vosges et le Fossé rhénan. In: Debelmas J (ed) Vieux massifs et grands bassins sédimentaires. Géologie de la France, Doin, 1, pp 63–104

Théobald N, Vogt H, Wittmann O (1977) Néotectonique de la partie méridionale du bloc rhénan. Bull BRGM 4:121–140

Théobald N, Ménillet F, Lougnon J, Bonnet C, Fluck P (1978) Notice explicative de la carte géologique de la France à 1/ 50 000, Feuille Neuf-Brisach, 0378N. BRGM, Orléans, 44 p

Vogt H (1992) Le relief en Alsace. Etude geomorphologique du rebord sud-occidental du fosse rhenan. Oberlin, Strasbourg, p 239

Wernert P (1957) Stratigraphie paléonthologique et préhistorique des sédiments quaternaires d'Alsace, Achenheim. Mém Serv Carte Géol Alsace Lorraine 14:259

15 Karstic valleys, Vineyards and Palafittic Settlements of the External Jura

Marie-Françoise André

Abstract

The Jura massif, which gave his name to the Jurassic period, comprises a folded mountainous eastern part (internal Jura) and a hilly and tabular western part, which is the focus of this contribution. The external Jura is a part of the charming French countryside displaying both original geomorphological landscapes as well as distinctive local products. It is especially the case of the fold-and-thrust belt overriding the Bresse graben, which bears the Jura vineyards mainly growing on Triassic and Liassic marls and providing the unique Château-Chalon 'yellow wine'. By contrast, the Jura plateaus include karstic plateaus with picturesque valleys called 'reculées' and glacially formed lakes. During the Holocene, these lake shores were inhabited by Neolithic cultural groups representative of the earliest European agrarian societies, who built pile-dwelling settlements inscribed in 2011 by UNESCO as World Heritage Sites.

Keywords

Jurassic period • Thrust plane • Reculée • Jurassian ice-cap • Neolithic palafittic sites • Yellow wine

15.1 Introduction

The geomorphological landscapes of the western Jura display vertical cliffs in Upper and Middle Jurassic limestones overlying more gentle slopes in Liassic marls covered with scree deposits (Fig. 15.1). On the geological map, the whole Jura massif appears as a blue crescent (Fig. 15.2), and the spatial extent of the Jurassic outcrops explains that Alexandre Brongniart gave the name of the Jura massif to the Jurassic period in 1829. The internal Jura is a folded medium-altitude mountain range characterised by conformable relief with high anticlines and low synclines (*relief jurassien*) looking like those of the external Bauges prealpine massif (Chap. 18). By contrast, the external Jura displays hills and plateaus painted by Gustave Courbet, which are characterised by vineyards associated with compartmented fold-and-thrust structures, scenic karstic valleys called 'reculées' and glacially formed lakes displaying well-preserved Neolithic pile-dwelling settlements known as 'palafittes'.

15.2 Geographical and Geological Setting

The crescent-shaped Jura mountain range (in blue in Fig. 15.2), which is 300 km long and 70 km in its central part, culminates at 1,723 m at the Crêt de la Neige. It is surrounded by the Rhine graben and the Paris basin to the north, the Bresse graben to the west and the Alpine belt and the Swiss molassic basin to the southeast. The Jura is divided into the eastern internal Jura and the western external Jura (Chauve 1975). The internal Jura, which is often called the *Jura plissé* (folded Jura) or the *Haute Chaîne* ('high range'), appears in dark blue in Fig. 15.2. It is characterised by its large strike-slip faults and folded structures giving rise to classical conformable

M.-F. André (✉)
Laboratory of Physical and Environmental Geography – GEOLAB, Université Blaise Pascal/CNRS, Clermont-Ferrand, France
e-mail: m-francoise.andre@univ-bpclermont.fr

M. Fort and M.-F. André (eds.), *Landscapes and Landforms of France*, World Geomorphological Landscapes, DOI 10.1007/978-94-007-7022-5_15,

Fig. 15.1 The Jurassic limestone cliffs and scree slopes of the 'reculée' of Baume-les-Messieurs, a landscape *Leitmotiv* of the external Jura (Photo M.F. André)

relief with high anticlines and low synclines (*relief jurassien*). These are the realm of fir forests and dairy farming for cheese production. The external Jura comprises two subunits. To the east, plateaus ranging from 500 to 750 m in altitude are resting upon subhorizontal Jurassic limestones (in medium blue in Fig. 15.2); they are characterised by karstic landscapes and glacially formed lake basins. To the west, a fold-and-thrust zone (in light blue in Fig. 15.2) culminates between 300 and 400 m. This compartmented zone, which is known as the *Jura des faisceaux,* overthrusts the Bresse graben and displays limestone-capped hills and marly depressions planted with vineyards between Salins-les-Bains and Saint-Amour (Fig. 15.2).

15.3 From the Triassic Lagoons to the Holocene Peat Bogs: A Multistage History

Geological presentations of the Jura massif provided by Chauve (1975) and Bichet and Campy (2008) allow to reconstruct the regional geological history. This starts with the deposition in Triassic lagoons of saliferous marly sediments, including the Keuper variegated marls. These deposits, which overlie the Hercynian basement rocks, contain well-preserved remains of the oldest 'French' dinosaur, *Plateosaurus* sp., first discovered in 1862 near Poligny; the Triassic salt has been exploited since the Neolithic period. During the Jurassic, marine conditions prevailed, first with the deposition of Liassic grey marls interbedded with a 10 m limestone layer containing fossil oysters (*Gryphaea arcuata*), used as decorative 'marble stone'. Then, the Middle and Upper Jurassic were dominated by limestone sedimentation in shallow tropical waters, including coral reef islands.

The Cretaceous period was a transition phase, with both the penetration of the Paris basin 'chalk sea' and the first compressive episode due to the interaction of the African and Eurasian tectonic plates. Then came the Oligocene distensive phase, with the formation of the Bresse graben, which is a part of the west European rift system including the Rhine graben. Whereas the western edge of the Bresse graben displays classical fault steps bearing the Burgundy vineyard, its Jurassian eastern edge displays an original compartmented fold-and-thrust zone (Fig. 15.3) as revealed by a series of

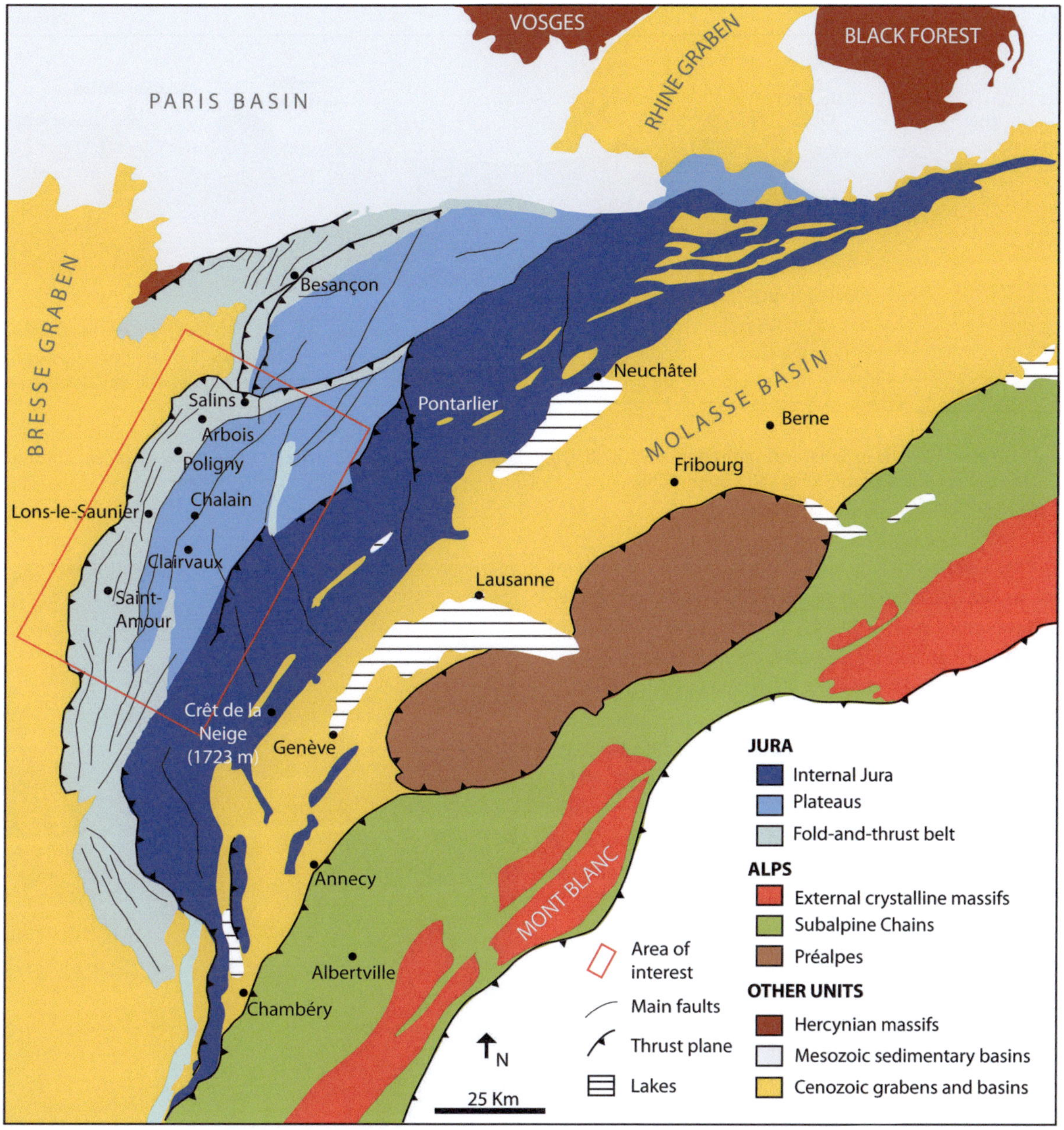

Fig. 15.2 Geological sketch of the Jura massif (in *blue*) and surrounding units (E. Roussel, modified from Chauve 1975)

deep cores (up to >1,000 m) carried out in the 1950–1960s for oil exploration. The western part of the Jura massif, up to 10 km wide, has slipped westwards, forming an external fold-and-thrust belt overriding the Bresse graben (Caire 1963; Chauve et al. 1988); and this 'décollement', favoured by a Triassic saliferous 'soap layer', is thought to have occurred slowly, from the Late Miocene (Pontian) to mid-Pliocene. It was followed by an erosional phase resulting in a dissected landscape with alternating limestone hills and marly depressions, known as the *Vignoble* ('vineyard') in the region of Lons-le-Saunier. This area was partly reached by the outlets of the ice-cap during the so-called 'Riss' penultimate glaciation (c. 347,000–125,000 BP), whereas the more internal plateaus were also carved by the Würmian glaciers, which left glacially formed lakes, partly filled by peat sediments during the Holocene.

15.4 The Jura Plateaus and Their Picturesque Karstic Landscapes

The Jura plateaus, which have developed upon subhorizontal limestones, display karstic features such as sinkholes and lapiaz, which are either covered by mosses in forested environments or vegetation-free in dry environments colonised by scattered xerophilous bushes (*Buxus sempervirens*

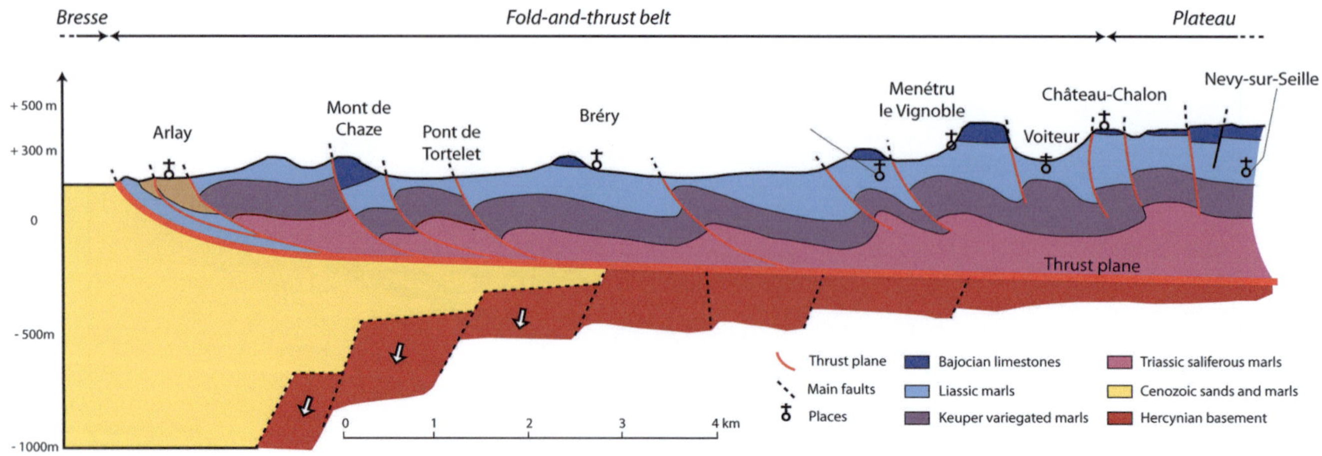

Fig. 15.3 Cross section of the external part of the Jura massif showing the fold-and-thrust belt overriding the Bresse graben, north of Lons-le-Saunier (E. Roussel, modified from Bichet and Campy 2008)

and *Juniperus communis*). Endokarst includes a number of caves such as the Moidons cave near Arbois and the Baume-les-Messieurs caves near Lons-le-Saunier, first explored in 1610. They are inhabited by large bat colonies: 28 of the 34 bat species present in France are found in the Jura massif. Due to the limestone substrate, surface rivers, such as the Doubs and Loue rivers, are few and communicate through a network of underground galleries: on 11 August 1911, during the dramatic fire of the Pernod distillery near Pontarlier, 600,000 l of absinthe and spirit were poured in the Doubs river to avoid explosion, and 2 days later, a pronounced absinthe smell was reported at the Loue springs, 15 km northwards!

Some of these exsurgence springs supply water to cities like Poligny, and most of them are located at the bottom of limestone cliffs of karstic valleys. It is the case of the Hérisson 'reculée', a scenic site with a series of cascades, which is protected since 2002 (Fig. 15.4). The word 'reculée' comes from a local Jura name, 'culée', given to the cirque-like head of such valleys. These 'reculées' are a *Leitmotiv* of the western edge of the Jura plateaus. Many of them were initiated by the Oligocene formation of the Bresse graben, and their evolution involves a combination of processes such as subaerial regressive erosion, roof collapse of underground galleries and (sub) glacial processes.

The Baume-les-Messieurs 'reculée', near Lons-le-Saunier (Fig. 15.5), which is the most scenic of the Jura reculées, displays a network of 200-m-deep valleys exploiting fault lines, with subvertical cliffs in Bajocian limestones, scree slopes underlain by Liassic marls and localised rockfall deposits of possible seismic origin. This 'reculée' provides the flora with a diversity of microenvironments, which explains that both Mediterranean and Boreal species can be found in this blind valley (Chevassus and Chevassus 1973). Apart from this floristic peculiarity, the Baume-les-Messieurs 'reculée' is a suitable location to assess the drastic

Fig. 15.4 One of the Hérisson waterfalls, at the *bottom* of a limestone cliff which is part of a karstic valley called 'reculée', a protected site since 2002 (Photo M.F. André)

Fig. 15.5 The 'reculée' of Baume-les-Messieurs in the 1910s (top) and in the 2010s (bottom). Note the change of vegetation cover induced both by agricultural decline and climate warming. The houses in the foreground are located near the Dard karstic springs and the entrance of the Baume-les-Messieurs touristic caves (Photos: ancient postcard, collection M.F. André (top) and recent photography by M.F. André (bottom))

contemporary changes in vegetation cover (Fig. 15.5a, b): the post-Little Ice Age bare landscapes of the 1910s, due both to cold conditions and grazing pressure, have been replaced by densely wooded vegetation almost hiding the springs with travertine deposits and the cave entrance in the valley bottom.

15.5 The Quaternary Glacial Legacy and Postglacial Palaeoenvironmental Record

15.5.1 The Jurassian Ice-Cap and Its Lacustrine Margins

Since the work of Penck and Brückner in 1909, two morainic complexes have been identified at the periphery of the Alps: the external Riss moraines and the internal Würmian moraines. These two generations of moraines are also present in the Jura massif, but there has been a long controversy about their origin (local or allochthonous). Although Agassiz had distinguished as early as 1843 the moraines deposited by local Jurassian glaciers from the ones brought by Alpine glaciers, several scientists assumed until the 1960–1970s that the Jura massif was too low for a mountain to have generated its own glaciers. The importance of local glaciers has been definitively established by Campy and colleagues since the 1980s (e.g. Campy 1985; Campy and Arn 1991; Buoncristiani et al. 2002): during the Last Glacial Maximum (LGM), the central part of the Jura massif, which is the highest in altitude, was entirely covered by an elongated local ice-cap culminating at 1,800 m, with west-flowing short glacial tongues, 200 m thick, channelised by the 'reculées' (Fig. 15.6). The geometry of the Würmian tongues during the LGM (20,000 BP) was reconstructed from the shape of the front moraines which are better preserved than the external moraines left by the penultimate 'Riss' glacier located 20 km westwards; this 'Riss' ice-cap is assumed to have been 200 m thicker than the Würmian one.

At the western foot of the Jurassian ice-cap, a depression called *Combe d'Ain* was occupied by a series of proglacial lakes (Fig. 15.6), partly filled with deltaic deposits ('Gilbert deltas') displaying the classical threefold complex associating topsets, foresets and bottomsets (Fig. 15.7). The bottom of the depression has been filled with silty laminated lake deposits ('rythmites' or *varves*).

15.5.2 Jura Lake Sediments and Peat Bogs: Archives for Palaeoenvironmental Reconstructions

The central part of the Jura plateaus is occupied by a number of lakes like the *Petit Maclu* lake, northeast of Clairvaux (Fig. 15.8), which is a part of the seven lake region, nicknamed 'Little Scotland' by Charles Nodier in his *Voyages pittoresques et romantiques dans l'ancienne France*. Most of these lakes are due to glacial overdeepening and/or river damming by front moraines. Some have become peat bogs, and both lake and peat sediments have been cored in the Jura massif in order to reconstruct the Holocene history of climate and environment. The lake-level fluctuations, which have been reconstructed over the last 12 millennia based on multi-proxy analyses, have been shown by Magny to be controlled by solar and orbital forcing (Magny 2004). The Holocene history of vegetal landscapes was reconstructed by Richard using palynological analyses (Richard 1994; Richard and Gauthier 2007) who showed that during the Holocene climatic optimum, the Jura forest was composed of oaks, elms and lime trees (instead of the present-day beeches, spruces and fir trees) and that agricultural practices started in the western Jura plateaus as early as 5500 BC.

15.6 The Neolithic and Bronze Age Palafittic Settlements of the External Jura, Part of a Group of Lacustrine World Heritage Landscapes

The Chalain and Clairvaux sites, which are located on the plateaus of the external Jura (see location on Fig. 15.2), are among the 111 Neolithic and protohistoric pile-dwelling settlements which have been inscribed in 2011 by UNESCO on the World Heritage list. This group of palafittic sites from 5000 to 500 BC is scattered in and around the Alpine belt in six countries (France, Switzerland, Italy, Germany, Austria and Slovenia). They are characterised by the remarkable preservation of remains (including textiles, tools, weapons and wooden wheels and piles) due to waterlogged conditions. These lacustrine sites provide an exceptionally detailed information about the daily life, means of transport, construction techniques and trade routes of cultural groups which are representative of the earliest European agrarian societies.

At Chalain and Clairvaux, the first archaeological findings have been made between the 1830s and the 1900s, including spectacular findings such as the one of a 9.35-m-long dugout canoe dating back to the Late Bronze Age. Since the 1970s, in-depth multidisciplinary studies have been carried out at Chalain and Clairvaux, combining archaeological excavations and experimentation (Fig. 15.9), ethnoarchaeological approaches, lake sediment cores, palynological analyses and dendrochronological and ^{14}C datings (Pétrequin and Pétrequin 1988; Pétrequin 1999; Magny 1993, 2004; Richard 1994; Richard and Gauthier 2007).

Based on detailed studies of 32 sites in Chalain and 18 sites in Clairvaux, it appears that these dwelling sites have been repeatedly occupied over four millennia, with a major occupation period between the fifth and the third millennium BC. Both environmental drivers (lake lowstands) and more importantly cultural drivers (e.g. defensive strategies) account for the chronology of occupation. The Neolithic and Bronze

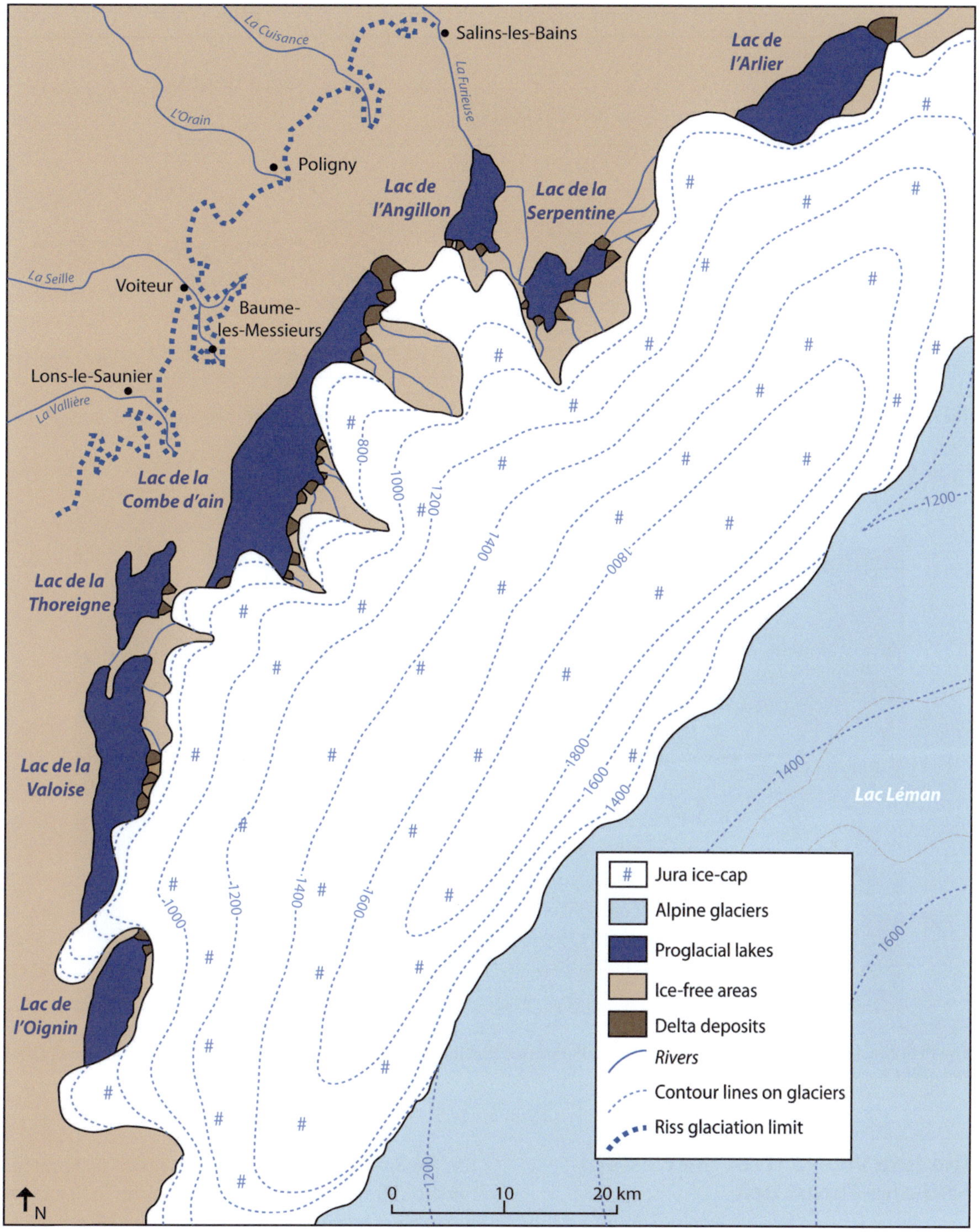

Fig. 15.6 Extent of the Würmian Jura ice-cap during the Last Glacial Maximum, about 20,000 years ago (E. Roussel, simplified from Campy in Hantzpergue and Bichet 2007)

Age pile dwellings were located on islands or peninsulas surrounded by wetlands and/or enclosed by fences. The duration of occupation of each site (10–20 years on average) corresponded to the duration of an agricultural cycle, and the size and density of villages were controlled by demographic pressure. These cultural groups, whose main activity was agropastoralism (including cheesemaking), were not autarkical societies. They were inserted within long-distance circulation networks and trade routes, as demonstrated by the occurrence of jadeite axes and flint blades originating from the Mont Viso and the Touraine region up to more than 300 km from the Jura plateaus.

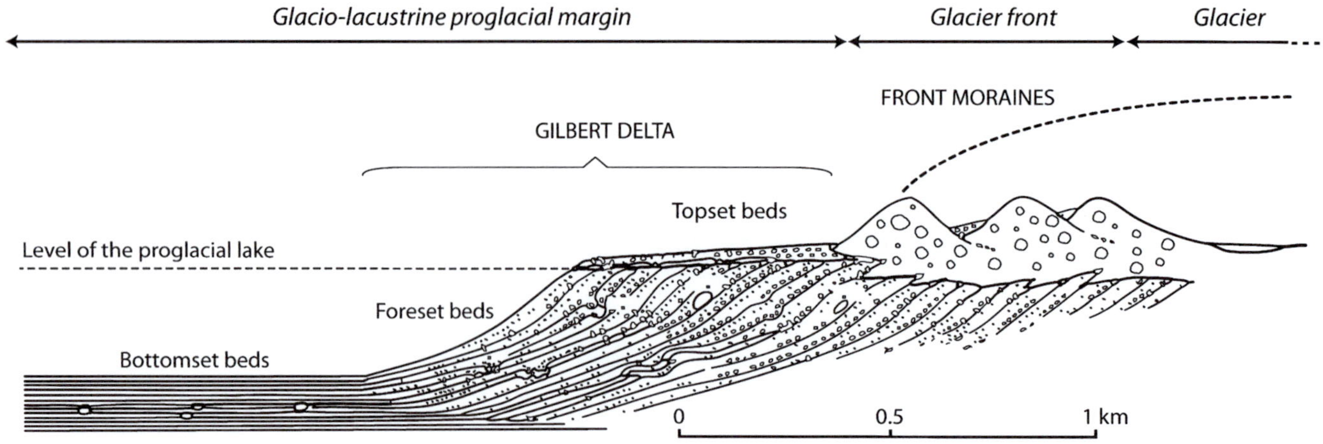

Fig. 15.7 The sedimentary complex of the lacustrine margin of the Würmian Jura glacier in the Combe d'Ain region (E. Roussel, modified from Campy 1985)

Fig. 15.8 The Petit Maclu lake, one of seven lakes of glacial origin located in 'Little Scotland', a picturesque region of the Jura plateaus depicted in 1825 by the poet Charles Nodier (Photo M.F. André)

15.7 The Jura Vineyards of the Western Fold-and-Thrust Belt

The Jura vineyard region stretches from Salins to Saint-Amour along the fold-and-thrust belt of the external Jura overriding the Bresse graben (see Figs. 15.2 and 15.3). Although the main thrust scarp is facing west, the tectonic compartmentalisation induces a diversity of aspects including south which is a particularly suitable slope aspect for vine cultivation. Most vineyards grow on marly slopes covered or not by scree deposits derived from the limestone scarps (Fig. 15.10). Disintegrated Triassic variegated marls and Liassic grey marls are undoubtedly first-class substrates for the Jura vineyards: they are easily penetrated by roots, remain dry at the surface and keep the moisture deep in the ground. Like the Champagne, Alsace and Burgundy vineyards, the Jura wine region belongs to the northern and continental French vineyards, experiencing harsh winters and dry summers. Initially, the end of harvest times was delayed until November to obtain the highest level of grape ripeness, but climate warming currently allows to harvest the grapes between early September and mid-October.

The Jura wine region is a very small vineyard (1,800 ha) but a first-quality one: 90 % of the Jura wines are classified as AOC (*Appellation d'Origine Contrôlée*). They include the famous Château-Chalon 'yellow wine' (*vin jaune*) which is a

Fig. 15.9 Original piles of the Lac de Chalain palafittic site in the foreground, with the reconstruction of a Neolithic pile dwelling in the background (Photo P. Pétrequin/CRAVA)

Fig. 15.10 Vineyard of Château-Chalon on a south-facing slope in Toarcian grey marls below a Bajocian limestone cliff. In the foreground, cereal field on alluvial deposits of the Seille river (Photo M.F. André)

Table 15.1 Grape varieties of the Jura vineyard and their attributes

Grape varieties	Percentage of Jura planting (%)	Favourable substrates	Wine characteristics	Accompanying dishes
Savagnin	17	Keuper variegated marls and Liassic grey marls including Toarcian *schistes carton*	Powerful and distinctive dry white and yellow wines (*Vins jaunes*), golden to amber, with unique long-lasting taste, walnut and mineral flavour	*Foie gras* Comté cheese Lobster *à l'américaine* (with spicy sauce) Chicken with morels in wine cream sauce
Chardonnay (= melon)	50	All substrates including scree slopes below limestone cliffs	Dry and fruity white wine, light yellow, more distinctive when blended with Savagnin in a balanced white wine called *Côtes du Jura Tradition*	Poultry Fish Seafood Mont d'Or cheese
Poulsard	18	Keuper variegated marls including bottom-slope cold soils	Aromatic and *gouleyant* rosé to light red wine, often displaying an original light orange 'onion skin' colour ('coral wine')	Grilled meat Morteau sausage Sweet and sour dishes
Trousseau	5	Gravelly well-drained warm soils	Powerful and distinctive red wine, dark red to purple, very tannic, with red fruit aromas	Red meat Game
Pinot noir	10	Nonspecific substrates (ubiquitous variety)	Red wine, more mineral than the Pinot noir from Burgundy, often used with Poulsard and Trousseau to make assemblage wines	Beef bourguignon Rabbit stew Eggs in red wine sauce (*oeufs en meurette*)

Partly based on Campy in France (2002, p. 154–159)

very distinctive wine, to be tasted with walnuts and Comté cheese to fully appreciate its unique taste. This is obtained by storing the wine after fermentation for 6 years and 3 months in oak or chestnut barrels and by allowing it to evaporate through a yeast veil (*Mycoderma vini*) before bottling it in customised 62 cl bottles known as *clavelins*. The oldest *clavelin* of 'yellow wine' tasted in the 1990s dated back to 1774 and had not lost its flavour; a similar bottle was sold for 57,000 euros in 2011. Beside Château-Chalon, which represents only 2.5 % of the surface area, the three other Jura *Appellations d'Origine Contrôlée* are L'Etoile (4.5 %), Côtes du Jura (42 %) and Arbois (51 %). This last AOC produces a significant quantity of red wine, in contrast to the three others. Overall, 80 % of the Jura wines are white wines.

Five grape varieties are grown in the Jura region and their attributes are described in Table 15.1. Whereas Chardonnay and Pinot noir have been imported from Burgundy in the Middle Ages, three grape varieties are typical of the Jura vineyard: Savagnin, which is used to make the famous 'yellow wine', and Poulsard and Trousseau, grown to make distinctive rosé and red wines. Some of these varieties are used to make specific wines, used as aperitifs or dessert wines, that include: the *Vin de paille*, sweet wine made with raisins dried on beds of straw until Christmas and sold in half-bottles (37.5 cl); the *Macvin*, in production since the fourteenth century, which is the Jurassian version of the *Pineau des Charentes*, and the *Crémant du Jura*, a very pleasing sparkling wine.

15.8 Conclusion

Although not as spectacular as the Alpine peaks or the Polynesian lagoons, the external Jura is part of the charming French countryside which provides long-lasting memories due to a unique blend of scenic landscapes and cultural heritage. The picturesque 'reculées' offer the same white to ochre colour of the Jurassic limestone in their vertical cliffs and their Medieval abbey churches hidden in the valley bottoms. The Jura lake region displays in winter icy landscapes evoking the glacial past of the Jura massif and allows to swim in summer near the Neolithic pile-dwelling settlements dating back to the Holocene climatic optimum. The thermal spas of Salins-les-Bains and Lons-le-Saunier keep using the Triassic salted waters, and geomorphologically oriented itineraries are proposed to visitors in the Jura massif (Marcel and Quesne 2011). Last but not least, the Jura wine route follows the fold-and-thrust belt overriding the Bresse graben. In this area in early February, the most vibrant wine festivity of the French wine regions, the *Percée du vin jaune*, is organised, an event not to be missed.

Acknowledgements The author is most grateful to Erwan Roussel (Maison des Sciences de l'Homme of Clermont-Ferrand) who adapted Figs. 15.2, 15.3, 15.6 and 15.7; to Michel Magny (Chrono-Environnement, Besançon) who provided updated scientific information; and to Pierre Pétrequin and CRAVA (Centre de Recherche Archéologique de la Vallée de l'Ain) who kindly allowed reproduction of Fig. 15.9.

References

Bichet V, Campy M (2008) Montagnes du Jura. Géologie et paysages. Néo-Editions, Besançon

Buoncristiani JF, Campy M, Nicoud G (eds) (2002) Le front würmien à la périphérie du Jura et des Alpes – expression sédimentaire et morphologique, chronologie. Livret-guide de l'excursion annuelle de l'AFEQ, 8–11 mai 2002

Caire A (1963) Problèmes de tectonique et de morphologie jurassiennes. Société Géologique de France, Paris, pp 105–158 (Livre à la mémoire du Professeur Pierre Fallot)

Campy M (1985) Dynamique d'une marge glaciaire au maximum würmien – La Combe d'Ain (Jura). Bulletin de l' Association Française d'Etude du Quaternaire 2–3:65–74

Campy M, Arn R (1991) The Jura glaciers: paleogeography in the Würmian circum-Alpine zone. Boreas 20:17–27

Chauve P (1975) Jura – Guides géologiques régionaux. Masson & Cie, Paris

Chauve P, Martin J, Petitjeanseq E, Sequeiros F (1988) Le chevauchement du Jura sur la Bresse. Données nouvelles et réinterprétation des sondages. Bulletin de la Société Géologique de France 8:861–870

Chevassus L, Chevassus P (1973) Fleurs du Jura. SAEP, Colmar-Ingersheim

France B (ed) (2002) Grand atlas des vignobles de France. Solar, Paris

Hantzpergue P, Bichet V (eds) (2007) Jurassique ... Jura. Métamorphoses d'un paysage. Centre Jurassien du Patrimoine, Lons-le-Saunier, Société Jurassienne d'Emulation, Porrentruy

Magny M (1993) Une nouvelle mise en perspective des sites archéologiques lacustres: les fluctuations holocènes des lacs jurassiens et subalpins. Gallia Préhistoire 35:253–282

Magny M (2004) Holocene climatic variability as reflected by mid-European lake-level fluctuations, and its probable impact on prehistoric human settlements. Quat Int 113:65–79

Marcel P, Quesne D (2011) Jura. Guides géologiques. Omniscience BRGM éditions, Montreuil

Pétrequin P (1999) Lake dwellings: archaeological interpretation and social perception. In: Stone PG, Planel P (eds) The constructed past. Experimental archaeology, education and the public. Routledge, London/New York, pp 217–228

Pétrequin P, Pétrequin AM (1988) Le Néolithique des Lacs. Préhistoire des lacs de Chalain et de Clairvaux (4000–2000 av. J.-C.). Editions Errance, Collection des Hespérides, Paris

Richard H (1994) Indices polliniques d'une néolithisation précoce sur le premier plateau du Jura (France). C R Acad Sci Paris 318(Ser. II):993–999

Richard H, Gauthier E (2007) Bilan des données polliniques concernant l'âge du Bronze dans le Jura et le nord des Alpes. In: Mordant C, Richard H, Magny M (eds) Environnements et cultures à l'âge du Bronze en Europe occidentale. éditions du CTHS, Paris, pp 71–87

Glacial Imprint on the Main Ridge of the Vosges Mountains

16

Jean-Luc Mercier

Abstract

In this paper we provide an example of a widely studied landscape where conflicting observations in geomorphology lead to opposite interpretations. The central Vosges Mountains consist of a multi-tilted fault block, covered by an ice cap during each Quaternary cold period. After the Last Glacial Maximum, deglaciation began on the main ridge, leaving periglacial formations on low-relief highland or "old surface" and young valley glaciers or cirque glaciers with their associated moraines. The bald summital part of the Vosges Massif, which is known as the Hautes Chaumes, is of disputed origin. Whereas the occurrence of typically alpine soils supports a primary origin of heathland and meadows at some places, recent archeobotanical studies indicate that most of these vegetal communities result from deforestation since the Bronze Age period. The protection of this natural heritage is organized at two levels, a nature park (*Parc Naturel des Ballons des Vosges*) in an integrated rural and natural landscape, which is visited each year by five million persons, and four nature reserves, which guarantee the overall preservation of the landscapes and associated threatened fauna and flora. On the western gentle slope of the Vosges Mountains, a geomorphic feature corresponding to the maximum extension of the Würmian glaciers (the Noirgueux end moraine) benefits from a specific protection.

16.1 Introduction

As an intraplate aborted ocean, the Rhine graben is limited on the west side by the Vosges Mountains and on the east side by the Black Forest. The Vosges Mountains are a midlatitude (48°N, 7°E) granitic Hercynian shield lifted up since the Mio-Pliocene time. As other rift margins, the Vosges are characterized by a very pronounced dissymmetry, a long "gentle" slope to the west and a steeper one on the east side. This organization has strongly influenced the development of the valley glaciers and river network (André 1991; Flageollet 2002). The axis of the range strikes NNE (N110°); the elevation ranges from 1,424 m in the south to the 700 m plateau in the north. Moreover, the main ridge is shaped into a smooth convex upland.

J.-L. Mercier (✉)
Université de Strasbourg, 3 Rue de l'Argonne,
67083 Strasbourg, France
e-mail: jlm@unistra.fr

During the Quaternary, this mountain range, situated in the 48–53°N corridors between the former Scandinavian and Alpine ice sheets, was completely covered by a local ice cap. It represents a common geomorphological situation in Central Hercynian Europe.

A specificity of this small mountain is the interest by academics since 170 years to this landscape including the upper flat area known as the *Ballons* or *Hautes Chaumes* and the collection of glacial cirques on the east and west sides (Fig. 16.1). During a century, the upper part of the Vosges has been described as – or derived from – an "old planation surface," the post-Hercynian or pre-Triassic peneplain. The aim of this chapter is to describe the main feature of this landscape, that is, the contrast between the so-called old smooth main ridge and the recent glacial cirques and moraines of valley glaciers on both sides of the massif. If the more recent cirques are hanged in the upper valleys, the majority of them are associated to deeply incised ones. This area has suitable characteristics for paleoweathering, paleosurface, and glacial studies.

M. Fort and M.-F. André (eds.), *Landscapes and Landforms of France*, World Geomorphological Landscapes,
DOI 10.1007/978-94-007-7022-5_16, © Springer Science+Business Media Dordrecht 2014

Fig. 16.1 Facing west, the smoothly undulated main ridge. Typical landscape of *Ballons* and *Hautes Chaumes* in the upper Fecht valley from *right* to *left*, the Kastelberg (1,350 m), the Rainkopf (1,305 m), and the Rothenbachkopf (1,318 m) in a porphyroïde granite. The Altenweiher cirque, the lake (930 m), is installed on a moraine and a *roche moutonnée* (Credit: Y. Kohler)

16.2 The Vosges Mountain Main Ridge: The *Ballons*

Along 80 km from the south culminating at the Grand Ballon at 1,424 m to the north at Tête des Faux reaching 1,219 m (Fig. 16.2), the topography of the crest is smoothly undulated (Fig. 16.1). The continuity of this ridge associated to the convexity or the *Ballon* shape of the summits is strongly opposed to the vertical glacial cirques walls.

16.2.1 The Ridge Around "Le Hohneck"

In the center of the ridge, the altitude is continuously high associated with a small variability; Le Hohneck (1,363 m) is the highest point of a 25 km crest, the other summits range within 144 m. The central ridge is a kind of Hercynian granitic horst lifted in altitude with its Mesozoic sedimentary cover. The erosion during the Cenozoic destroyed the Jurassic and Triassic deposits and exhumed the buried pre-Triassic topography. The latter has been eroded on the sides of the blocks, leaving the central *Ballon*-shaped part as a remnant of the paleotopography. This evolution gave rise to multi-paleosurface theories.

From north to south and orthogonal to the main ridge, faults and fractures individualized block units which evolved as *Ballons* separated by erosion gaps or incipient gaps. At a smaller scale, each *Ballon* has a slightly different shape, according to the effects of petrography, tectonics, and associated fracture patterns. Avoiding monotony each summit can be perceived differently, with more or less aesthetic pleasure, according to everyone's sensitivity (Fig. 16.3).

Recent work on the Ringelbach experimental catchment issued from a 150 m drilling core facies analysis shows a paleoweathering record on the upper part of the granite buried below the Triassic sandstones and below the post-Hercynian paleosurface. The weathering observed in this drilling core has a thickness of 100 m (Baltassat et al. 2005; Wyns 2012). This ante-Triassic weathering mantle produced specific horizons: arenaceous residuals on the upper part, planar fissures, and fractured granites. First observed in the Rhine graben (Carozza 2013), these resulting weathered levels have been used as proxies of the pre-Triassic topography to infer the location of paleotopography, as well as to highlight the influence of tectonics on the tilted blocks.

Fig. 16.2 The Vosges main ridge: Grand Ballon, Hohneck, and Tête des Faux, between two cities, Colmar and Gérardmer. Image Google Earth

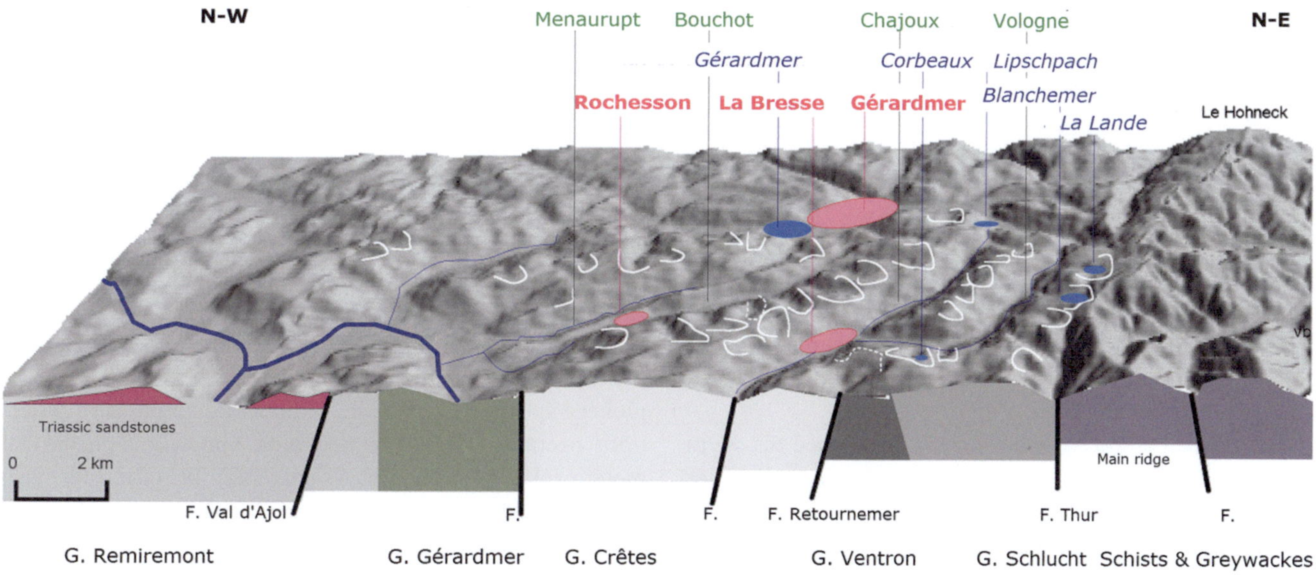

Fig. 16.3 South of main ridge and western side (Lorraine) in the Hohneck area (*F* fault, *G* granite, *Bold pink* cities, *Blue italic* lakes, Green: valleys). The blocks are steps from Paris Basin to the horst. The glacial cirques are facing west; on the horst they are carved on opposite sides. The Grande Pile peat bog (Woillard 1978) is on the far southwest of this area (Adapted from Flageollet 2008)

The Vosges' topography has a strong influence on the local climate, being the first main orographic barrier from the Atlantic; the annual precipitation increases slowly from the Paris Basin to the main ridge with a mean annual rainfall of 2,000–2,200 mm over half a year. The reverse occurs on the eastern lee side with fast decreasing precipitations toward the Rhine graben with a gradient of 150–200 mm/km due to the foehn effect resulting from the steep N–S topography and westerly winds. The main ridge is cold, windy, and foggy. The annual mean temperature is 5 °C at 1,200 m and exceeds

Fig. 16.4 Snow patches and vegetation of the main ridge. Facing south, in the center, the Hohneck (1,363 m), the bald effect is illustrated by the density of the vegetation: (*A*) *Chaumes* heather, blueberry, and broom, (*B*) stunted beech, (*C*) upland beech, and (*D*) beech-fir mixed forest (Credit: Y. Kohler)

10 °C during 4 months only; the frost occurs 160 days/year; these local edaphic conditions limit the vegetation; the wind was noticed by the first botanists (Mappi 1742) and later (Jenik 1958; Carbiener 1963) proposed the concept of anemo-orography to describe the canalized winds and interpret the spatial distribution of the vegetation of the *Hautes Chaumes* on the main ridge.

But the same processes have similarly shaped the asymmetry of the landscape; the wind blowing from the west accumulated snow in leeward positions with resulting firn and glacier masses progressively transforming the east-side depression into cirques.

16.2.2 The Main Ridge: A Bald Mountain of Disputed Origin

Hautes Chaumes is the local name for the summit, encompassing the local ecology, including petrography, geomorphology, and the vegetation, as well as postglacial and human impact on these mountains. On gentle slopes, the vegetation of the *Chaumes* mainly consists of a mosaic of heathland with low shrubs (heather, blueberry, and broom), grasses (matt grass, fescue, and other Poaceae), and moorland (peat mosses and Cyperaceae).

The bald effect is emphasized by the contact between the *Chaumes* vegetation and upland beech forest (beech and pioneer species, like maple, willow, mountain ash), where trees become stunted, loaded with epiphytic lichens and mosses (elfin woodland or *Krummholz*). In some sites, peat bogs tend to be colonized by Norway spruce.

The vegetation is a consequence of geomorphology, soil, climate, and man. During many years, the convex *Ballons* of the main ridge were described as upper and lower *Chaumes* according to altitude or primary vs. secondary *Chaumes* according to natural vs. anthropic origin. Did the forest ever colonize this upper part of the Vosges Massif, or are these bare landscapes of the *Hautes Chaumes* of primary origin? The occurrence of cryptopodzolic rankers, evolving typically under alpine meadows, which are found in some parts of the *Chaumes,* support their primary origin (Fig. 16.4; Guillet 1972).

For a long time, the *Chaumes* of the upper area (Gazon de Faing, Gazon Martin) were considered as primary and the *Chaumes* below 1,250 m (Haut Fourneau, Tanet, Wurzelstein) as secondary (Carbiener 1963). Recent historical, biogeographical, and anthracological works have provided new insights in this question (Goepp 2007). Some arguments show that deforestation begun during the Bronze Age and that heathland is resulting from anthropic activity. Following this author, all summits of the area would have been grazed during several centuries before this activity disappeared in recent decades. This explains why, from Col de la Schlucht to Tanet, the summits harbor forest with pioneering species like Norway spruce, mountain ash, and maple, from the edge of the beech high forest (Wlodarczyk 1992). These recent woodlands have replaced previous heathland and meadows, which have almost disappeared from La Schlucht to Missheimle. At Haut Fourneau and Tanet, Norway spruce and mountain ash are firmly established over the *Chaumes*, preparing the new naturally forested area. All in all, some *Chaumes* may have retained their primary nature until now in very limited areas, but most of them are certainly anthropogenic to some extent.

16.3 The Glacial Cirques

Because climate is dependent on topography, we hypothesize that climatic processes acting on vegetation occurred also during the cold periods of Pleistocene. The present-day position of the permanent snow line is above the Hohneck (1,363 m) while the lowest snow line position was approximately at 900 m during the last glaciation. The main consequence of the snow accumulation is the incision and deposition in the valleys. The western valley glaciers were 30–40 km long; the maximum extension of the Würmian glaciers (the Noirgueux end moraine) benefits from a specific protection. On the eastern side, the end moraines are located at 5–10 km from the main ridge. On the east and west sides, 23 and 19 cirques are respectively located along the main ridge.

16.4 The Eastern Cirques

The smooth convex upper summits and bare mountains *Ballons* contrast strongly with the steepness of the limiting hillslopes and east-facing cirque headwall. Two examples to illustrate the incision of the valley originated from the main ridge. After only 4 km the Thur River at Wildenstein flows 670 m down the Schweiselwasen (main ridge). In the upper Fecht valley at Erbersch, the flat valley bottom is situated at 3 km from the Kastelberg (main ridge).

The principal evidences of the Pleistocene Vosges Mountain glaciation are cirques, steep-sided valleys, hanged valleys, overdeepened basins, glacier accumulations, and fluvioglacial sediments. The valleys are mainly controlled by morphostructural features (Flageollet 2008). The dissected valleys are characterized by abrupt slopes of 500–700 m. This glacial erosion is due to the *excess* of snow accumulation on the lee side and by ice movement. The lower parts of valleys were occupied by the ice and end moraines.

Many glaciologists have been intrigued by the size of the moraines in the Vosges: "Why these huge moraines for such a small mountain?" Some new observations have recently clarified the point. The question is not related to climate, latitude, or petrography but could be probably reduced to the origin of the accumulated glacial sediments. To accumulate, the morainic sediments had to be easily removed and transported. Two hundred million years after their production, the three-level weathering mantles have provided sands, pebbles, and boulders. In the valleys, the till deposits are known as water reservoir since the nineteenth century; the thickness of the glacial and fluvioglacial sediments can reach up to 40 m in the Fecht valley near Metzeral, 30 m at Stosswihr, 20 m at Munster, and 20–60 m at Turckheim.

The youngest glacial cirques associated with periglacial forms are located in altitude around the *Ballons*. Several valley closures below the ridge show features of a remodeling by nivation processes. In some of these nivation cirques, accumulation characters, altitude, and valley exposition suggest the occurrence of firn glaciers.

Among a series of beautiful cirques, such as those of lac Blanc (Fig. 16.5), lac des Truites (Fig. 16.7), Missheimle (Fig. 16.8), lac Vert, and lac Noir, some were occupied by peat bogs and some of them have been turned into artificial lakes. The lac Blanc cirque is more impressive by its size, the cirque headwall, and the moraine closing the lake. The lac Blanc and lac Noir are easily reached by car and were recognized in the 1870s by Charles Grad as suitable for hydroelectricity, it will be done 60 years later. Since the 1930s, *lac Blanc* is flowing into *lac Noir* situated 1 km and 120 m below during the day, producing electricity, and during the night water is refueled into the upper lake.

16.5 The Western Cirques

With its valleys, vales, and cirques, the western slope is the counterpart of the eastern side. Due to the rifting, on this side, the topography is less carved, the glacial valleys are less incised, and the slopes are smoother. This results in the past and nowadays a greater accessibility to the main ridge. The mean altitude of the western slope is greater than the eastern one; the extension of the ice cap was also greater, reaching 30–40 km, and the deglaciation left a great variety of moraines, peat bogs including the "Grande pile" (Woillard 1978), rock bars, and contributed

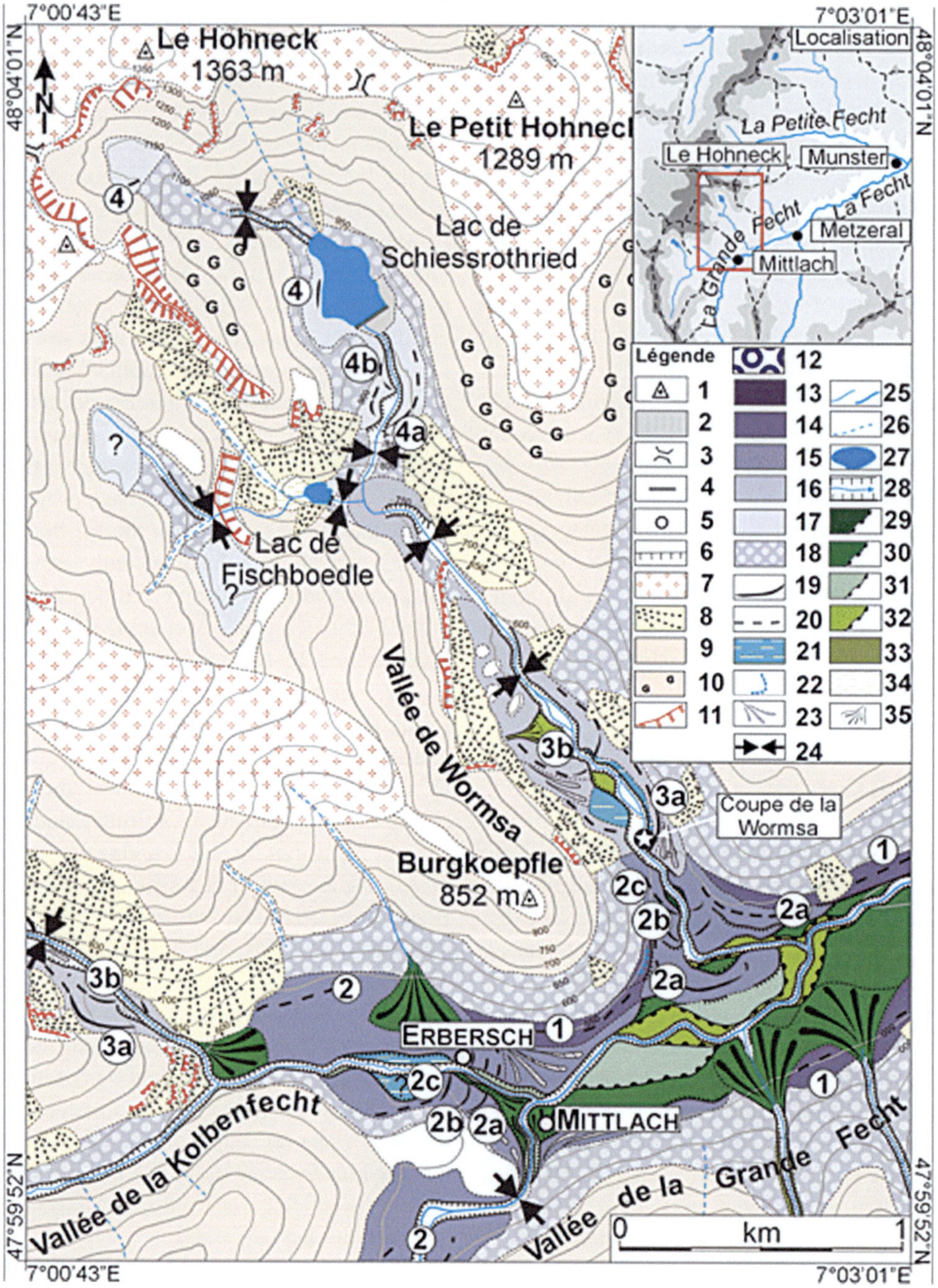

Fig. 16.5 Map of the Wormsa valley field (Vosges, France). *1* Summit, *2* backfill, *3* pass, *4* dam, *5* locality, *6* bank, *7* Grus, *8* bouldery talus, *9* undetermined or polygenic periglacial deposits, *10* gelifluxion, *11* steep slope, *12* till (Riss), *13* till (middle Würm), *14* stage 1 till, *15* stage 2 till, *16* stage 3 till, *17* stage 4 till, *18* reworked till, *19* morainic ridge, *20* discontinuous morainic ridge, *21* fluvioglacial and glaciolacustrine deposits indicating ice retreat, *22* ice contact valley-side terrace or kame delta, *23* proglacial fan, *24* rock bar, *25* river, *26* ephemeral stream, *27* lake, pond, *28* present-day riverbed, *29* fluvioglacial terrace T1a, *30* fluvioglacial terrace T1b, *31* fluvioglacial terrace T1c, *32* post-glacial terrace, *33* torrential deposits, *34* peat bog, and *35* alluvial or torrential fan (Adapted from Andréoli et al. 2006)

to the beauty and diversity of the landscape. In this area, human activity deeply transformed the natural environment by dams, artificial lakes, electric hydropower, ski resorts, snow gun towers, and ski lifts. Nowhere else in the massif is the impact of human activities on wildlife so visible (Fig. 16.6).

16.6 Deglaciation of the Cirques and Main Ridge

The local deglaciation, which is documented by tors, erratic blocks, moraine boulders, and *roches moutonnées,* is similar to other deglaciations in Central Europe (Bourlès et al. 2004).

Fig. 16.6 Western side (Lorraine), one of the more impressive ski runs at Artimont-Belle Hutte ski resort, on the land dependant of the municipality of La Bresse. Snow will have to cover these slopes to hide the heavy destruction of the natural environment aimed to serve the needs of private interests. It clearly shows how difficult it can be to share a common space (Credit: J-L Mercier)

Fig. 16.7 The "lac des Truites" and cirque, on a *roche moutonnée* (48°06′12 N, 7°04′25 E, 1,110 m): sample RM Lt03 gives an exposure age of (9.5 ± 1.7 ^{10}Be ka; Mercier et al. 1999; Mercier and Jeser 2004)

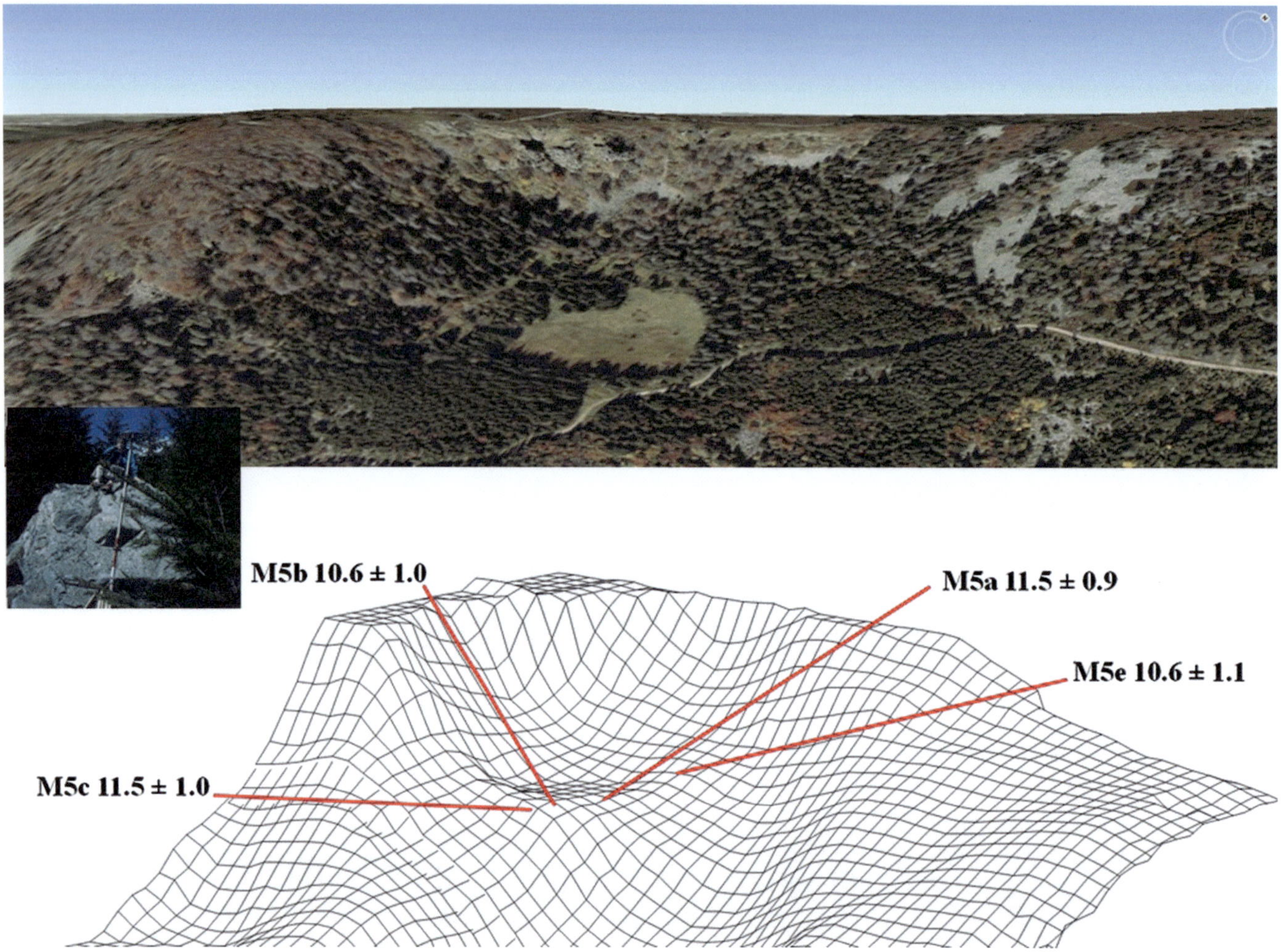

Fig. 16.8 Missheimle (48°04′31 N, 7°02′25 E), a cirque limited by two moraines. Four erratic boulders (1,102 m) (Insert: student for scale). Boulders are dated (^{10}Be ka) from the Younger Dryas (Mercier et al. 1999; Mercier and Jeser 2004; Image Google Earth)

After the Last Glacial Maximum, the deglaciation was not simply a smooth warming, because cold events also occurred. The ^{10}Be datings of moraine boulders provide exposure ages attributed to the Younger Dryas ((Fig. 16.8, Mercier et al. 1999). During the Early Holocene, decaying glaciers occurred in the Vosges Mountains, especially in the upper cirques and within shallow depressions of the main ridge above 1,100 m. A peat bog has developed in the La Lande valley (924 m) after closure by the *Belle Hutte* end moraine; the lower part of the peat has been dated with four ^{14}C dates ranging from 8,300 ± 50 to 6,600 ± 40 years in the Boreal (Walter-Simonnet et al. 2008) and an indirect older age for the *Belle Hutte* end moraine. Exposure age of deglaciated *roches moutonnées* (Fig. 16.8) was obtained at 9.5 ± 1.7 ^{10}Be ka (Mercier et al. 1999; Mercier and Jeser 2004), and a cold event was evidenced at 8.3 ± 1.1 ^{10}Be ka in the Wormsa valley (Fig. 16.5, cf. supra) (Figs. 16.7 and 16.8).

Permanent ice and firn fields possibly persisted in relief positions suitable for snowdrift accumulations until as recently as 5,000–6,000 years ago (Mercier and Jeser 2004). Other snow-ice patches were still present (Mappi 1742), i.e., during the Little Ice Age and near the Kastelberg in 1860 (Wahl and David 2007), and also persisted in 1978 for more than a year.

16.7 Conclusion

With a bit of irony, some authors explained that the number of scientific works and disputes is related to the quality of the landscape; if this assertion proves true, the beauty of the Vosges landscape is well ranked. It is vital to preserve these unique features: the *Hautes Chaumes*, the upland beech forests, the old maple forests on scree slopes, the endemic floristic association encountered on the upper part of the cirque headwall, and the peat bogs in the cirques. At higher altitude, the conservation of the *Hautes Chaumes* and the upper part of the main ridge is in peril because of climatic change. The invasive plants in the valleys and the lower parts of slopes threaten the autochthonous flora and modify the natural landscape. The winter snow leisure industry, an unsustainable development activity in a natural paradise, should be reconsidered in the realm of climate change and conservation of a natural heritage.

The legacies of the past did not have the same weight in geomorphology and in phytogeography. The climatic factor has been a major driver for the evolution of both the relief and the vegetation. The landscape encountered in the Vosges main ridge acted as a strong amplifier on the climate, the vegetation cover as well as land use by rural population. The burden of the past is extremely strong in geomorphology, as noticed with the main ridge, its shape and extension, its discussed and disputed origin, and its cirques and moraines. It is less pronounced on the more flexible vegetation cover able to adapt to climate variability and human activity. These natural processes are fully expressed where human footprint is limited as in the *Hautes Vosges*. The overall preservation of the natural heritage of the Vosges Massif is currently ensured by the existence of a Nature Park (*Parc Naturel des Ballons des Vosges*) and four Nature Reserves.

Acknowledgemnts An early draft of this chapter was improved with useful discussion and collaboration with Prof. M.F. André (Geomorphology), Dr A. Fort (Physics), Dr J.M. Walter (Botany), and Dr L. Wilmé (Geomorphology).

References

André M-F (1991) L'empreinte glaciaire dans les Vosges. Presses Universitaires de Nancy, Nancy (coll. «Etudes Géographiques»)

Andréoli R, Rosique T, Schmidt M, Carozza J-M (2006) La dernière phase glaciaire du haut bassin de la Fecht (Vosges, France): dynamique de l'englacement et chronologie relative de la déglaciation. Géomorphologie: Relief, Processus, Environnement 1:23–36

Baltassat J-M, Legchenko A, Ambroise B, Mathieu F, Lachassagne P, Wyns R, Mercier J-L, Schott JJ (2005) Magnetic Resonance Sounding (MRS) and resistivity characterization of a mountain hard rock aquifer: the Ringelbach catchment, Vosges massif, France. Near Surf Geophys 3(4):267–274

Bourlès D, Braucher R, Engel Z, Kalvoda L, Mercier J-L (2004) Deglaciation of the Giant Mountains indicated by ^{10}Be dating. In: Drbohlav D, Kalvoda J, Vozenilek V (eds) Czech geography at the dawn of the millennium. Czech Geographic Society, Olomouc, pp 25–39

Carbiener R (1963) Les sols du massif du Hohneck – Leurs rapports avec le tapis végétal. In: Association Philomatique Alsace Lorraine (ed) Le Hohneck, Aspects physiques, biologiques et humains. Association Philomatique Alsace Lorraine, Strasbourg, pp 103–154

Carozza J-M (2013) The Vosgian-Alsatian side of the Rhine Graben: a unique, tectonically controlled and manmade landscape. In: Fort M, André M-F (eds) Landscapes and Landforms of France. Springer, Dordrecht, pp 139–148

Flageollet J-C (2002) Sur les traces des glaciers vosgiens. CNRS éditions, Paris

Flageollet J-C (2008) Morpho-structures in the Vosges massif. Géomorphologie: Relief, Processus, Environnement 2:75–86

Goepp S (2007) Origine, histoire et dynamique des Hautes-Chaumes du massif vosgien. Déterminismes environnementaux et actions de l'Homme. Thèse Strasbourg, Université Louis Pasteur, http://scd-theses.u-strasbg.fr/1455/

Guillet B (1972) Relation entre l'histoire de la végétation et la podzolisation dans les Vosges. Thèse Etat Sc. Nat., Université de Nancy I, 112 p

Jenik J (1958) Die Wind und Schneewirkungen auf die Pflanzengesellschaften im gebirge Belansky Tatry. Vegetatio 8(2):130–135

Mappi M (1742) Historia Plantarum Alsaticarum posthuma opera & studio. In: Dulseckeri JD (ed) Amsterdam, 335 p

Mercier J-L, Jeser N (2004) The glacial history of the Vosges mountains. In: Ehlers J, Gibbard PL (eds) Quaternary Glaciation – extent and chronology, part I: Europe, vol 2a, Development in Quaternary Science. Elsevier, Amsterdam, pp 113–118

Mercier J-L, Bourlès D, Kalvoda J, Vergne V, Braucher R, Paschen A, Raisbeck GM, Yiou F (1999) ^{10}Be dating of moraines and roches moutonnées at low altitude in continental Europe during the Holocene. Geophys Res Abstr 1:576

Wahl L, David PM (2007) Les névés temporaires des Hautes Vosges. In: AFEQ Compte Rendu excursion juin Lorraine, Luxembourg, Rhénanie-Palatinat, Sarre, pp 343-345

Walter-Simonnet A-V, Bossuet G, Develle A-L, Begeot C, Ruffaldi P, Magny M, Adatte T, Rossy M, Simonnet J-P, de Beaulieu J-L, Vannière B, Thivet M, Millet L, Regent B, Wackenheim C (2008) Chronologie et spatialisation de retombées de cendres volcaniques tardiglaciaires dans les massifs des Vosges et du Jura, et le plateau suisse. Quaternaire 2(19):117–132

Wlodarczyk D (1992) Evolution du couvert végétal sur les hautes chaumes du Tanet, Mém. Maîtrise UFR Géographie, Université Louis Pasteur, Strasbourg

Woillard G (1978) Grande Pile peat bog: a continuous pollen record for the last 140,000 years. Quat Res 9:1–21

Wyns R (2012) Etude géologique du cadre structural et des forages du bassin versant de recherche du Ringelbach (Soultzeren, Haut-Rhin). BRGM/RP-56540-FR, 123 p

On the Roof of Europe: High-Altitude Morphodynamics in the Mont Blanc Massif

17

Philip Deline and Ludovic Ravanel

Abstract

The Mont Blanc Massif, the highest of the main external crystalline massifs of the Western Alps, is renowned for its extensive glacier cover, steep granite rockwalls and vertiginous peaks.

Bordered by populated French, Italian and Swiss valleys, the massif has attracted both tourists and scientists over several centuries. Numerous sites are remarkable associations of landforms, for example, (a) the Aiguilles de Chamonix, a superb alignment of arêtes and peaks reaching 3,500 m a.s.l. which dominates the valley of Chamonix; (b) the Mer de Glace, the largest French glacier, which has been a major centre for tourism and glaciology since the eighteenth century; and (c) the Glacier du Miage, one of the largest debris-covered glaciers in the Alps.

Since the Last Glacial Maximum, the landscape of the massif has changed significantly and its morphodynamics is evidently still very active. Glacier retreat accelerates, and permafrost degradation has triggered more rockfalls. Nevertheless two million people are still attracted each year by the 'top of Europe'.

Keywords

Western European Alps • High-mountain morphodynamics • Glacier • Permafrost • Rockfall

17.1 Introduction

The Mont Blanc Massif has worldwide renown as 'the top of the Western Europe (4,810 m a.s.l.)', and owing to the Montenvers rack railway (1,913 m) and the Aiguille du Midi cable car (3,842 m), it is more than just a tourist attraction. At a crossroads between Savoy (France), the Aosta Valley (Italy) and Valais (Switzerland), it is an area of great geomorphological, human, historical and cultural interest. Since the mid-eighteenth century, scholars visited the massif, fascinated by an exceptional landscape, a considerable relief and huge ice masses. They studied its structure, its diverse landforms and especially its glaciers. Today, owing to global climate change (mean annual air temperature has increased by 2 °C in Chamonix since 1934), the dynamics of those landforms is changing rapidly. Thus, the Mont Blanc Massif is a major site for the study of the relationships between climate and morphodynamics in high mountain areas. After an overview of the massif, we will describe three emblematic sites that combine these aesthetic, historic and scientific aspects. Then the landscape evolution will be considered on different space and timescales, ending with various aspects of its remarkable tourism development over three centuries.

17.2 Geographical and Geological Setting

The European Alps formed during Tertiary collision of the European and the African continental margins. Ongoing collision resulted in a re-exhumation of these parts of the European basement which formed before ~20 Ma external crystalline massifs such as the Mont Blanc Massif (Glotzbach

P. Deline (✉) • L. Ravanel
EDYTEM Lab, Université de Savoie, CNRS,
Le Bourget-du-Lac, France
e-mail: philip.deline@univ-savoie.fr; ludovic.ravanel@univ-savoie.fr

M. Fort and M.-F. André (eds.), *Landscapes and Landforms of France*, World Geomorphological Landscapes,
DOI 10.1007/978-94-007-7022-5_17, © Springer Science+Business Media Dordrecht 2014

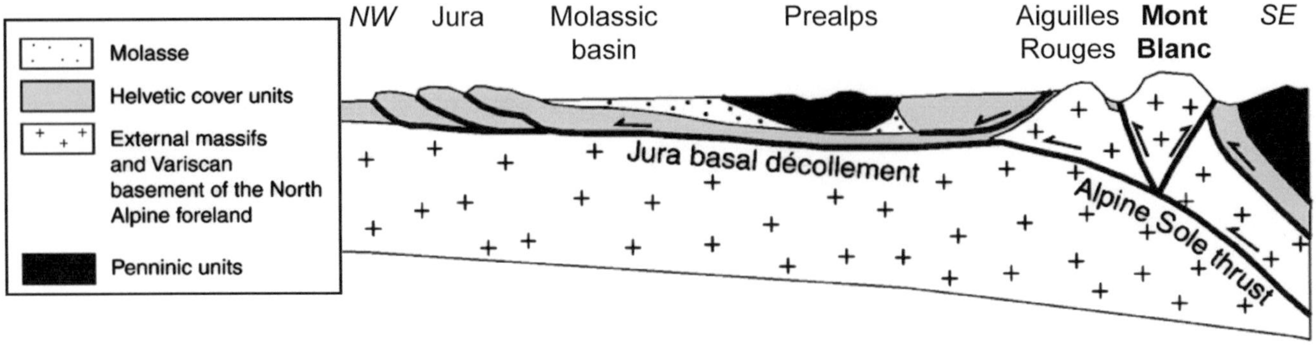

Fig. 17.1 Cross section of the NW Alps (Modified after Glotzbach et al. 2011)

et al. 2011). Thrusting subsequently propagated to the foreland along the Alpine sole thrust located below the massif and the Jura basal decollement (Fig. 17.1).

Shared between France, Italy and Switzerland, the massif has an area of about 550 km². About 30 % of this is covered by glaciers which are bordered by steep rockwalls having a relative relief of 500–1,000 m (Fig. 17.2). Delimited by deep peripheral valleys, it culminates at 4,810 m a.s.l. in a nearly 40-m-thick ice cap, and many of its fractured granite faces and ridges stand well above 3,000 m a.s.l. (Fig. 17.3).

The Mont Blanc Massif is formed by a Variscan granitic batholith intruded into the gneissic basement which also formed the main part of the much lower Aiguilles Rouges massif lying to the NW of the valley of Chamonix. Subvertical faults and shear zones determine the distribution of the large spurs of massive granite and the main couloirs. Bounded by a wide shear zone in the NW and a thrust in the east and SE ('pop-up' structure: Fig. 17.1), the massif displays a cross-range asymmetry (Fig. 17.2). Gentler slopes on the NW side are covered by six of the eight largest glaciers of the massif which are nourished by the dominant westerly winds and protected by a shaded northern aspect: Trient, Tour, Argentière, Mer de Glace, Bossons and Tré-la-Tête (Fig. 17.2); the SE flank is generally very steep, with small glaciers and high rockwalls.

The asymmetry is also reflected in climate of the massif. The mean annual air temperature (MAAT) is +7.2 °C in Chamonix (1,040 m a.s.l.) and +10.4 °C in Courmayeur (1,220 m a.s.l.). The average annual precipitation is 1,288 mm in Chamonix and 854 mm in Courmayeur. Precipitation is about three times higher at 3,500 m a.s.l. than at Chamonix (Vincent 2002). The MAAT at Aiguille du Midi (3,842 m a.s.l.) was −8.4 °C during the period 2008–2010 and with a July average lapse rate of 0.6 °C 100 m^{-1} on the French side and 0.75 °C 100 m^{-1} on the Italian side.

The high elevation of the massif is responsible for an extensive cryosphere. Twelve glaciers are longer than 5 km², whereas the present glacier surface area on the French side (102 km² in 2008) represents 37 % of the French Alpine glaciers. The steady-state equilibrium line altitude was in the range 3,000–3,150 m a.s.l. during 1971–1990 (Zemp et al. 2007), but is probably around 3,200 m currently. Steep rockwalls towering above these glaciers are permafrost-affected. Finally, the massif has been affected by intense glacial erosion from the Last Glacial Maximum (Coutterand 2010) to the Little Ice Age as indicated by glacial trimlines and deposits.

With 12,500 inhabitants, the Chamonix valley is relatively densely populated, whereas settlement is sparser on the Swiss and Italian sides. Nevertheless, these areas have major summer and winter tourism, and more than two million tourists visit the Chamonix valley every year. Both main sides of the massif have major transport axes connecting northern and southern Europe: 1.85 million vehicles per year (2011) pass through the Mont Blanc Tunnel. A third is trucks. The combination of past and present glaciations, steep slopes, fractured rockwalls and strong relative relief results in high rates of geomorphic activity along the sides of the massif; therefore, its bounding valleys are hazardous.

17.3 Remarkable Landforms

We have selected three emblematic sites of the Mont Blanc Massif: the Aiguilles de Chamonix, a superb and famous set of sharp peaks and arêtes that tower above the Chamonix valley; the Mer de Glace, the largest French glacier and one of the most frequently visited glaciers in the world; and the Glacier du Miage, located on the Italian side of the massif, the size, entrenched tongue and large supraglacial debris cover of which give it the appearance of a Himalayan glacier.

17.3.1 The Aiguilles de Chamonix: Sentinels of the Valley

The Aiguilles de Chamonix form a part of the NW side of the Mont Blanc Massif (Fig. 17.4). They correspond to a 5-km-long sequence of peaks from the Mer de Glace to the

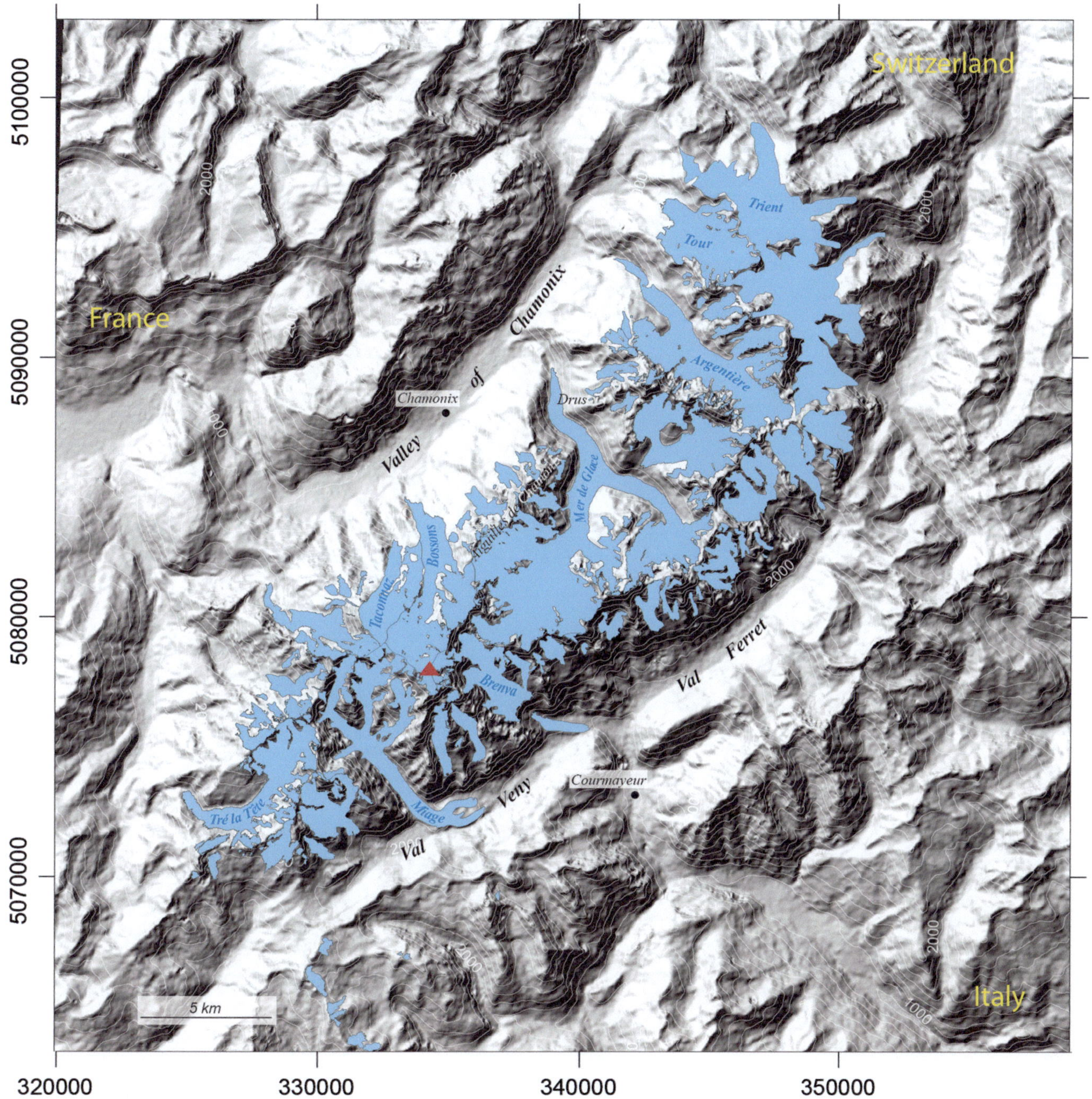

Fig. 17.2 The Mont Blanc Massif straddling France, Italy and Switzerland. Present glacier extension is in *blue*; *red triangle:* Mont Blanc summit (From Le Roy 2012, modified)

Glacier des Bossons, regularly exceeding 3,500 m a.s.l., cut into panels by two main sets of subvertical joints and faults (N0° and N40°–60°). Consequently, the north side shows high (over 1,000 m), steep and fractured rockwalls. All visible from Chamonix, the eponymous needles contribute greatly to the overall attractiveness of the valley. The most famous and the highest is the Aiguille du Midi (3,842 m a.s.l.), which has been accessible by cable car since 1955. Three small glaciers (Pélerins, Blaitière and Nantillons) are flowing at the foot the north side of the Aiguilles, whereas the Vallée Blanche/ Glacier du Géant is well developed on a large part of the other side (Fig. 17.2). Some of the rockwalls are ice- or snow-covered with hanging glaciers. The 50 unweathered rockfall scars which mark the faces and the amount of debris covering all three small glaciers in retreat at the foot of the slope indicate a strong geomorphic activity of this area.

Fig. 17.3 View of the Mont Blanc Massif from the SW. Glacier de Tré-la-Tête in the *centre*; Val Montjoie and Valley of Chamonix in the *lower* and *upper left* corners, respectively; Val Veny in the *upper right* corner

Fig. 17.4 The north side of the Aiguilles de Chamonix. From *left* to *right*: Aiguilles des Grands Charmoz (3,445 m a.s.l.), de Blaitière (3,522 m), du Plan (3,673 m) and du Midi (3,842 m) with the string of hanging glaciers (Photo: J. Malbert)

17.3.2 The Mer de Glace: Three Centuries of Tourist Attraction

The Mer de Glace *sensu lato* is a c. 30-km^2 compound valley glacier occupying the central area of the massif. It flows from the Mont Blanc du Tacul (4,248 m a.s.l.) and forms the Glacier du Géant, a very large accumulation zone between the Aiguille du Midi and the Dent du Géant (4,013 m a.s.l.) where, every year, more snow accumulates in winter than snow and ice melt in summer. When the Glacier du Géant develops a tongue, this is called the Glacier du Tacul. Several glaciers are tributaries of the Tacul upstream from its confluence with the Glacier de

Fig. 17.5 Mont Blanc Massif, viewed from the east (elevation in m a.s.l.). Ogives form in the Géant icefall and develop on the *left side* of the tongue of the Mer de Glace. A supraglacial debris cover (the Veine noire) lies on the *right side*, the expansion or retraction of which is related to the mass balance of the glacier (Photo: J.-F. Hagenmuller)

Leschaux. Downstream from this confluence, a 5-km-long and 0.5–0.9-km-wide tongue that terminates at an elevation of 1,500 m a.s.l. forms the Mer de Glace *sensu stricto*. Since the end of the 1940s, the large Glacier de Talèfre (7.6 km^2) is no longer connected to Leschaux.

The Mer de Glace is well known for its ogives (Fig. 17.5), studied by J. Forbes in the 1840s. These arcuate alternating light and dark bands have apices which point down-glacier. They form at the Géant icefall. Each pair of bands results from the deposit of rocky dust in the icefall crevasses during a summer season, followed by snowfall in winter. Both types of ice are deformed downstream from the icefall because of the difference of velocity between the centre (thicker ice) and the margins of the ice flow (higher friction).

The right part of the tongue is called the Veine noire (i.e. Black Vein), a debris-covered ice stream comprising 1.2 km^2 (35–40 %) of the area of the tongue (Fig. 17.5). It is dominated by the right lateral moraine which exceeds 220 m in height.

The Mer de Glace is also well known for its artificial ice cave annually visited by tens of thousands of people since 1946. The entrance to the cave is farther to reach every year because of the fast lowering of the glacier surface (4 m per year during the period 2000–2008: Berthier and Vincent 2012). A consequence of this is that the cave managers have to insert each year some extra steps to the already 400-step-high staircase.

17.3.3 The Glacier du Miage: A 'Himalayan' Glacier

The Glacier du Miage is a 10.5-km^2 and 10-km-long compound valley glacier fed by four steep tributaries. Originating on the Mont Blanc summit, its lower section is an 8-km-long, gently sloping tongue, partly entrenched in a deep and straight valley. When reaching the Val Veny, its arcuate course is channelized by huge lateral moraines up to 200 m high, before forming three divergent moraine-bounded lobes, the lowest terminating at *c.* 1,730 m a.s.l. The ice-marginal Miage Lake is contained within a morainic amphitheatre formed by a staircase-like sequence of breach-lobe moraines. It is located on the right flank of the glacier and was drained at irregular intervals at least 15 times in the last century, that in 2004 being the last (Fig. 17.6). The Miage's distal part (4 km^2) is totally debris-covered. Together with the

Fig. 17.6 The Glacier du Miage, one of the largest debris-covered glacier in the Alps, mainly supplied by rockfall and rock avalanche. The Plan Combal (*low left* corner) results from the filling of a palaeolake formed by the damming of the Val Veny by the glacier and its lateral moraine (Photo: M. Le Roy)

Unteraargletscher (Bernese Oberland), Miage is the largest debris-covered glacier in the Alps. Its size, steep slopes, entrenchment, debris cover and marginal and supraglacial lakes give a Himalayan look to the glacier.

Twenty to forty centimetres thick in average, the supraglacial debris cover consists of several dozen units distinguished by lithology, fabric, particle size, roundness and distribution, all of which have resulted from rockfall or rock avalanche. Flow-parallel units derived from deposits onto the glacier accumulation area have been transported englacially. Irregular units with hummocky topography derived from deposits onto the ablation zone were transported supraglacially. More than ten rockfalls and small rock avalanches were deposited onto the Miage during the last century (Deline 2009).

17.4 Evolution

17.4.1 Glacier Fluctuations in the Mont Blanc Massif Since the LGM

While one of the fronts of the Last Glacial Maximum (LGM) ice cap was close to Lyon, the ice surface was towered by many high nunataks above 2,400 m a.s.l. all around and within the Mont Blanc Massif. As suggested by the preserved trimlines, the ice was 750 m thick in the Montenvers area (Coutterand 2010).

The valley remained entirely glaciated up to the end of the Older Dryas (c. 15 kyr). During the Younger Dryas (12.8–11.5 kyr), the by-then individualized glaciers (Mer de Glace, Bossons and Argentière) readvanced (Fig. 17.7). The front of the Mer de Glace reached the present centre of Chamonix and calved icebergs into a lake dammed by the front of the glacier formed by the confluence of the Bossons and Taconnaz.

Two readvances of the Glacier du Tour took place at the beginning of the Holocene at 9.9 and 8.4 cal. kyr, but there is no geomorphological evidence for any glacial advance in the massif until the Neoglacial period, i.e. around 5 kyr BP (Le Roy 2012). Researches in the Miage morainic amphitheatre and the Plan Combal infill indicate that moraine sets of the first glaciation correspond to the main Neoglacial phases distinguished in other alpine regions, starting with the damming of the Val Veny valley around 4.8 cal. kyr BP (Deline and Orombelli 2005). The right lateral moraine of the Mer de Glace *s.s.* contains a large amount of subfossil wood which has been used to identify 13 fluctuations of this glacier during the Neoglacial, from 3.6 cal. kyr to 1350 AD (Le Roy 2012).

Fig. 17.7 Reconstruction of the past extension of the glaciers in the valley of Chamonix (corresponding to the Younger Dryas) by E. Viollet-le-Duc (Glacier des Bois et vallée de Chamonix, Aiguille du Dru, Aiguille Verte; drawing, watercolour and gouache, 1874 [290×695 mm])

During the Little Ice Age (LIA), the front of the Mer de Glace advanced seven times between 1550 and 1852 AD (Nussbaumer et al. 2007), generally farther downstream than during any other cold phases of the Neoglacial.

After the LIA, the glacier area of the massif strongly decreased: the loss for the period 1850–2000 was 43, 24 and 23 % on the Italian, Swiss and French sides of the Massif, respectively, with an acceleration of the shrinkage in the three last decades (Deline et al. 2012). The decrease of the Mer de Glace s.s. thickness was between 40 and 185 m during the period 1939–2001 (Nussbaumer et al. 2007). The post-LIA retreat was first driven by a winter precipitation decrease during the end of the LIA, then by a temperature increase from the beginning of the twentieth century which accelerated in the last few decades, especially during summers (Vincent 2002).

Because of the reduced ablation due to their extensive debris cover, recent fluctuations of the Miage and the Glacier de la Brenva were much smaller than those of debris-free glaciers. For instance the retreat of the Miage front is less than 300 m since 1860, compared with 2,300 m at the Mer de Glace (Fig. 17.8).

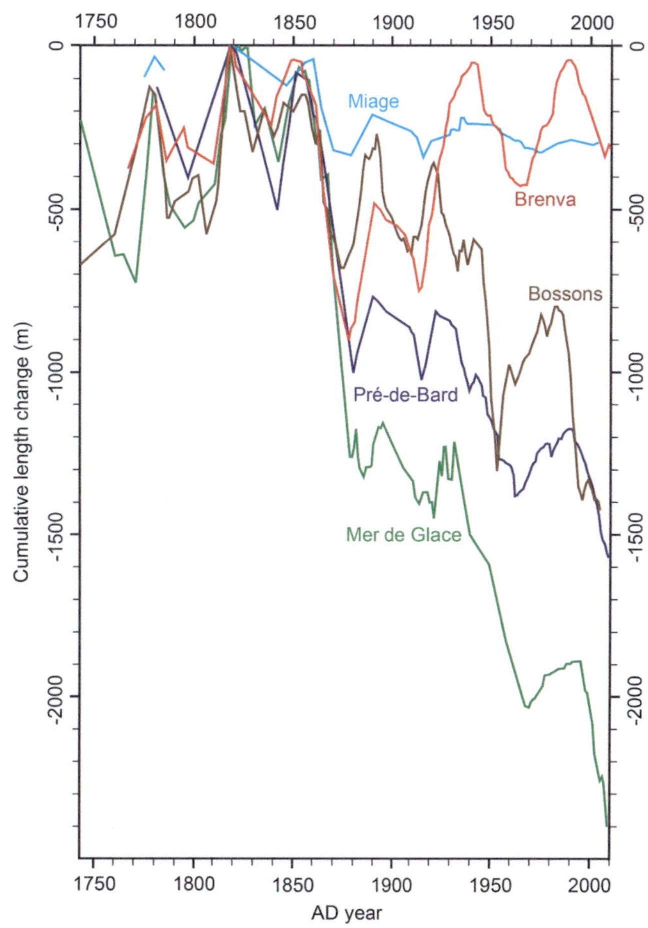

Fig. 17.8 Cumulative length variations of five glaciers of the Mont Blanc Massif since the eighteenth century (See Deline et al. 2012 for data sources)

17.4.2 Rockwall Morphodynamics Since the End of the Little Ice Age

1920 and 1997 Brenva Rock Avalanches

These two $2–3 \times 10^6$-m^3 rock avalanches detached from very steep rockwalls located above 3,500 m a.s.l. The low-friction

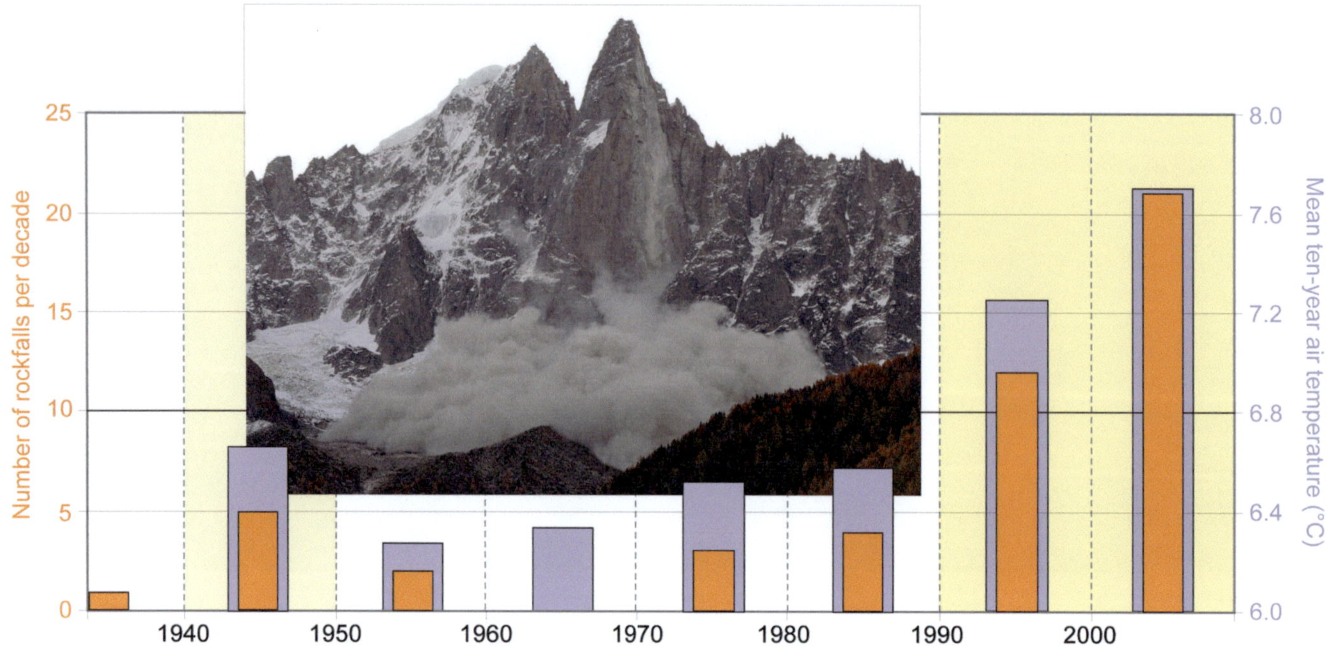

Fig. 17.9 Air temperature in Chamonix (1,040 m a.s.l.) and rockfall occurrence on the west face of the Drus and the north side of the Aiguilles de Chamonix since the 1940s. Insert: the Drus rockfall of October 2011 (43,000 m^3; Photo: J. Malbert)

surface offered by the Glacier de la Brenva (one of the largest of the Massif) caused excessive run-out distances, and both avalanches reached the valley floor. Because of the protection against the ablation due to the 1920 rock deposit onto the glacier surface, the Brenva advance that started in 1913 lasted atypically until the early 1940s (Fig. 17.8).

The twentieth-century Brenva rock avalanches might be related to the degradation of rockwall permafrost at depth, an effect attributable to global warming, although historical and geomorphological evidences suggest that several rock avalanches occurred during colder phases of the Late Holocene (e.g. at 2,500 BP, 1,540 BP, fourteenth century AD, 1760s AD), whereas the largest known rock avalanche in the massif (Triolet) occurred in 1717 AD.

Rockfall Evolution at the Drus and the Aiguilles de Chamonix Since the 1860s

The relationship between rock instability frequency and warm periods is more obvious at a different scale. Photo comparison and field work on the west face of the Drus (Ravanel and Deline 2008) and the north side of the Aiguilles de Chamonix (Ravanel and Deline 2011) identified 58 rockfalls in these areas since the termination of the LIA. Only two of this series occurred before the mid-twentieth century (1905 and 1936), whereas the two main clusters of rockfalls are at the end of the 1940s and, especially, the two last decades with 71.5 % of the series, i.e. in close correspondence with the warmest periods (Fig. 17.9). Permafrost degradation is probably the main factor in explanation of this recent increase in the rockfall frequency (Ravanel and Deline 2011).

Present Rockfall Frequency in the Massif

Among 242 rockfalls surveyed in the central area of the Mont Blanc Massif during the period 2007–2011, 115 have been precisely dated: 95 % occurred during the hottest months of the year, mostly during the warmest periods of these months. Even more striking, the rockfall frequency in 2003 has been exceptional in the central area of the massif (152 rockfalls) as in the whole Alps, while the 0 °C isotherm remained above 3,000 m a.s.l. for three full months (June–August) except for 5 days (Ravanel et al. 2011).

Among other elements, this strong correlation between rockfall occurrence and warmest periods at the timescale of year supports the hypothesis of the relationship between permafrost degradation and rockfall (Ravanel 2010).

To confirm this relationship, studies of the rock temperature at the surface and inside the rockwalls were started in 2005, especially at the Aiguille du Midi. Many surface sensors and three 10-m-deep boreholes equipped with a chain of 15 thermistors are combined with repeated geophysical (ERT) surveys, a wireless system to measure crack movements and temperature modelling to study the permafrost in high-Alpine steep rock faces.

Fig. 17.10 Lithography Vue de la Mer de Glace et de l'Hôpital de Blair, du Sommet du Montanvert dans le mois d'aoust 1781 by K.L. Hackert

17.5 Cultural Value and Touristic Promotion

17.5.1 From the 'Invention of the Mont Blanc'...

Tourism in the valley of Chamonix started with the visit of Montenvers by English and Swiss travellers (Windham and Pococke, Martel) in the 1740s, when Mont Blanc became recognized as the highest European mountain (Fig. 17.10). Studies of glaciers and related landforms started earlier in the Mont Blanc Massif than in any other alpine areas: a paragon of the Enlightenment scholars for a long time, Saussure was very active during the last third of the eighteenth century. He first visited Chamonix in 1760 and encouraged the conquest of the summit of Mont Blanc, attained in 1786 by Paccard and Balmat. In 1788, he made observations during his 2 weeks at the Col du Géant (3,371 m a.s.l.), on the present boundary between France and Italy.

After Saussure, many investigations took place on the French side of the massif (Forbes, Tyndall, Vallot, Lliboutry, Vivian and Reynaud) in pioneering studies of glaciology. The study of the permafrost in the massif was much more recent. This probably results from the minor importance of the massif for Switzerland (permafrost studies began in the early 1970s in other Swiss massifs), the absence of large rock glaciers in the massif and a limited interest in mountain permafrost from Italian and French scientists until recently. The massif has thus been the object of much research, including issues outside of geology, geomorphology and glaciology. For example, a first Refuge des Cosmiques (3,613 m a.s.l.) located close to the Aiguille du Midi was built by L. Leprince-Ringuet in 1943 in order to study cosmic rays – this was destroyed by a fire in the 1980s.

17.5.2 ...to the Main Natural Touristic Site in the Western Alps

With 785,000 visitors in 2010, the Aiguille du Midi (Fig. 17.11a) was the 7th noncultural touristic site in France, whereas Montenvers (8th) and Brévent (19th) were visited by 720,000 and 455,000 people, respectively. The physical characteristics of the Mont Blanc Massif explain its interest for tourists and scientists but also, of course, mountaineers. As mountaineering became an established and popular sport, the Compagnie des Guides de Chamonix was created

Fig. 17.11 The Aiguille du Midi. (**a**) Poster (1956) praising the 'highest cable car in the world' built between 1951 and 1955 (Payot collection); (**b**) Gaston Rébuffat climbing the south face of the Aiguille du Midi in the end of the 1950s (Compagnie des Guides collection)

in 1821. The 'mountaineering golden age' began with the conquest of the main peaks of the massif: Whymper, Mummery, Ravanel, Charlet, Bonatti, Rebuffat (Fig. 17.11b) or Desmaison participated in the strongest challenges of their times and strengthen the aura of the Chamonix valley and the massif. In the early twentieth century, Payot introduced skis, and Chamonix hosted the first Winter Olympic Games in 1924.

17.6 Conclusion

Located close to densely populated European areas and easy to access with cable cars on all its sides and within, the Mont Blanc Massif has been a focus of attention for a long time. People can discover the high mountain environment not only as an impressive landscape but also as a geomorphologically very active alpine area. As far as the French are concerned, the Mont Blanc Massif is one of their most important natural and cultural heritage sites. However, controversy about attendance levels and facilities has so far prevented its acceptance onto the World Heritage List of UNESCO, an application first made in 2000.

Acknowledgements M. Fort and M.-F. André are warmly acknowledged for their encouragement to write this chapter. Many thanks are due to P. Migon for his suggestions and to P. Walsh and Y. Battiau-Queney for their careful proofreading of the English language.

References

Berthier E, Vincent C (2012) Relative contribution of surface mass-balance and ice-flux changes to the accelerated thinning of Mer de Glace, French Alps, over 1979–2008. J Glaciol 58(209):501–512

Coutterand S (2010) Etude géomorphologique des flux glaciaires dans les Alpes nord-occidentales au Pléistocène récent. Du maximum de la dernière glaciation aux premières étapes de la déglaciation. PhD thesis, Université de Savoie

Deline P (2009) Interactions between rock avalanches and glaciers in the Mont Blanc massif during the late Holocene. Quat Sci Rev 28(11–12):1070–1083

Deline P, Orombelli G (2005) Glacier fluctuations in the western Alps during the Neoglacial as indicated by the Miage morainic amphitheatre (Mont Blanc massif, Italy). Boreas 34(4):456–467

Deline P, Gardent M, Magnin F, Ravanel L (2012) The morphodynamics of the Mont Blanc massif in a changing cryosphere: a comprehensive review. Geografiska Annaler Ser A Phys Geogr 94:265–283

Glotzbach C, van der Beek PA, Spiegel C (2011) Episodic exhumation and relief growth in the Mont Blanc massif, Western Alps from numerical modelling of thermochronology data. Earth Planet Sci Lett 304:417–430

Le Roy M (2012) Reconstitution des fluctuations glaciaires holocènes dans les Alpes occidentales. Apports de la dendrochronologie et des datations par isotopes cosmogéniques produits in situ. PhD thesis, Université de Savoie

Nussbaumer SU, Zumbühl HJ, Steiner D (2007) Fluctuations of the Mer de Glace (Mont Blanc area, France) AD 1500–2050. Part I: The history of the Mer de Glace AD 1570–2003 according to pictorial and written documents. Z Gletsch Glazialgeol 40(2005/2006):5–140

Ravanel L (2010) Caractérisation, facteurs et dynamiques des écroulements rocheux dans les parois à permafrost du massif du Mont Blanc. PhD thesis, Université de Savoie

Ravanel L, Deline P (2008) La face ouest des Drus (massif du Mont-Blanc): évolution de l'instabilité d'une paroi rocheuse dans la haute montagne alpine depuis la fin du petit âge glaciaire. Géomorphol Relief Processus Environ 4:261–272

Ravanel L, Deline P (2011) Climate influence on rockfalls in high-Alpine steep rockwalls: the North side of the Aiguilles de Chamonix (Mont Blanc massif) since the end of the Little Ice Age. Holocene 21(2):357–365

Ravanel L, Allignol F, Deline P, Bruno G (2011) Les écroulements rocheux dans le massif du Mont-Blanc pendant l'été caniculaire de 2003. In: Lambiel C, Reynard E, Scapozza C (eds) La géomorphologie alpine: entre patrimoine et contrainte. Actes du colloque de la Société Suisse de Géomorphologie. Géovisions 36:245–261

Vincent C (2002) Influence of climate change over the 20th century on four French glacier mass balances. J Geophys Res 107:D19. doi:10.1029/2001JD000832

Zemp M, Hoelzle M, Haeberli W (2007) Distributed modelling of the regional climatic equilibrium line altitude of glaciers in the European Alps. Glob Planet Change 56:83–100

18 In the Folds of the Earth: French Prealpine Geomorphological Landscapes

Fabien Hobléa

Abstract

The French Prealps constitute the Western margin of the Alpine range. They are composed of about 20 distinct massifs. Their geomorphological landscapes are typical of sedimentary folded relief, but the combination of several physical and human factors introduces nuances and diversity between the Northern and Southern Prealps as well as between the Prealpine *s.s.* and Subalpine massifs. In 2012, the Prealps host the four French members of the European and Global Geoparks Network. One of them, the Bauges Massif Regional Nature Park, is taken as the example of archetypal subalpine folded and karstified relief, famous for its perched synclines, its caves, and its karst waters which supply the two largest natural lakes in France. At last, Mont Granier, in the Chartreuse Massif, is presented as one of the major geomorphosites of the Northern Prealps, which offers many scientific interests regarding the rock failure of 1248, karst development, and associated paleontological remains of cave bears.

Keywords

French Prealps • Geomorphological landscape • Folded relief • Karst, geomorphosite • Geopark

18.1 The French Prealps: The Western Margin of the Alpine Range

The generic name *Préalpes* was popularized by the famous French geographer Raoul Blanchard in his monumental work *Les Alpes occidentales* (the Western Alps), composed of seven volumes published between 1938 and 1956 (Blanchard 1938–1956). It refers to all the mountainous massifs forming the sedimentary folded external margin of the Alpine arc (Fig. 18.1).

The French, Western Prealps extend from the Chablais in the north to the Prealps of Nice in the south, striking along the curvature of the Alpine range (Fig. 18.2). They are made of sedimentary rocks that consist mainly of alternating limestone strata, resistant to erosion but usually karstified, overlying softer, marly layers. Selective erosion has exploited these contrasts in hardness to generate a leitmotiv landscape represented by a typical morphological binomial of limestone cliffs overlooking marl slopes more or less vegetated or covered by scree.

Hence, the Prealps present a certain landscape homogeneity. They are nicknamed "cow mountains" because of their relative low elevation with respect to the Alpine Central Massifs (most of the summits do not reach 2,500 m a.s.l.) that induces the presence of extended forests and meadow areas.

The Prealps are nevertheless remarkable for the infinite range of landscape nuances that distinguish each of their massifs (Fig. 18.3). This variety, which is also expressed in the landforms and specific features of the relief, is largely due to the combination of geological, climatic, and human parameters.

F. Hobléa (✉)
EDYTEM, UMR CNRS – Université de Savoie,
Le Bourget du Lac, France
e-mail: Fabien.Hoblea@univ-savoie.fr

M. Fort and M.-F. André (eds.), *Landscapes and Landforms of France*, World Geomorphological Landscapes,
DOI 10.1007/978-94-007-7022-5_18,

Fig. 18.1 Winter facing northeast aerial view showing the Northern subalpine chains alignment (here Chartreuse and Bauges) bordering the central crystalline massifs (including Mont Blanc visible on the *right* in the background) (Photo Didier Monin)

18.2 Between Geology and Geography: The Factors of the Prealpine Landscape Diversity

18.2.1 Geologist's Prealps Are Not Geographer's Prealps!

From a geological point of view, the word "Prealps" is used to distinguish the sedimentary massifs which are in external position in the Western Alps but formed by allochthonous thrust nappes coming from the internal alpine zone and now separated from their region of origin located to the east of the "Penninic Front" (a major overlap that delineates the internal and external alpine domains) (Schmid et al. 2004). In the French Alpine range, only the Chablais Massif in the north (which continues in Switzerland) and the Embrunais-Ubaye nappes in the south correspond to this definition. All the other massifs of the external alpine zone belong for the geologist to the subalpine domain (*chaînes subalpines*). They are formed by the sedimentary cover of the crystalline Central Alps. This cover has been detached from the crystalline basement (décollement surface) and moved to the west during the uplift of the Central Alps. It has also been strongly shortened by folding and overlapping (Fig. 18.4).

The set has in common to be composed of sedimentary rocks of marine origin and Mesozoic age (Fig. 18.5) mainly formed during the phases of extension (190–120 My) then subduction (120–60 My) of the Alpine Domain; they have gradually and mainly been structured during the Cenozoic era through successive tectonic phases in the context of continent–continent collision that induced several décollement surfaces, a counterclockwise rotation and an E-W displacement of the sedimentary cover totalizing 5–30 km from the south to the north in the subalpine domain (Gratier et al. 1989) and over distances of several tens or even hundreds of kilometers for the Prealpine nappes (Coward and Dietrich 1989).

The current morphostructural landscape is largely inherited from the final phase of major shortening in the late Miocene (Tortonian around eight million years), from the Pliocene epeirogenic uplift and river incision, then the Quaternary glaciations that began around two millions of years.

18.2.2 The Northern Prealps Are Not the Southern Prealps!

The Drôme valley marks the separation between the Northern and the Southern Prealps (dotted line on the Fig. 18.2). This

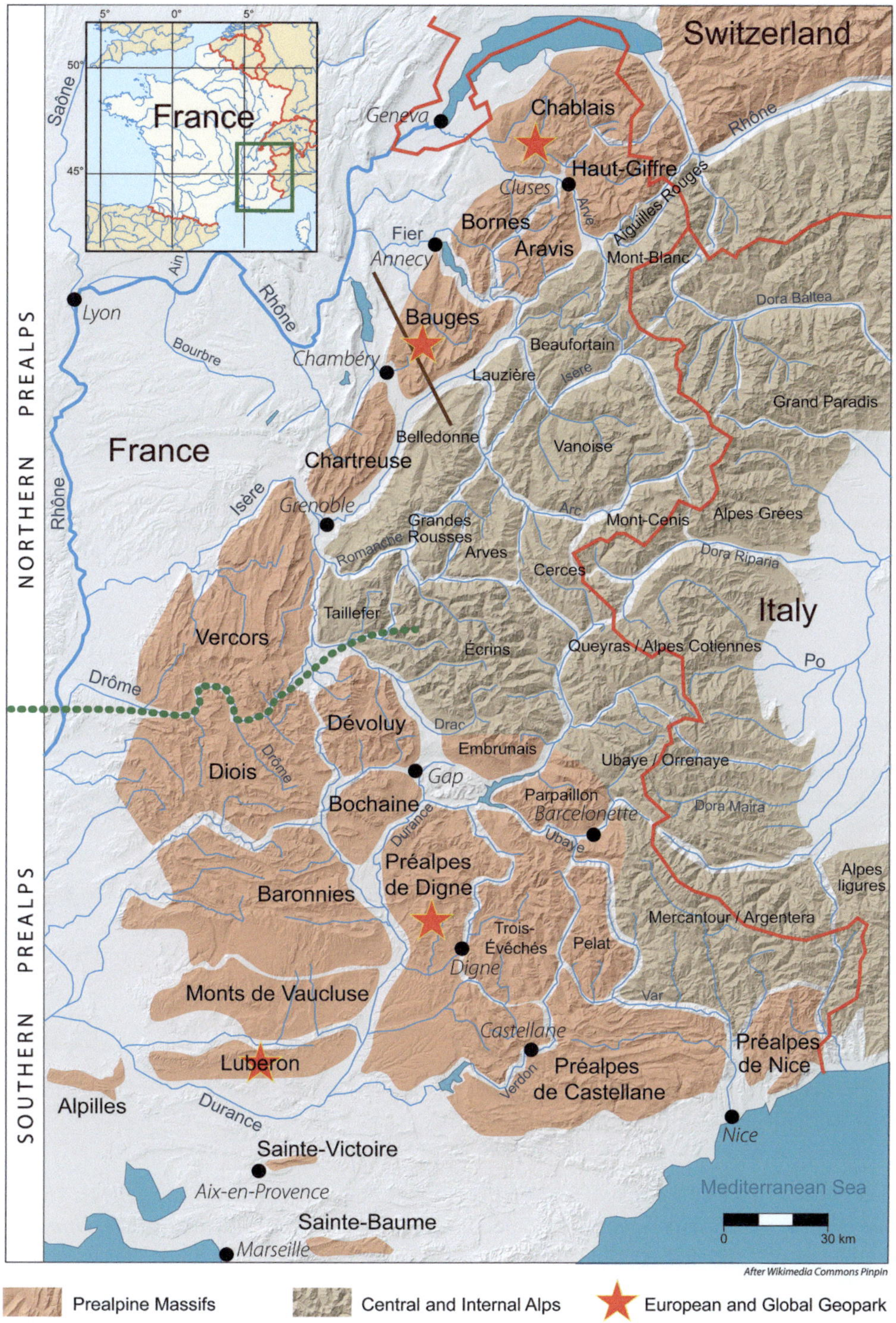

Fig. 18.2 Location of the Prealpine massifs on the map of French Alps (the *straight line* indicates the trace of the geological cross section Fig. 18.4)

boundary can be explained by both geological and bioclimatic factors.

The Northern Prealps, from Chablais to Vercors, form a narrow band of well-individualized massifs clearly separated from each other by transverse valley of tectonic origin. Each massif is composed of folded structures parallel and stretched in a NNE-SSW axis (Fig. 18.1), due to the "Alpine tectonic stage" that began during the Oligocene. They are armed with thick and massive Jurassic and Cretaceous limestone beds (Fig. 18.5, left). The Urgonian reef limestone facies (Barremian–Aptian, around 125–115 million years) can reach 400 m thick. They are shaped by karst processes and they constitute the highest cliffs of most of the summital ridges. All the folds are at least overturned and often over-

Fig. 18.3 Northern Prealpine landscape (**a**) vs. Southern Prealpine landscape (**b**) (Photos Fabien Hobléa). (**a**) The ridge of subhorizontal Jurassic limestones of the Tenneverge Peak (2,989 m a.s.l.) in the Haut-Giffre Massif and (**b**) cluse of Sisteron and the vertical strata of Jurassic limestone of the Montagne de la Baume (1,147 m a.s.l.)

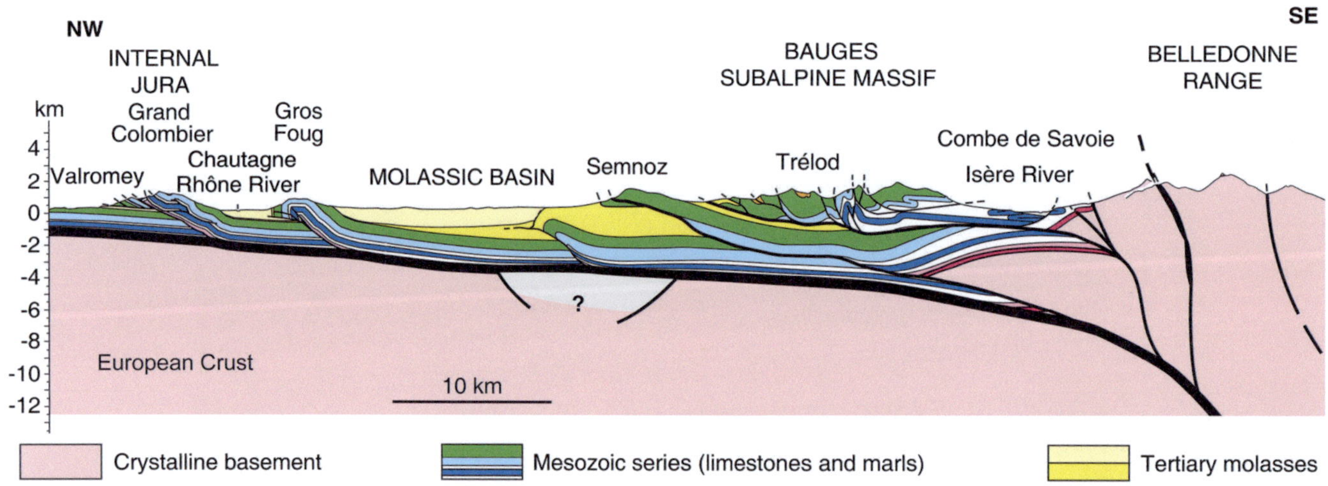

Fig. 18.4 Geological cross section through the external Alpine domain (see location on Fig. 18.2). (After Tardy et al., modified)

thrust westward. The western part of the massifs is usually lower and less eroded than the eastern part where the folded relief is inverted (highlands and perched synclines).

The landscapes of the Northern Prealps have been strongly influenced by the glacial periods. The present relict glaciers are tiny and rare, confined to summits above 3,000 m a.s.l. in the Haut-Giffre massif (Fig. 18.3a) at the boundary with Switzerland (Mt Ruan and Mt Buet). But the current mountain climate is still harsh and wet due to the barrier effect against the oceanic fluxes from the west. The dominant vegetation is one of the beech/fir mountain forest and of the subalpine stage (mountain pine, moor, and lawn). The rural landscape is mainly influenced by the silvopastoral activities, with cattle turned to milk and cheese. After a period of population decrease due to the world wars, the Northern Prealps stayed away from the industrialization of the Alpine Valleys. They possess only a few major ski resorts compared to the *Grandes Alpes*. These are territories that have remained preserved at the gates of major cities (Geneva, Annecy, Chambéry, Grenoble). Some of them became regional nature parks: Vercors (1970), Chartreuse (1995), and Bauges Massif (1995) in which the population is now increasing.

The Southern Prealps, from Diois to Nice Prealps, spread widely in a double row of massifs less individualized than their northern neighbors. The merging of east-west structural directions, namely, "Pyreneo-Provençal" (from the end of the Cretaceous) and north-south Alpine directions (due to the later "Alpine tectonic stage") generated a particular style of folding. This gives anticline domes with very steep slopes (as in Sisteron) and circular (e.g., Céüse mountain) or east-west elongated

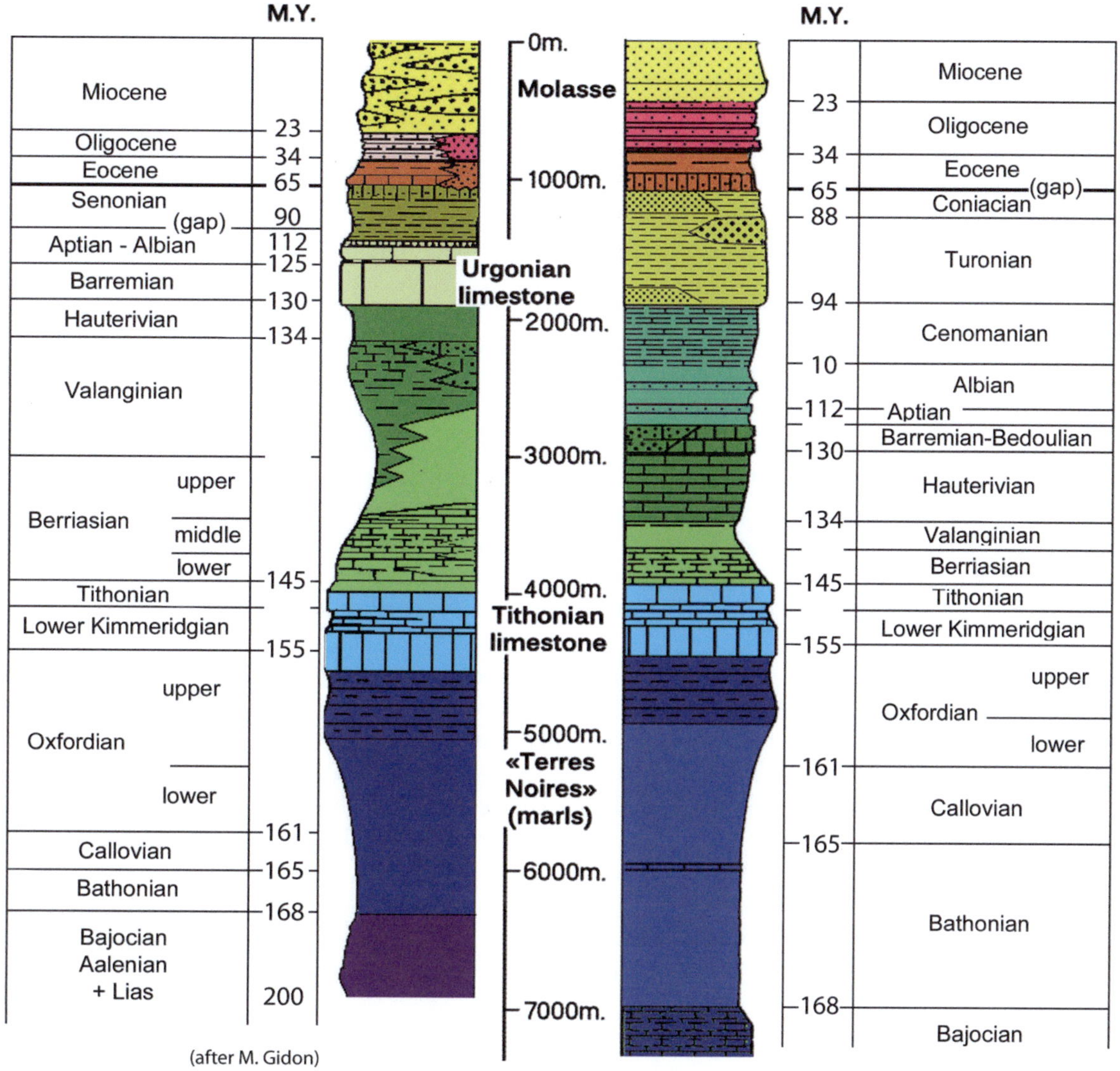

Fig. 18.5 Comparison between Northern (*left*) and Southern (*right*, Vocontian domain) subalpine lithostratigraphic series

synclines (the famous Forêt de Saou), often evolved into inverted reliefs. Between Vercors and the Mounts of Vaucluse, on the site of a deep basin of the ancient Alpine Sea ("Vocontian basin"), the Urgonian limestone facies disappears because of a lateral variation of facies in the Barremian–Aptian stages (Fig. 18.5, right). The landscape is here marked by outcrops of thick marly layers (Dogger stages). This is the area of the *Terres Noires* ("black terrains"), dissected into gullies (or *roubines* in local language) that remind the American "badlands" landscapes, especially as the climate here is under Mediterranean, semiarid influences. Further south, in the Mounts of Vaucluse, the Luberon, the Castellane, and the Nice Prealps, reef limestones appear again, arming long crests stretched from west to east like Mont Ventoux or the Lure Mountain. The vegetation, adapted to the summer drought, stands in stark contrast to one of the Northern Prealps: pines and larches at greater altitude and Mediterranean species in the lower belt. Aridity is often reflected in the landscape by bare slopes. There, erosion is exacerbated by the Mediterranean climatic conditions characterized by intense precipitation events succeeding to summer drought. In altitude, periglacial processes shape a landscape that is today unglaciated and was less affected by Quaternary glaciation than the Northern Prealps (at the Würmian maximum, the Durance glacier did not flow downstream from the rock bar of Sisteron) (Fig. 18.3b). Having experienced the same depopulation as the Northern Prealps, the Southern Prealps remained fairly deserted. Local people make their living from extensive sheep farming and diffuse tourism. The Luberon became a regional nature park in 1977 and the Verdon in 1997, and the *Préalpes d'Azur* (eastern part of the Castellane Prealps) obtained the same recognition on March 30, 2012.

If these six Prealpine regional nature parks are protecting and enhancing their natural and cultural heritage, some Prealpine territories highlight their geoheritage through a Geopark approach.

18.2.3 The Prealps: The Land of the French Geoparks

If the reliefs of the Prealps are relatively young, the different lithostructural conditions between Prealpine (*s.s.*) and Subalpine massifs on the one hand and changing morphogenic conditions between Northern and Southern Prealps on the other hand induce a real diversity of geomorphological landscapes.

It is therefore not a coincidence that the Prealpine area (*s.l.*) is the home of the four French territories members of the European and Global Geoparks Network (in 2012, Fig. 18.2). The geological reserve of Haute-Provence, located in the surroundings of Digne-les-Bains, pioneered the concept of Geopark in 2000. The Luberon Regional Nature Park became Geopark in 2004. Recently, two Geoparks were created in the Northern Prealps: the Bauges Massif Geopark in September 2011, belonging to the subalpine area, and the Chablais Geopark in March 2012, archetype of the Prealpine thrust nappes.

The concept of Geopark distinguishes territories highlighting their geoheritage in development policy. The approach aims to raise awareness among both the local people and the visitors about the links between man and geosphere and to enhance these links as a basis for a local sustainable way of development. In the Prealps, we find evidences of such relations dating back to the Magdalenian (15000 BC) and the Mesolithic periods (10000–5000 BC), when caves and rock shelters were used as home by the first Alpine inhabitants (Bintz and Griggo 2011).

In the context of an agroforestry–pastoral economy, local people began quite early to domesticate the driving force of the water and dig tunnels in the mountains to extract various mineral resources like flint, limestone, slate, iron, lignite, and silica sand.

Today, the Prealpine massifs, which are facing the climate change issue, rediscover this old heritage and use it for geotouristic activities, as it is considered as a good support for their diversification and sustainable development.

This is particularly the case for the four Geoparks, territories revealing the wealth and the diversity of the Prealpine morphological landscapes as well as the modalities of the geoheritage management and promotion. Each Prealpine Geopark boasts about a wide geodiversity. But each of them must assert its own geoheritage identity through some key themes: fossils and folded relief for the Haute-Provence Geopark; a 150-million year geological history and thousands years of relations between man and mineral resource for the Luberon; thrust nappes, water, and glacial heritage for the Chablais and, for the Bauges, folded subalpine reliefs, water, and karst.

18.3 The Bauges Massif: A Representative Geopark of the Northern Subalpine Reliefs

18.3.1 The Bauges Massif Regional Nature Park

The Bauges Massif belongs to the "Prealps of Savoie." The massif covers an area of 850 km^2 and culminates at 2,217 m a.s.l., at the top of Mount Arcalod. It is bordered by the two largest natural lakes in France: the Bourget Lake in the foreland (molassic basin) at its western side and the Annecy Lake in a transverse valley at its northern side. The massif is bordered by the Isère valley at the eastern side and by the transverse valley of Chambéry at the southern side (Fig. 18.6). This is a Subalpine massif belonging to the Delphino-Helvetic geological alpine zone. Its stratigraphical series shows mainly an alternation of Mesozoic limestones and marls crowned by the Urgonian reef limestone (Figs. 18.5, left, and 18.7). Cenozoic and Quaternary deposits complete the series. The Bauges Massif Regional Nature Park, which is the administrative structure of the Geopark, has identified 54 geosites to preserve and promote through educational, cultural, and touristic activities (Peisser and Renau 2010). At least 45 of them can be considered as geomorphosites. They are representative of several geotypes

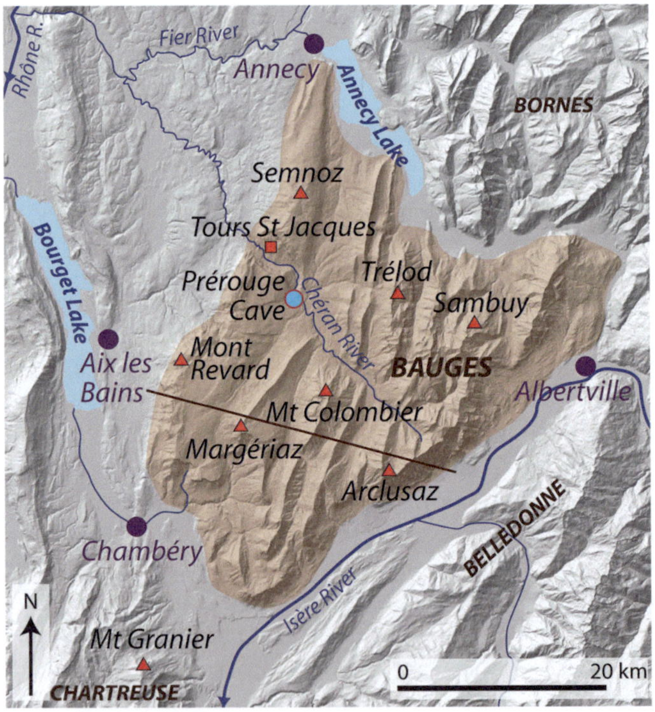

Fig. 18.6 Location of the Bauges Massif (in *brown*) and the quoted Geosites (the *straight line* indicates the trace of the geological cross section Fig. 18.7)

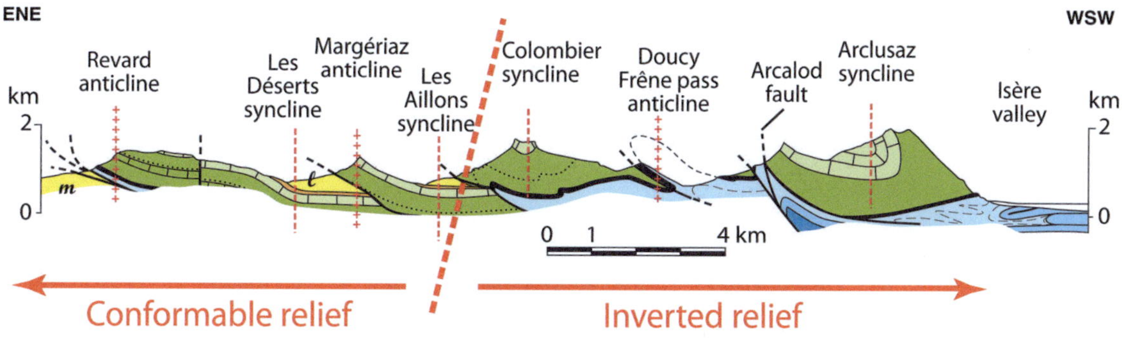

Fig. 18.7 Morphostructural cross section of the Bauges Massif (see location Fig. 18.6 and caption Fig. 18.5) (After Rampnoux 2008; Peisser and Renau 2010)

(including the landforms and the morphogenic processes): morphostructural landscapes, karst, hydro-morphological features, gravity movements, and processes. They are promoted in and outside the Geopark through different kinds of media: classical in situ interpretative panels but also innovative numerical tools (smartphone applications). Managing and promoting geomorphosites supposes to question and build on fundamental and local knowledge which sometimes lack or are obsolete. This motivates further research integrating the current tools and paradigms in geomorphology, introducing a new look at old scientific representations of the landforms, as, for example, with the question of the evolution of the folded relief or one of the tectonic mechanisms involved in the formation of the transverse valleys.

18.3.2 Up Hills and Down Dales: From Conformable to Inverted Relief

As their subalpine neighbors, the Bauges Massif is slipped off from the bedrock of the central crystalline alpine massifs. It is so overthrusted to the west over the molassic foreland. The morphostructural landscape consists of parallel sub-meridian ranges and crests. The geological cross section (Fig. 18.7) shows that the massif can be divided in two equal parts showing the two main landscapes types in such a folded structure: conformable or "Jurassian-like" reliefs in the western part and inverted or "Prealpine-like" relief in the eastern part.

The summits of the western anticline mounts are accessible by car (Mt Revard, 1,550 m a.s.l.; Mt Semnoz, 1,600 m a.s.l.) or cable car (Mt Margériaz, 1,845 m a.s.l.) and constitute impressive scenic lookouts.

The inverted relief of the eastern part gives a very famous landscape marked by perched synclines evoking a flotilla of stone ships named Dent d'Arclusaz (the flagship, 2,041 m a.s.l., Fig. 18.8), Mt Colombier (2,045 m a.s.l.), Trélod (2,181 m a.s.l.), and Sambuy (2,198 m a.s.l.). The Dent d'Arclusaz is a reference in this geomorphotype and appears in several geographical and Earth Sciences textbooks (Blanchard 1938, 1947; Debelmas 1979). This confers to this landform a high educational value. But the reasons why such a morphostructural duality between the western and eastern parts of the Bauges Massif exists are still a matter of debate (Figs. 18.5, left, and 18.7).

18.3.3 The "Tours Saint Jacques": A 10,000-Year Slip on a Structurally Controlled Slide

The Chéran river is the main watercourse of the Bauges Massif, crossing it from east to west. Just before the river exits out of the massif, three curious rock towers called "Tours Saint Jacques" appear on the upper hillslope of the right bank (Fig. 18.6). These towers correspond to slipped masses of Valanginian limestone derived from the above cliffs and affected by a slow down-sliding movement which began to develop some 10,000 years ago, after the deglaciation (Fig. 18.9). The movement is still active, with a significant velocity varying from 0.2 to 2 cm/year (Marconnet 1999). These values are important for managing the vulnerability of the village located 300 m below. New studies supported by the Geopark demonstrate that this movement did not develop along an erosional slope as initially believed, but rather on a structural slope.

For the visitor, the Tours Saint Jacques seem motionless and constitute above all a very aesthetic natural monument located in a scenic place where an interpretative panel has been installed. The towers are also present on numerous pictures, post cards, posters, etc. A few climbing routes exist and the place is now famous among the slackliners. The towers also have a historic value since Pliny the Elder already mentioned them in his *Historiae Naturalis*. The Tours Saint Jacques are so a unique and symbolic element of the local geoheritage. They introduce to the theme of natural hazard, also represented in the Bauges Geopark by other kinds of mass movement as active landslides and mudflows (Les Garins-Nant des Granges) or large collapses (Plainpalais).

Fig. 18.8 The famous perched syncline of the Dent d'Arclusaz, the "flagship" of a flotilla of "stone boats" visible in the background (Photo Claude Kerckhove)

18.3.4 The Karst Landscape and the Prérouge Cave: Unveiling the Mysteries of the Underground Bauges

Karst processes and landforms are very developed in the Bauges Massif and are highlighted by the Geopark. If the karst surface is quite spectacular and easy to show and explain, it is a landscape very sensitive to heavy equipments like ski resorts and quarries. In the center part of the Bauges Massif, on the Margériaz mountain, the *tannes et glacières* (holes and iced caves) geotrail explains to the hikers the mountain karst features: different kind of karren (or dissolution grooves) and dolines (dissolution hollows), former glaciokarst features, and caves. It also explains the interactions with past and present human activities, especially regarding the past extraction of ice from caves, and also the landscape damages and water pollution problems linked with the presence of a ski resort.

However, it is not easy to show and promote the underground landscape of the Bauges Massif, which develops along more than 200 km of karst networks. They are indeed not very accessible, often difficult or dangerous, also for cavers. The Bauges Massif counts only one show cave, included in the tourist site of "Seythenex cave and waterfall" on the northern side of the Geopark. But the largest explored karst networks are developed in the western part of the massif, benefiting from extensive outcrops of Urgonian limestone on medium-altitude plateaus. The *Montagne de Bange–Prérouge* system is the most important (55 km length; 860 m of denivellation). As many other karst springs of the Bauges Massif, the outlet is located in the Chéran valley (Fig. 18.6), at the Prérouge spring and mill. The permanent spring is a siphon but the overflow outlet is the famous Prérouge cave (Fig. 18.10a): it is not too difficult, yet a dangerous cave to visit because of flooding hazard (Fig. 18.10b).

Upstream in the cave, there are other siphons which block the cavers. It is a pity because beyond is a fabulous underground landscape, renowned among the most beautiful in the Alps: the Fitoja's Chamber and its incredible "soda-straws" stalactites (which exist also in the show cave of Choranche in the Vercors Massif) (Fig. 18.10d).

Fig. 18.9 The Saint Jacques Towers: masses of Valanginian limestones sliding slowly along a marly slope in the Chéran valley (Photo PNR Massif des Bauges)

To overcome these difficulties of management and promotion of the cave landscape, the Geopark, in partnership with scientists and filmmakers, fosters the development of innovative tools such as 3D movies full arch projected in a mobile inflatable "GeoDome" or also the "GeoVision" concept, system of monitoring and visualization in real time of the geomorphic and hydrologic events, and processes using a new generation of camera associated with laser scanning measurements. The system is experimented since 2012 for the study, prevention, and educational and geotouristic promotion of the spectacular and dramatic floods of the Prérouge cave.

18.4 Mont Granier in the Chartreuse Massif: A "Broad Spectrum" Geomorphosite

Other Prealpine massifs and geosites, like Mt Ventoux or the Devoluy in the Southern Prealps, or Vercors and Chartreuse more to the north, are also of great geomorphological interest.

A convincing example is the 3-km^2-wide, imposing karst plateau of Mont Granier, culminating at 1,933 m a.s.l. at the northern end of the Chartreuse Massif (see location Fig. 18.6). It is a geomorphosite of primary importance, because of the number and the diversity of its heritage interests.

18.4.1 A Formative Event: The Big Disaster of 1248

The breathtaking north face of the Mont Granier, nearly 1,000 m high, is no more than the giant scar of one of the greatest and famous collapses/landslides that have occurred in the Alps (Fort et al. 2009) (Fig. 18.11a, b). The event occurred on the night of November 24–25, 1248 (Guillomin 1937). The deposit of 500 million m^3, which spread over 11 km at the foot of the mountain (Fig. 18.11c), swept the city and the villages that were in its path, causing about 1,000 deaths (Guillomin 1937; Guilleré 1999). This disaster attributed to a divine punishment led to an important site of pilgrimage dedicated to the Black Madonna, still very much alive today in the religious sanctuary of Notre Dame de Myans. The limestone debris of the rock avalanche spread over glacial siliceous terrains and are now colonized by vineyards producing the famous white wine *Abymes* (a word meaning "collapse" in Old French) (Fig. 18.11a).

Fig. 18.10 Prérouge cave: (**a**) gate of the cave, (**b**) flood of February 16, 2006, (**c**) underground deep helical pothole, and (**d**) "soda-straws" stalactites in the Fitoja's Chamber (Photos Fabien Hobléa (a–c) and CDS 73 (d))

18.4.2 Mont Granier: A "Hollow Tooth", Very Significant for Paleogeographic Reconstruction

With more than 90 km of galleries explored, including an interconnected network of 55 km, the Granier cave system is a good model of polyphased multilevel karst network (Fig. 18.11d). For the first time in a French cave, cosmogenic nuclide dating of karst fillings has been done, showing that the upper levels of galleries were already in existence more than four million years (Hobléa et al. 2011). This cave level is now dry and located above 1,670 m a.s.l., but during Pliocene, it

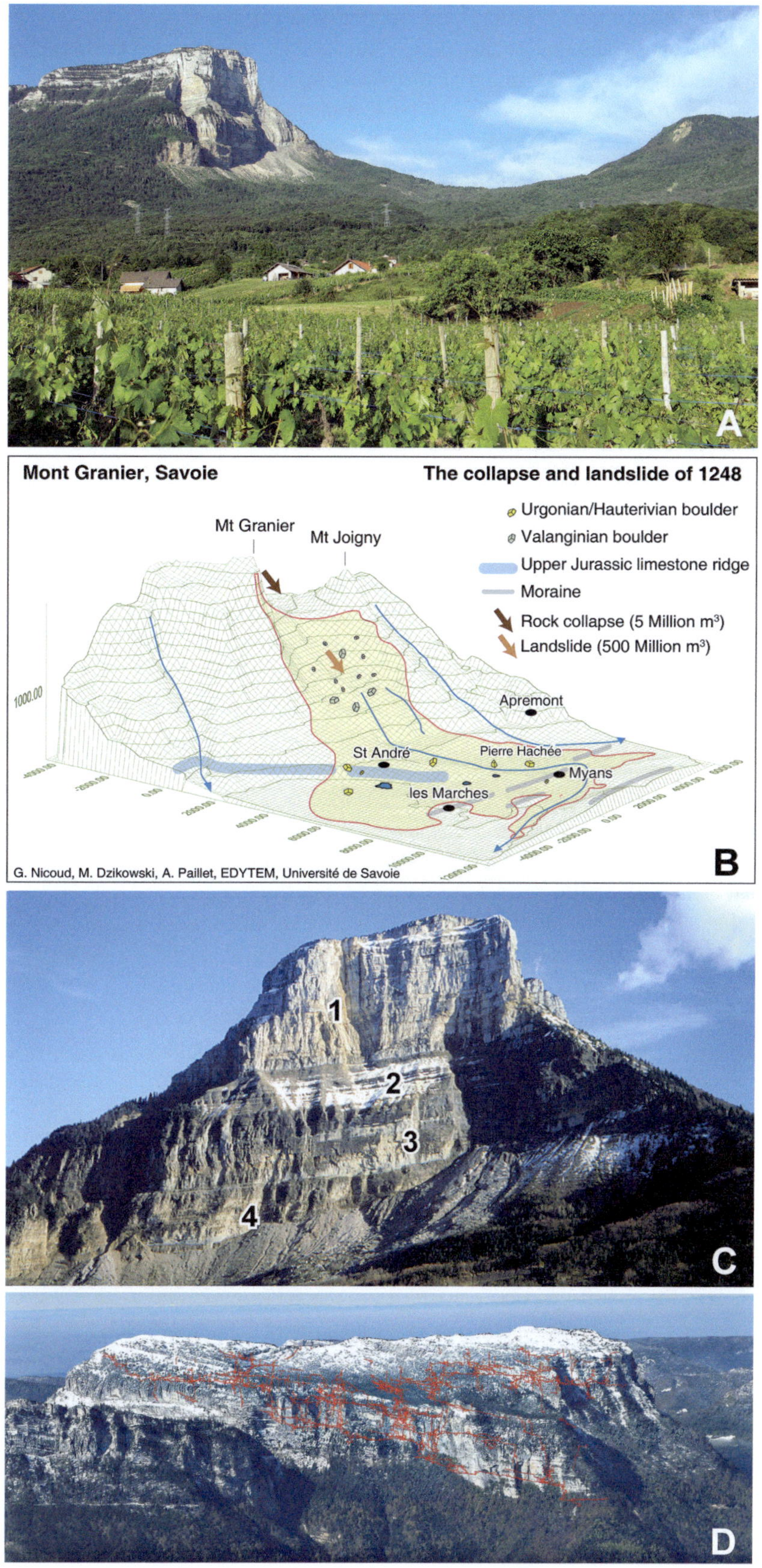

Fig. 18.11 Mont Granier geomorphosite: (**a**) The Abymes wineyard is located on the rock avalanche deposit coming from the north face of the Granier in 1248 (background); (**b**) 3D modeling of the collapse and landslide of 1248; (**c**) the Granier north face, a natural geological cross section created by the collapse of 1248 (*1* Urgonian limestone, *2* Hauterivian marls, *3* Valanginian limestone, *4* Berriasian marls); and (**d**) the Granier plateau (view of the eastside) with its multilevel karst network (*red stokes*) (Sources: Fabien Hobléa (a and b); Gérard Nicoud, Marc Dzikowski, André Paillet (c); Claude Kerckhove, Robert Durand, Jacques Nant (d))

was a major drain collecting underground water at the base level. This means that at that time Mont Granier was not a mountain but a low karstified plateau evolving under tropical conditions, as revealed by the analysis of clays found in the karst fillings (Hobléa 2001).

18.4.3 The Balme à Collomb Cave: House of the Cave Bears

In the Balme à Collomb cave, belonging to this old level of galleries perched at 1,700 m a.s.l., a very important paleontological site with thousands of bones and skeletons of cave bears (*Ursus Spelaeus*) was discovered in 1988. ^{14}C dating attested that the cave was occupied by the bears for hibernation from 45 to 24 Ky BP (Philippe 1999). After 10 years of excavations and multidisciplinary studies, the cave had to be closed by a reinforced door because of its vulnerability to looting. Nowadays, the cave is rarely open, only every 2 years during the prehistoric festival celebrated in the village of Entremont-le-Vieux at the foot of Mont Granier, where an original museum of cave bear was inaugurated in 2002. It has received 180,000 visitors in 10 years.

18.5 Conclusion

Between unity and diversity, the French Prealpine morphological landscapes are the perfect illustration of folded sedimentary reliefs linked to the alpine tectonics. They have recorded more than 200 My of the Alpine cycle history, from marine to glacial phases. Limestone cliffs overhanging marls slopes constitute a leitmotiv in the landscape. Karst features are very developed and some Prealpine massifs shelter karst networks among the largest and deepest in the world (more than 50 km long and 1,000–1,700 m deep).

If the Prealpine bedrock provides mineral resources still more or less exploited, the Prealpine geomorphological landscapes are today also perceived as heritage of high scientific interest but also of cultural, historical, and aesthetic value. The Prealpine geomorphosites are now protected and promoted in most of the massifs, especially in the four French Geoparks, all located in Prealps. Indeed, these are at the core of rebuilding relationships between man and the geosphere.

Acknowledgments The author thanks very much for their contribution to the writing and the illustration of the text: Dr. Stéphanie Gallino-Josnin, Mr. Jean-Luc Desbois, and Mrs. Carole Vanesse from the Massif des Bauges Regional Nature Park and Geopark; Mr. Pierre Renau and Dr. Carine Peisser (CalcEre); Mr. Robert Durand, Didier Monin, and Jacques Nant (CDS 73), Mr. André Paillet, Ms. Estelle Ployon, Dr. Nathalie Cayla, and Dr. Sylvain Coutterand (lab Edytem); special thanks to Prof. Monique Fort (University Paris-Diderot) and Prof. Marie-Françoise André (Université Blaise Pascal-Clermont-Ferrand) for their trust as coeditors of this book.

References

Bintz P, Griggo C (2011) Climats et premiers peuplements des Alpes du Nord françaises: des derniers chasseurs aux premiers paysans (15 000 à 5 000 ans av. JC.). English edition: Climate and early settlements of the French Northern Alps: From the last hunters to the first farmers (15 000 to 5 000 BC). Revue de primatologie [on line], 3|2011, document 13. URL: http://primatologie.revues.org/789; doi:10.4000/primatologie.789

Blanchard R (1938) Les Alpes occidentales. Tome 1: Les Préalpes Françaises du Nord. Arthaud, Paris

Blanchard R (1947) Les Alpes occidentales. Tome 2: Les cluses préalpines et le sillon alpin. Arthaud, Paris

Blanchard R (1938–1956) Les Alpes occidentales. Arthaud, Paris

Coward M, Dietrich D (1989) Alpine Tectonics – an overview. In: Coward MP, Dietrich D, Park RG (eds) Alpine Tectonics, vol 45, Geological Society Special Publication. Geological Society, London, pp 1–29

Debelmas J (1979) Découverte géologique des Alpes du Nord. BRGM, Orléans

Fort M, Cossart E, Deline P, Dzikowski M, Nicoud G, Ravanel L, Schoeneich P, Wassmer P (2009) Geomorphic impacts of large and rapid mass movements; a review. Géomorphologie: Relief Processus Environnement 1:59–74

Gratier JP, Ménard G, Arpin R (1989) Strain-displacement compatibility and restoration of the Chaînes Subalpines of the western Alps. In: Coward MP, Dietrich D, Park RG (eds) Alpine tectonics, vol 45, Geological Society Special Publication. Geological Society, London, pp 65–81

Guilleré C (1999) Espace et population après l'effondrement du Granier. In: L'éboulement du Granier et le sanctuaire de Myans. Proceedings of the Myans conference, 5–7 June 1998. Académie de Savoie. Documents, 2e série. Tome I, Chambéry, pp 159–168

Guillomin A (1937) Les Abîmes de Myans (Savoie). Revue de Géographie Alpine, Grenoble XXV(1):583–606

Hobléa F (2001) Genesis of the karstic network of the mount Granier (Chartreuse, Northwest Prealps, France). In: Häuselmann P, Monbaron M (eds) Cave genesis in the Alpine Belt. Proceedings of the 1st workshop for Alpine Speleogenesis, Habkern, Switzerland, 10–13 September 2000. UKPIK 10/2001, pp 29–39

Hobléa F, Häuselmann P, Kubik P (2011) Cosmogenic nuclide dating of cave deposits of Mount Granier (Hauts de Chartreuse Nature Reserve, France): morphogenic and paleogeographical implications. Géomorphologie: Relief Processus Environnement 4:57–68

Marconnet J-F (1999) Evolution des escarpements calcaires instables: quantification des mouvements actuels par G.P.S. et inclinométrie dans les Alpes du Nord (France). PhD thesis, University of Savoie

Peisser C, Renau P (2010) Aspiring Bauges Subalpine Geopark. Application dossier European Geopark Network. Parc Naturel Régional du Massif des Bauges, Le Châtelard

Philippe M (1999) Le Granier, un haut lieu pour la connaissance de l'ours des cavernes. In: L'éboulement du Granier et le sanctuaire de Myans. Proceedings of the Myans conference, 5–7 June 1998, Académie de Savoie, Documents, 2e série. Tome I, Chambéry, pp 103–120

Rampnoux JP (2008) Contexte stratigraphique, lithologique et structural des massifs subalpins des Bornes et des Bauges. Localisation des aquifères karstiques et circulation des eaux souterraines. In: Hydrogéologie et karst au travers des travaux de Michel Lepiller. Proceedings of the annual meeting of the French Committee of the International Association of Hydrogeologists, pp 61–70

Schmid SM, Fügenschuh B, Kissling E, Schuster R (2004) Tectonic map and overall architecture of the Alpine orogen. Eclogae Geol Helv 97:93–117

19 The Lower Ardèche River Karst Landscapes and Caves (Lower Rhône Valley): Unique Morphologies Induced by the Messinian Salinity Crisis

Ludovic Mocochain and Stéphane Jaillet

Abstract

The lower Ardèche and Cèze valleys are famous for their touristic sites. Besides their dramatic karstic landscapes such as the natural arch of the Pont d'Arc or the deep canyon of the Ardèche River, they conceal prehistoric caves among which the famous Chauvet Cave discovered in 1994. These landscapes of high plateaus, canyons, and caves were formed by a catastrophic geologic event, the Messinian Salinity Crisis (MSC), which affected the entire Mediterranean Basin and the morphological evolution of the surrounding mountains and plateaus. More specifically, the Saint-Remèze Plateau displays some good evidences to demonstrate that the Mediterranean Sea was completely desiccated. The MSC was also responsible for the downcutting of the Ardèche and Rhône canyons in less than 100,000 years. This evolution was followed by the formation of deep karst drainages. The MSC ended by a sudden reflooding of the basin, characterized by a rapid rise of the Mediterranean base level responsible for large river aggradations. It also provoked the development of a *per ascensum* karst adaptation characterized by (1) large galleries in the Saint-Marcel and Chauvet caves and (2) huge chambers in the Aven d'Orgnac Cave. Nowadays, more than one million tourists visit the area every year, attracted by its natural biotopes and sites, by its caves like Saint-Marcel and Aven d'Orgnac. In the future, the Chauvet Cave, where the oldest paintings in the World were found, will hopefully be nominated as a UNESCO World Heritage.

19.1 Introduction

The emblematic Ardèche canyon is one of the most spectacular karstic landscape in France (Fig. 19.1). Twenty-nine kilometers long and over 300 m deep, it results from the karstic nature of the Cretaceous limestone that composes the Saint-Remèze Plateau. It is visited each year by more than one million tourists, eager to explore and practice canoeing, hiking, climbing, or caving, and enjoying natural landscape and wilderness. This is also the place where prehistoric men left signs of their presence in several caves, the most famous being the Chauvet Cave, where wall paintings (underground rock art) are dated from the Aurignacian Period (30,000–35,000 years) and are among the oldest of humanity.

For more than a century, the origin of the Ardèche canyon downcutting together with the age of the caves have been debated issues: tectonics versus eustatism, Quaternary age or older, etc. There are many different interpretations, but none of them is able to integrate all fields of observations. Until recently, the Messinian Salinity Crisis (the complete desiccation of the Mediterranean Sea) was not solicited as a possible explanation for the karst formation around the Mediterranean

This chapter is dedicated to the memory of Georges Clauzon (1937–2013). He worked tirelessly and passionately over a scientific career spanning 40 years, giving the scientific community a greater understanding of Mediterranean landscapes through his reconstructions of the impacts of the MSC.

L. Mocochain (✉)
Université Pierre et Marie Curie, Paris 06, UMR 7193, IsTEP, F-75005, France
e-mail: ludomocochain@gmail.com

S. Jaillet
EDYTEM, Université de Savoie, CNRS, Chambéry, France
e-mail: stephane.jaillet@univ-savoie.fr

M. Fort and M.-F. André (eds.), *Landscapes and Landforms of France*, World Geomorphological Landscapes,
DOI 10.1007/978-94-007-7022-5_19, © Springer Science+Business Media Dordrecht 2014

Fig. 19.1 The Gorges de l'Ardèche, a most spectacular karstic canyon cut into Urgonian limestones.The red arrow points out the Pont d'Arc arch near the Cirque d'Estre escarpments where the entrance of Chauvet cave is situated (purple arrow). Note the flat surface of the Saint-Remèze Plateau (Photo credit: S. Jaillet/Edytem)

Basin (Audra et al. 2004). In this chapter, we present the Messinian Salinity Crisis (MSC) and demonstrate the major consequences it has had on the formation and the evolution of the Ardèche landscapes.

19.2 Geological Setting

19.2.1 The Saint-Remèze and Mejanne-Le-Clap Plateaus

The Saint-Remèze Plateau and the Méjannes-Le-Clap Plateau areas are underlain by the Mesozoic limestones that cover the Cévennes basement on its eastern border. Before traversing these plateaus, the Ardèche and Cèze rivers flow out of the Cévennes Monts and then cross southward through Mesozoïc rocks. At Vallon-Pont-d'Arc, the Ardèche turns eastward and crosses the Urgonian Cretaceous limestones, along a spectacular canyon (Fig. 19.1). The Cèze crosses the same carbonate formation in a similar canyon (Fig. 19.2).

The basement of the Urgonian Saint-Remèze Plateau forms a large brachyanticline. It displays typical karstic features, such as poljes, dolines, shafts, and karren fields. The Saint-Remèze Plateau was shaped by weathering and erosional processes acting with variable intensity, which has led to the origin of several generations of geomorphic surfaces (Mocochain et al. 2006). The most important surface forms the largest part of the Saint-Remèze Plateau at a constant elevation of 400 m a.s.l. This upper surface is considered to be a wave-cut platform, developed during the Middle Miocene high-stand sea level (Besson 2005; Mocochain et al. 2006).

19.2.2 The Messinian Salinity Crisis and Pliocene Eustatic Cycle in the Mediterranean Sea

A major event occurred 5.9 Ma ago: the Mediterranean Sea became completely insulated from the World Ocean, due to the closing of the Gibraltar Strait. The sea level started

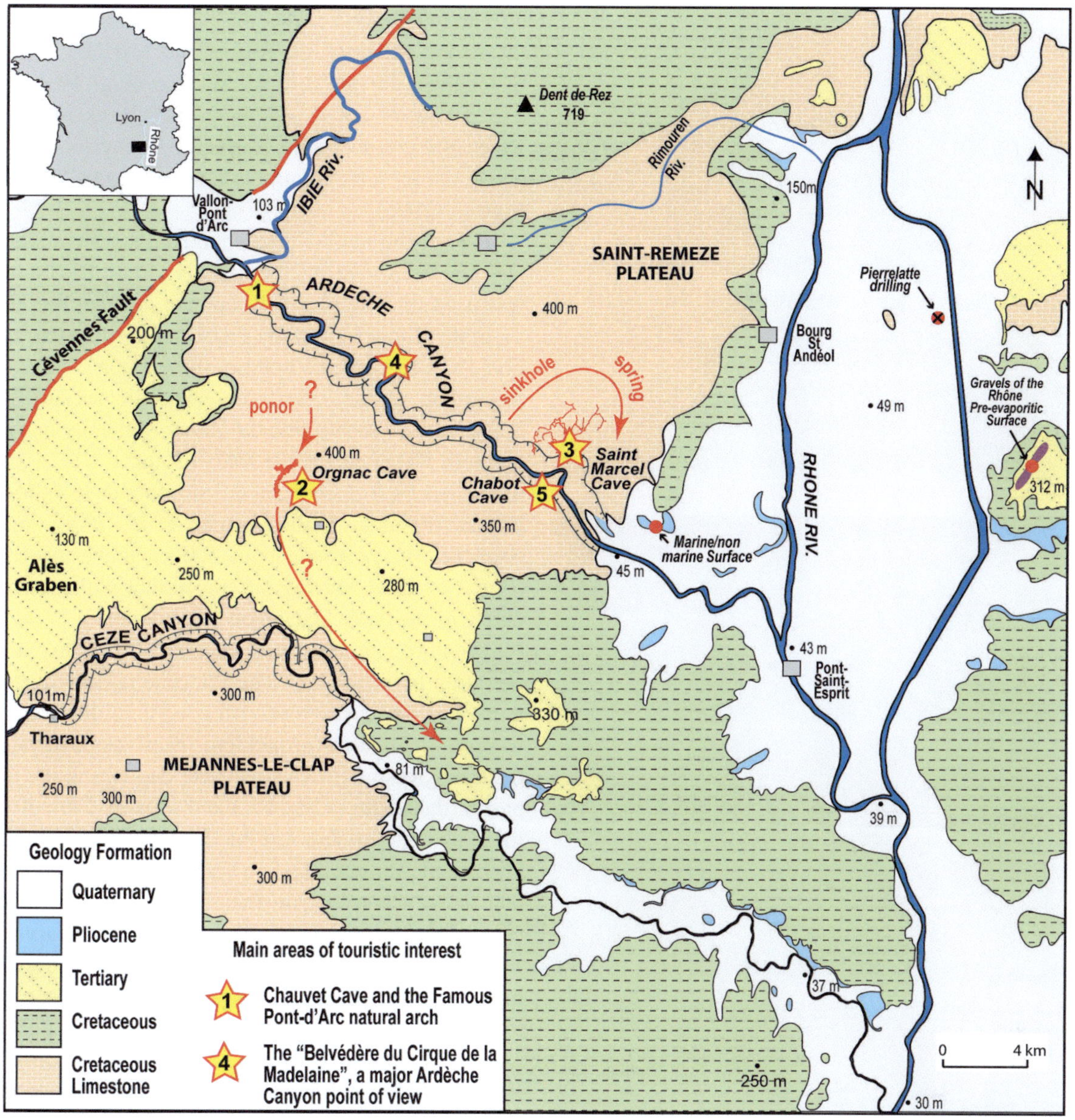

Fig. 19.2 Regional geological and geomorphological setting of the lower Ardèche and Cèze rivers. The Rhône River flows toward the Mediterranean Sea, located 150 km southward (From Mocochain et al. 2006, modified)

lowering, initiating the MSC eustatic cycle (Hsü et al. 1973). During this period, the sea level dropped by 1,500 m, leading to the dramatic desiccation of the Mediterranean between 5.9 and 5.32 Ma, then was followed by an extremely rapid transgression (Garcia-Castellanos et al. 2009).

The MSC caused considerable changes in the predominantly horizontal landscape, especially in carving huge Messinian canyons and forming deep karst drainages (Audra et al. 2004; Mocochain et al. 2006).

During the first stage and in response to the sea level collapse, a 1,000-m-thick evaporitic layer (salt) deposited in the old Abyssal plains under shallow water 5.6 Ma ago, whereas a deeply dissected erosional "surface" appeared on the emerged margins: the Messinian Erosional Surface (MES). This erosional surface can be traced onshore through deep canyons along the rivers called the Messinian canyons (MC). More specifically along the Northern Mediterranean continent, the Rhône MC developed up to the Jura Mountains and presently constitutes the most significant morphological evidence for the MSC (Fig. 19.3). Near its confluence with the Ardèche, the fossil Rhône MC is 300 m under its current position at 236 m below sea level (b.s.l.) (Clauzon 1982).

In the second stage, the MSC ended by another catastrophic event known as the "Pliocene reflooding," 5.32 Ma ago (Garcia-Castellanos et al. 2009). The MC were first flooded and then completely filled by sediment during the Pliocene (Fig. 19.3).

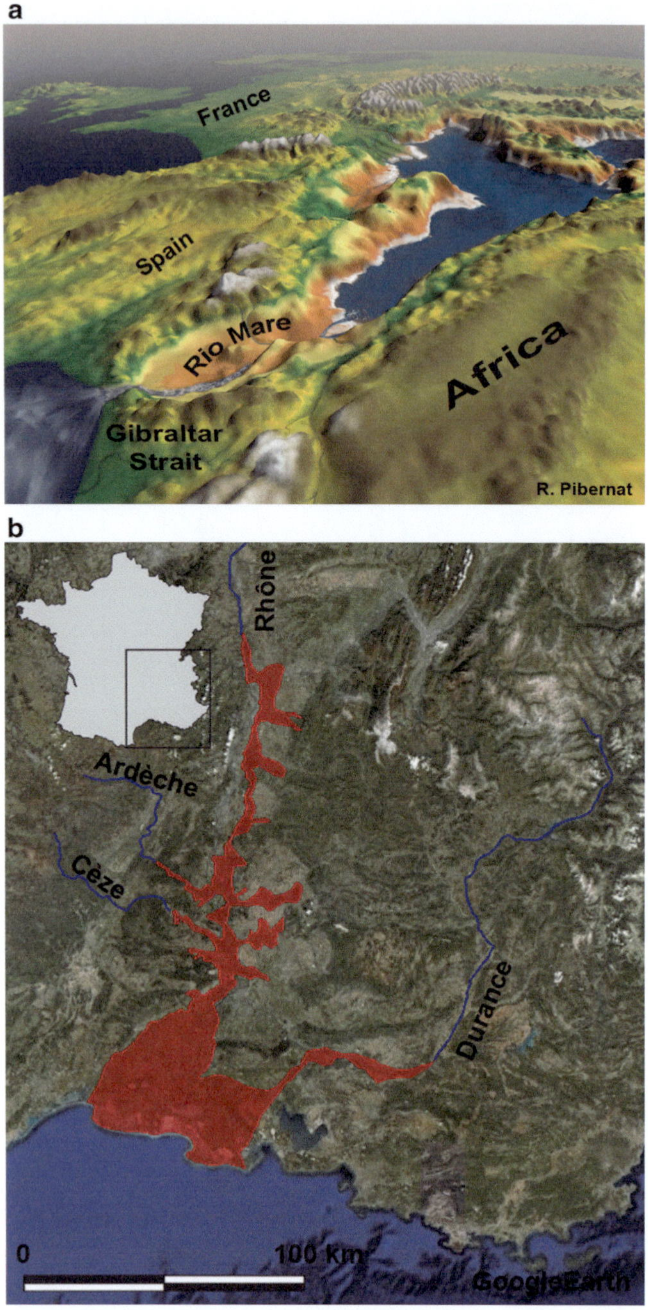

Fig. 19.3 (**a**) Catastrophic reflooding of the Mediterranean Sea: this scheme illustrates the huge Rio Mare River flooding the Mediterranean Basin from the Atlantic Ocean. (**b**) This transgression transformed the canyons of the Rhône and its tributaries into rias (*red area*), extending far off into the continent (From Clauzon 1982, modified)

19.2.3 Field Evidences of the Messinian-Pliocene Eustatic Cycle

The Mediterranean sea level fluctuations of this Messinian-Pliocene eustatic cycle have been reconstructed on the basis of four stratigraphic and geomorphological markers (i.e., Clauzon 1996).

The first marker is the pre-evaporitic surface corresponding to the extent of the Rhône and Ardèche alluvial plains just before the MSC, 5.95 Ma ago (Fig. 19.4-1; Clauzon 1982; Martini 2005). Along the left bank of the Ardèche River, the pre-evaporitic surface is today located at 330 m above sea level (a.s.l.) and is estimated to be 5.45 Ma old by Martini (2005).

The second marker corresponds to the Messinian canyons (MC) of the Rhône and the Ardèche rivers. The bottom of MC was used to recognize the maximal downcutting (Fig. 19.4-2). In the Rhône valley, the Pierrelatte drilling (location in the Fig. 19.2) reached the bottom of the Rhône MC 236 m b.s.l. According to the initial pre-evaporitic surface located at 330 m a.s.l., it means that the Rhône vertically incised its bed of about 570 m during the MSC (Clauzon 1982). The bottom of the current Ardèche canyon corresponds to its MC bottom, located 50 m a.s.l. (Mocochain et al. 2009), making the Ardèche MC perched about 300 m above the Rhône MC (Fig. 19.4-2), a striking difference that is due to the karstic nature of the Ardèche canyon (see below).

The third marker is the recording of the Pliocene sea level elevation by sediments. The Pliocene reflooding submerged these canyons and abruptly stopped their downcutting. They were suddenly transformed into "*rias*" (flooded valleys), and the transgression drove back river deposits at the uppermost part of these *rias* (Figs. 19.3 and 19.4-3). Fluvial sediments filled up the *rias* seaward due to the combination of a rapid horizontal progradation and vertical aggradation of their deltaic fans according to the Gilbert-type-fan-deltas mode (Fig. 19.4-3, 4; Clauzon 1996). The sedimentary process of progradation recorded a clear sedimentary transition, the marine/non-marine surface that fossilizes the sea level position (Fig. 19.5a). Close to the Saint-Remèze Plateau, the marine/non-marine surface is located at 130 m a.s.l. (Figs. 19.1 and 19.5a). The age of the marine/non-marine surface near Saint-Remèze Plateau is estimated to be 4.7 Ma (Mocochain et al. 2006).

The last marker is the Upper Pliocene surface that corresponds to an old alluvial plain and indicates the highest limit of the MC infilling around 2 Ma (Fig. 19.4-4). Perched about 200 m a.s.l. at the Ardèche/Rhône confluence, this surface is characterized by a large alluvial terrace composed of reddened quartz and quartzite pebbles (Fig. 19.5b), recording the last position of the rivers just before Quaternary downcutting phases.

During the Quaternary, the climatic-induced, glacio-eustatic cycles produced successive stages of downcutting and partial exhumation of the MC. This dynamics is recorded into fluvial deposits as alluvial terraces, staged between the Upper Pliocene surface at 200 m and the current level of rivers.

Noteworthy is the fact that since the Neogene time, the studied area was only affected by epeirogenetic uplift, without any local, differential displacement. Consequently, the

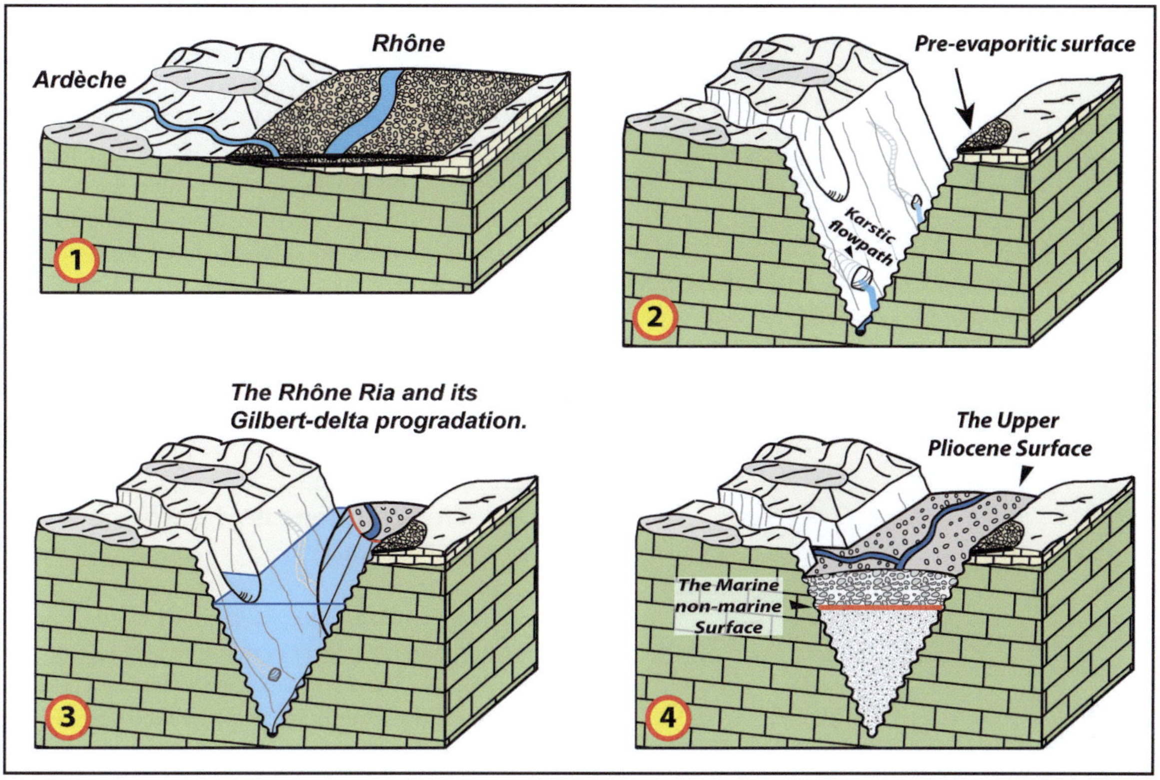

Fig. 19.4 Evolution of the Rhône Valley before and after the MSC till the end of Pliocene. (1) The Rhône River just before the initiation of the MSC. (2) The Rhône Messinian canyon during the MSC, with karstic flow path that corresponds to the Ardèche subterraneous river. (3) Reflooding of the Mediterranean Sea at 5.32 Ma. (4) The Rhône Valley after the filling process by deltaic sediments at the end of Pliocene (From Mocochain et al. 2006, modified)

elevation of the Messinian-Pliocene markers could be used to reconstruct the base-level evolution during the last 6 Ma and shows, after the large base-level drop during the MSC, several stages of base-level rise during the Pliocene, and finally several stages of base-level drop again during the Pleistocene.

Furthermore, the recognition of MSC event gave new clues to determine the Ardèche and Cèze canyons' age, and to better constrain the stages of caves formation.

19.3 When Did the Ardèche and Cèze Canyons Form?

The age of the Ardèche canyon formation has long been debated, due to the difficulty to date landforms. Three hypothesis were considered: (1) Alpine tectonics caused a regional uplift and the subsequent downcutting of the Ardèche canyon during the Upper Miocene; (2) the Ardèche canyon resulted from both glacial-interglacial cycles and sea level fluctuations during the Pleistocene; and (3) the formation of the Ardèche canyon is the consequence of the dramatic Messinian sea level drop (the hypothesis we defend here).

This latter theory was demonstrated by Mocochain et al. (2009), who showed that the Ardèche canyon was fossilized by marine Pliocene sediments in its downstream part, whereas in its upstream part, near Vallon-Pont-d'Arc city, an important alluvial deposit indicates that the canyon was completely filled. Mocochain et al. (2009) demonstrated that the only possible time to fill the canyon was the Pliocene, during the sedimentary aggradation in the Rhône Valley (see Fig. 19.4). This process is neither possible after the Pliocene because rivers started to downcut their new valleys, as attested by the Pleistocene alluvial terraces that are omnipresent in the region, nor before the Pliocene, as demonstrated by Martini (2005) who pointed out the presence of the Ardèche River on the plateau at 360 m a.s.l. before the MSC. Therefore, the only possible time for the Ardèche to incise its canyon is during the MSC. Near Tharaux village (Fig. 19.2), at the entrance of the Cèze canyon, remains of Pliocene filling similar to the ones encountered in the Ardèche canyon suggest the Ardèche and Cèze valleys had a similar evolution.

The relative shallow depth of Ardèche and Cèze MC compared to the Rhône MC is due to their karstic nature. Actually, scuba divers discovered a deep, flooded karst drainage system below the Ardèche and Cèze canyons. This flooded karst

Fig. 19.5 (**a**) Marine/non-marine surface marker. This marker shows a sea level record at the contact between horizontal gravel bed (alluvial deposits) and clinoforms (marine foreset bed deposits), presently at 130 m a.s.l. (**b**) Upper Pliocene surface marker, composed of fluvial quartzitic gravels at 200 m a.s.l. In the Rhône Valley the Upper Pliocene surface is famous for its vineyards, and the same gravels are also found underlying the Château-Neuf-du-Pape vineyards (Photo credit: L. Mocochain)

suggests that the Ardèche and Cèze rivers were formed by karstic flow paths under their canyons (Fig. 19.4-2) that stopped the downcutting of their valleys during the MSC (Mocochain et al. 2006).

One good example of the role played by the underground waters is the dramatic Pont d'Arc natural arch and promontory (Fig. 19.6). Formerly, the Ardèche River bypassed by a meander this promontory. Erosion by small underground drains allowed water to progressively enlarge the drains until a unique large drain was formed, so that the river left its former meander (the "Cirque d'Estre") to slip through this new channel that it has gradually expanded, hence passing under the actual huge arch (34 m high and 59 m wide at water level). This process is thought to have occurred during the Upper Pleistocene, some 400,000 years ago. The abandoned meander is exceptionally reactivated during exceptional floods, the last one dating back to September 1890.

19.4 The Uniqueness of the Karst Formation

The Saint-Remèze and Mejannes-Le-Clap plateaus are famous among the cavers community. The Saint-Marcel and Aven d'Orgnac caves are among the largest caves in France, but their main particularity remains their formation process, different from each other.

19.4.1 Saint-Marcel: A Multilevel Cave

The Saint-Marcel Cave is located on the downstream part of the Ardèche canyon (Fig. 19.2). This cave is 48 km long and 11 km of galleries are flooded under the level of the Ardèche River. It is formed by three horizontally superimposed levels (Figs. 19.7 and 19.8). The lower level develops between 50 m a.s.l. and 10 m b.s.l. This level is active and partly flooded. The middle (between 100 and 130 m a.s.l.) and upper levels (between 170 and 200 m a.s.l.) are not anymore active. The three levels are characterized by (a) horizontal galleries (Fig. 19.7) and by (b) the presence of chimney shafts, vertical conduits allowing communication between the different levels and exhibiting evidence of water drainage in upward direction (Fig. 19.7; Mocochain et al. 2006, 2011).

On the basis of the cave geometry and hydrodynamic process, we defend the idea that the Saint-Marcel Cave formation results from both the MSC and the subsequent canyon-filling process that occurred during the Pliocene (Fig. 19.4-3 and 4). The geometric argument points out that several levels of the Saint-Marcel Cave are at the same elevation as the stratigraphic and geomorphic markers, as described above (Fig. 19.8). Consequently, we correlated the lower level with the Ardèche and Rhône Messinian canyons, the middle level with the Pliocene high-stand sea level recorded by the marine/non-marine surface, and the upper level with the Upper Pliocene surface (Mocochain et al. 2006). In addition, one can observe the presence of chimney

Fig. 19.6 The spectacular Pont d'Arc natural arch bypasses the former meander of the Cirque d'Estre, presently occupied by vineyards. The blocked entrance of the Chauvet Cave is located below the cliff *(right)* (Photo credit: Françoise Prud'homme, with permission)

Fig. 19.7 *Left:* The Galerie des Peintres, a typical horizontal drain in Saint-Marcel Cave (Photo V. Ferrer, with permission). *Right:* A huge chimney shaft of Saint-Marcel Cave that connects two superimposed horizontal levels similar to the Galerie des Peintres (Photo P. Goossens, with permission)

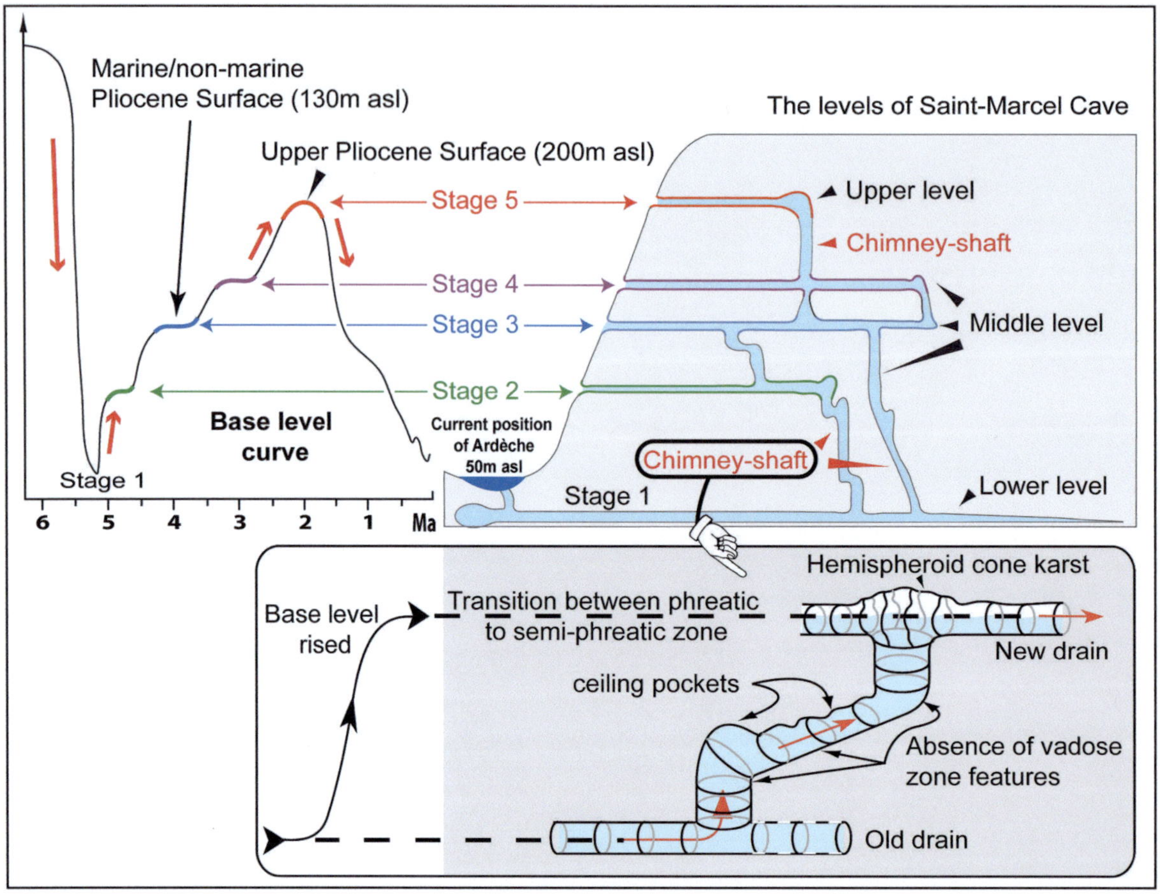

Fig. 19.8 During the Pliocene, the base level rose (*top left*) and the karstic water circulated through chimney shafts to find a new spring near the new base-level position (*top right*). The lower sketch illustrates the chimney shaft role in the formation and superposition of cave levels (The base-level curve is adapted from Tassy et al. 2013)

shafts that connect each level and where water flows vertically (Fig. 19.8).

Both hydrodynamic and geometric arguments suggest an upward, "per ascensum" formation from one level to another (Fig. 19.8). Recent studies based on cosmogenic dating (Tassy et al. 2013) reveal that the lowest cave levels are the oldest and that the highest cave levels are the youngest, which confirms the "per ascensum" formation of the cave levels in Ardèche.

19.4.2 Large Cave Formation by Paragenesis Process: Example of the Aven d'Orgnac Cave

The Aven d'Orgnac Cave is located between the Ardèche and Cèze rivers (Fig. 19.2). The cave is 5 km long and extends along a north-south direction. The Aven d'Orgnac Cave formation was different from the Saint-Marcel Cave. This network is a very large drain partly filled with (1) an important deposit of cave sediments, (2) several collapses of cave ceiling, and (3) huge stalagmites growing on sediment or collapses.

In the *Salles Rouges* Chamber (in the touristic part of the cave), a succession of bench walls indicates an upward erosional process by "paragenesis." This particular process of caving formation results from a major corrosion of the ceiling gallery with a simultaneous deposition that protects the floor of the gallery (Fig. 19.9). The paragenesis process, enhanced by a long-duration clay sedimentation favored by slow steady flow and a very large corrosion of ceiling galleries, caused the formation of such huge galleries in the Aven d'Orgnac Cave. Currently, when visiting the Aven d'Orgnac Cave, one can still observe two consequences of this erosional process: very high ceilings in huge galleries and several generations of bench walls on the galleries' banks (Fig. 19.9) that indicate an upward progress of corrosion that ultimately extended more than 100 m high.

This paragenesis process is controlled by base-level rise. The height of the Aven d'Orgnac Cave galleries proves that the base level rose by 100 m. Considering the regional setting, the process of post MSC reflooding that filled up the Cèze MC during the Pliocene is the only possible mechanism to explain this base-level rise (Jaillet et al. 2011).

Later on, and similarly to most of the caves in Ardèche, the final stage of the Aven d'Orgnac Cave evolution is the partial erosion of clay filling in the galleries. This process occurred during the Quaternary cold periods when the rivers incised their valleys. It was after these clay erosion episodes that the stalagmites began to grow (Fig. 19.10).

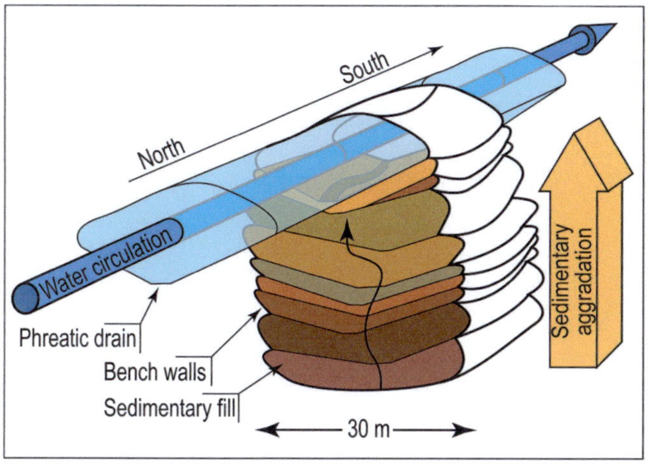

Fig. 19.9 Process of a drain during the Pliocene paragenesis. The succession of bench walls (in *brown color*) records the underground detrital filling that is now observable with the erosion of sediments (in *white color*) due to further lowering of sea level (From Jaillet et al. 2011, modified)

19.4.2 An Exceptional Karstic Landscape for Prehistoric Men

Prehistoric men were present in this scenic landscape. Indeed, several prehistoric and historic remains were found in the Ardèche gorges and caves. Among these remains, cave paintings are the most remarkable. Prehistoric art was first discovered in the Chabot Cave one century ago (see Fig. 19.2 for location). Compared to many others, the Chauvet Cave stands as an exceptional site for the prehistoric art knowledge. Located close to the 60 m-high natural arch of Pont d'Arc above the former bed of the Ardèche River, the Chauvet Cave displays on its walls an exceptional number of paintings in very good state of conservation, representing a large variety of animals (at least 13 species, with the predominance of rare themes such as felines and the woolly rhinoceros) (Fig. 19.11). It also exhibits animal and human footprints. More generally, the cave reveals unknown techniques in Paleolithic rock art, such as stump drawing.

Fig. 19.10 Spectacular speleothems in the tourist part of the Orgnac Cave. These stalagmites were formed upon deposits of ceiling collapse subsequent to erosion of detrital filling (Quaternary period) (Photo credit: S. Jaillet/Edytem)

Fig. 19.11 The cave paintings of the Chauvet Cave are among the oldest and most spectacular in the world. In the last room of the cave, the "panneau des lions" (Lion's panel) is probably one of the most impressive underground rock art ever found (Photo credit: S. Jaillet/Edytem)

Recent studies revealed human presence in the cave during the Aurignacian (30,000–35,000 years BP) and the Gravettian (c. 28,000–23,000 ya) periods. Using $_{36}$Cl cosmic-ray exposure ages, it was demonstrated that the cliff overhanging the Chauvet Cave had collapsed several times since 29 ka until definite sealing of the cavity entrance that prohibited the access to the cave (Delannoy et al. 2010) at least since the last 21 ka (Sadier et al. 2012). This dating confirmed that the Chauvet Cave paintings are the oldest and the most elaborated cave paintings ever discovered to date. This is why the Chauvet Cave is candidate to be claimed as a UNESCO World Heritage in the future!

19.4.4 A Natural Reserve Praised by Wilderness Lovers

The Pont d'Arc, the only case in France where a stone arch span over a still active river and the most spectacular short-cut meander in Europe (Fig. 19.8), is the starting point of many geotourism activities, all the more as the area is of very easy access from the right bank of the Rhône River.

Firstly, the Ardèche canyon is a playing ground for many outdoors activities: it is famous for hiking (with many breathtaking outlooks over the canyon), climbing, caving (speleology), and canoeing (with its 25 rapids, it is considered as the Mecca of canoeing!). Secondly, the area is praised for its wilderness assets: classified in 1980 as natural reserve, it is also a Natura 2000 site, ranked as a protected biotope by prefectoral authorities. It provides a habitat for endangered and protected species (Bonelli's Eagle, Egyptian Vulture, and pilgrim falcon). In addition, the area is rich in architectural patrimony: small, well-preserved villages such as Aiguèze or the "Maladrerie des Templiers," the old episcopal city of Viviers, or the archaeological remnants of Alba-La-Romaine.

Potential conflicts could be foreseen between human activities and tourism development on the one hand, and the respect of natural and human heritage on the other hand, especially after the Chauvet Cave discovery. There is a need for regulatory measures to keep the sites their wilderness and integrity. For example, access to Chauvet Cave has been prohibited right from the discovery, and a "facsimilé" project is planned in 2015 to open to the public to avoid damages to paintings as what occurred in Lascaux. Yet, despite the large number of visitors, the Ardèche River was able to preserve its integrity and remains today one of the wildest landscapes of France.

19.5 Conclusion

In the Ardèche and Cèze areas, the canyons and the caves are the consequence of a huge base-level oscillation over 600 m caused by the MSC, which directly influenced the landscape and karst evolution. First, the MSC caused the downcutting of the Messinian canyons and the formation of deep karst drainages. Then, during the Pliocene, the base level rose by steps due to the catastrophic reflooding of the Mediterranean Sea and river aggradation. The base-level rise process induced the formation of chimney shafts and huge chambers by paragenesis process in karst systems. The last phase occurred in the Quaternary with new stages of rivers downcutting which provoked a general exhumation of karst levels. Eventually, the river course evolution interplayed locally with karst evolution, resulting in unique landforms like the Pont d'Arc natural arch.

Humanity discovered more than 30,000 years ago this particular landscape made of huge canyons, high plateaus, and numerous caves. In the Ardèche area, more than a hundred of caves show paintings and rock art that tell a new story, the human story of this land. Nowadays, local population is still very much attached to this land, and thousands of tourists come every year to visit the numerous caves open to the public or to admire the Ardèche canyon and its exceptional natural, scenic landscapes.

Acknowledgments We are grateful to Monique Fort, Associate Editor, for many inspiring and instructive discussions, and for reading and improving this manuscript.

References

Audra P, Mocochain L, Camus H, Gilli E, Clauzon G, Bigot JY (2004) The effect of the Messinian Deep Stage on karst development around the Mediterranean Sea. Examples from Southern France. Geodin Acta 17–6:389–400

Besson D (2005) Architecture du bassin rhodano-provençal miocène (Alpes, SE France). Relations entre déformation, physiographie et sédimentation dans un bassin molassique d'avant-pays. PhD thesis. Ecole des Mines de Paris, p 358

Clauzon G (1982) Le canyon messinien du Rhône: une preuve décisive du "Desiccated deep-basin model " (Hsu, Cita, Ryan, 1973). Bull Soc Géol Fr 24:597–610

Clauzon G (1996) Limites de séquences et évolution géodynamique. Géomorphologie 1:3–22

Delannoy JJ, Sadier B, Jaillet S, Ployon E, Geneste J-M (2010) Reconstitution de l'entrée préhistorique de la grotte Chauvet-Pont d'Arc (Ardèche, France): les apports de l'analyse géomorphologique et de la modélisation 3D. Karstologia 56:17–34

Garcia-Castellanos D, Estrada F, Jimenez-Munt I, Gorini C, Fernandez M, Verges J, De Vicente R (2009) Catastrophic flood of the Mediterranean after the Messinian salinity crisis. Nature 462:778–781

Hsü KJ, Cita MB, Ryan WBF (1973) The origin of the Mediterranean evaporites. Initial Rep DSDP Wash XIII(2):1203–1231

Jaillet S, Sadier B, Hajri S, Ployon E, Delannoy JJ (2011) Une analyse 3D de l'endokarst: applications lasergrammétriques sur l'aven d'Orgnac (Ardèche, France). Géomorphologie 4:379–394

Martini J (2005) Etude des paléokarsts des environs de Saint-Remèze (Ardèche, France): mise en évidence d'une rivière souterraine fossilisée durant la crise de salinité messinienne. Karstologia 45–46:1–18

Mocochain L, Clauzon G, Bigot JY, Brunet P (2006) Geodynamic evolution of the peri-Mediterranean karst during the Messinian and the Pliocene: evidence from the Ardèche and Rhône Valley systems canyons, Southern France. Sediment Geol 188–189:219–233

Mocochain L, Audra P, Bigot JY, Clauzon G, Bellier O, Parize O, Monteil P (2009) The Messinian Salinity Crisis manifestations on landscape geodynamic (karst surface, river piracies, and cave levels): example of the Lower Ardèche River (Rhône Mid-Valley). Geomorphology 106:46–61

Mocochain L, Audra P, Bigot JY (2011) Base level rise and per ascensum model of speleogenesis (PAMS). Interpretation of deep phreatic karsts, vauclusian springs and chimney-shafts. Bull Soc Géol Fr 182(2):85–91

Sadier B, Delannoy JJ, Benedetti L, Bourlès D, Jaillet S, Geneste JM, Lebatard AE, Arnold M (2012) Further constraints on the Chauvet cave artwork elaboration. Proc Natl Acad Sci USA 109(21):8002–8006

Tassy A, Mocochain L, Bellier O, Braucher R, Gattacceca J, Bourlès D (2013) Coupling cosmogenic dating and magnetostratigraphy to constrain the chronological evolution of peri-Mediterranean karsts during the Messinian and the Pliocene: example of Ardèche Valley, Southern France. Geomorphology 189:81–92

The Lower Valley and the Delta of the Rhône River: Water Landscapes of Nature and History

20

Gilles Arnaud-Fassetta and Mireille Provansal

Abstract

Linking the Mediterranean to northern Europe, the lower Rhône valley is the major axis of communication in the south of France since the antiquity. The valley ends by a large deltaic plain rich in history and landscapes of international renown. It annually attracts thousands of tourists and nature lovers and artists through several festivals and exhibitions. Original water landscapes were built in the valley and at the contact between the river and the sea (delta). Their development over the centuries is the product of a complex evolution between the natural environment and human history. Numerous archaeological excavations reflect man–environment interactions, emphasising the strong constraint of fluviatile–deltaic environments, and how human societies, since the Greeks and the Romans, found some parades to mitigate the fluvial risk. The geomorphology of the landscapes is most of the time discrete, and we must learn to decipher the old and recent landforms derived from fluvial, lagoonal or marine environments. The landforms are now subject to different aggressions (erosion, pollution) and metamorphoses related to human activities combining with the natural dynamics: the protection of the "nature" has become a real challenge.

Keywords

Water landscape • Palaeoenvironment and geoarchaeology • Biodiversity • Fluvial and marine hazards • Cultural heritage • Lower Rhône valley

20.1 Introduction

The Rhône delta, the so-called Camargue, is one of the major Mediterranean deltaic plains. Known as one of the most extensive natural areas (1,700 km^2) in Western Europe, the Rhône delta is famous for its vast sandy beaches and amphibious landscapes, which constitute an ornithological reserve of global importance, with white horses and black bulls living in freedom. These large "natural" spaces and their mythical inhabitants, the "gardians", were celebrated by both movies and literature, while the Gypsies gather every year at Les Saintes-Maries-de-la-Mer for their "Great Pilgrimage".

The location at the interface between a large anthropised catchment and a "closed" sea sensitive to pollution created original landform/landscape units downstream from the city of Avignon. It is an environment of great touristic, ecological, historical and societal importance, but fragile and under threat in the near future.

This text invites the reader to discover the main features of the (palaeo)environment of both the lower valley and the delta of the Rhône River, through urban and rural water landscapes representative of the area.

G. Arnaud-Fassetta (✉)
UMR PRODIG 8586 & LGP 8591, Université Paris-Diderot, Paris, France
e-mail: gilles.arnaud-fassetta@u-pec.fr

M. Provansal
UMR CEREGE 7330, Aix-Marseille Université, Aix en Provence, France
e-mail: provansal@cerege.fr

M. Fort and M.-F. André (eds.), *Landscapes and Landforms of France*, World Geomorphological Landscapes,
DOI 10.1007/978-94-007-7022-5_20, © Springer Science+Business Media Dordrecht 2014

20.2 Geographical and Geological Setting

The alluvial plain of the lower Rhône valley (Fig. 20.1) is situated downstream of a large catchment (97,800 km^2). At Avignon, towering the entire valley from the left side, Mt. Ventoux (1,800 m) is the last signal of the Alpine mountains range. Between Avignon and Arles, the river flows across alluvial basins alternating with resistant geological structures (E–W limestone folds). The historical sites such as the mediaeval castles in Avignon, Beaucaire and Tarascon or the Roman town in Arles were built on these "hard points".

From the alluvial plain to the delta, longitudinal channel profile and water surface slope decrease drastically (from 0.07 to 0.003 m/km), enhancing a drop in stream power and reinforcing the alluviation. The mean 1920–2007 annual discharge at the Beaucaire station is 1,720 m^3/s, influenced by the Mediterranean inputs from the downstream tributaries (Ardèche, Durance, Gard). The mean annual value of total wash load, which is estimated *ca.* 6.7×10^6 t at Arles (Antonelli et al. 2008), has been reduced by more than 75 % over the last 150 years in relation with deep climatic and socio-economic transformations in the fluvial catchment (reduction of the torrential erosion and the frequency in major floods at the end of the LIA, rural depopulation, gradual withdrawal of agriculture, reafforestation and building of Alpine reservoir dams).

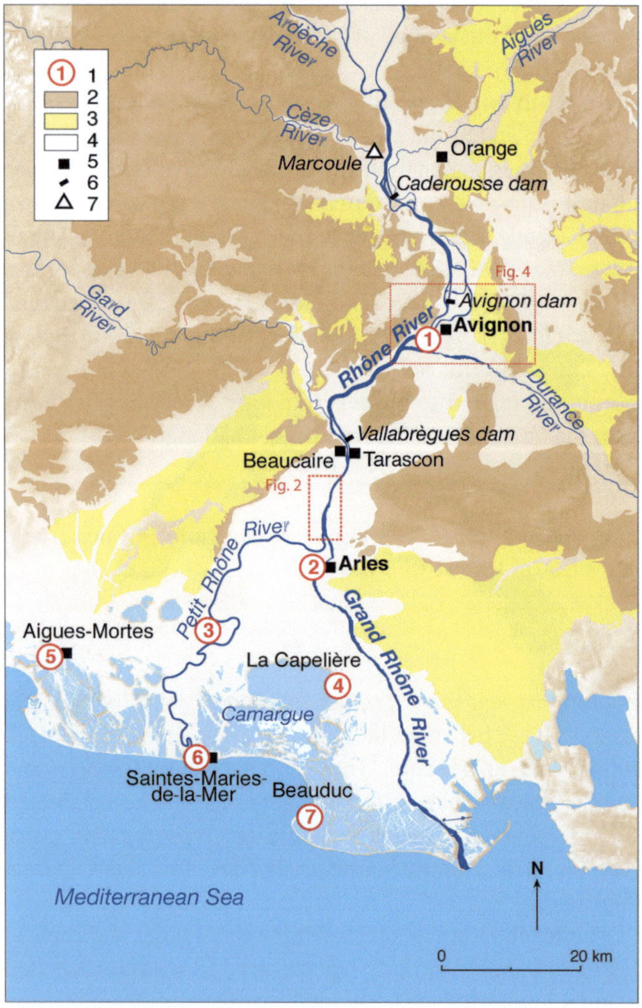

Fig. 20.1 The Rhône River in the lower valley and the delta. *1* Selected sites (sites 1–7 form the structure of Part 4, with 1 = Avignon, 2 = Arles, 3 = Petit Rhône, 4 = La Capelière, 5 = Aigues-Mortes, 6 = Les Saintes-Maries-de-la-Mer, 7 = Beauduc), *2* Resistant geological substrate, *3* Pleistocene terraces and fans, *4* Recent alluvium of the Rhône River and its tributaries, *5* Main cities, *6* Ddams managed by the National Company of the Rhône River (CNR), *7* Nuclear power plant of Marcoule (Credit: G. Arnaud-Fassetta and M. Provansal)

At the apex of the "Camargue" delta, the Rhône River splits into two branches, the Grand Rhône (50 km) and the Petit Rhône (60 km), which export respectively 90 and 10 % of the discharge. The Grand Rhône, with a sub-straight channel, supplies an active prodelta in the Mediterranean. Its sediment plume plays an essential role in nutrient and contaminant inputs in the marine environment up to the Pyrenees. But the deltaic coastline is generally retrograding, and its lobed shape is due to the predominant action of marine erosion.

20.3 Evolution and Landforms

Between the slopes formed by the outcrop of the geological substrate or alluvial terraces, the alluvial and delta plains of the Rhône River include several landscape units. They are inherited from the recent geomorphological history and hydraulic structures that have modified the dynamics of fluvial forms and determine the spatial extent of current risk.

20.3.1 The Alluvial Plain, Between Avignon and the Deltaic Diffluence

Downstream of Avignon, the structural constraint reduced the extension of the alluvial plain. Between Beaucaire and Arles, the river pattern was controlled by a late-glacial palaeochannel incised into the Pleistocene deposits and joining a palaeo-coastal fringe located at – 120 m under the present sea level. During the Holocene, the Rhône River gradually infilled this palaeo-valley by adjusting its longitudinal profile to both the sea-level rise and mouth progradation advance. In alluvial basins, archaeological data showed several stages of braiding expansion, associated with rising groundwater, during periods of flood events and high detrital supplies (e.g. during the Little Ice Age, LIA; Bruneton et al. 2001). Alternatively, meandering periods correspond to a good natural drainage. Urban sites are located on highest points, usually geological bedrock outcrops (see sites 4.1. Avignon and 4.2. Arles).

Fig. 20.2 Some palaeochannels of the lower Rhône River. (**a**) *Lône des Pêcheurs*. At the kilometric point (Kp) 306.5, this channel was active until the early twentieth century. Its cross-section area was then reduced by aggradation of a large sandy bar on the riverbank. (**b**) *Lône Pillet* (Kp 272–276). This old channel was closed in 1870 by a longitudinal dyke then cross thresholds. Today, the palaeochannel is often flooded downstream but only inundated by moderate flood events upstream (Photographs by M. Provansal)

The Rhône channel is actually located at the top of a large alluvial ridge, which dominates the Holocene floodplain of several metres: breaches opened in its riverbanks during flood events that flooded very large areas. Since the end of the nineteenth century, changes in natural or socio-economic conditions (end of LIA, slowing down in mountain agriculture), then hydraulic structures (continuous unsubmersible dykes, groynes, compartments and barred arms (Fig. 20.2)

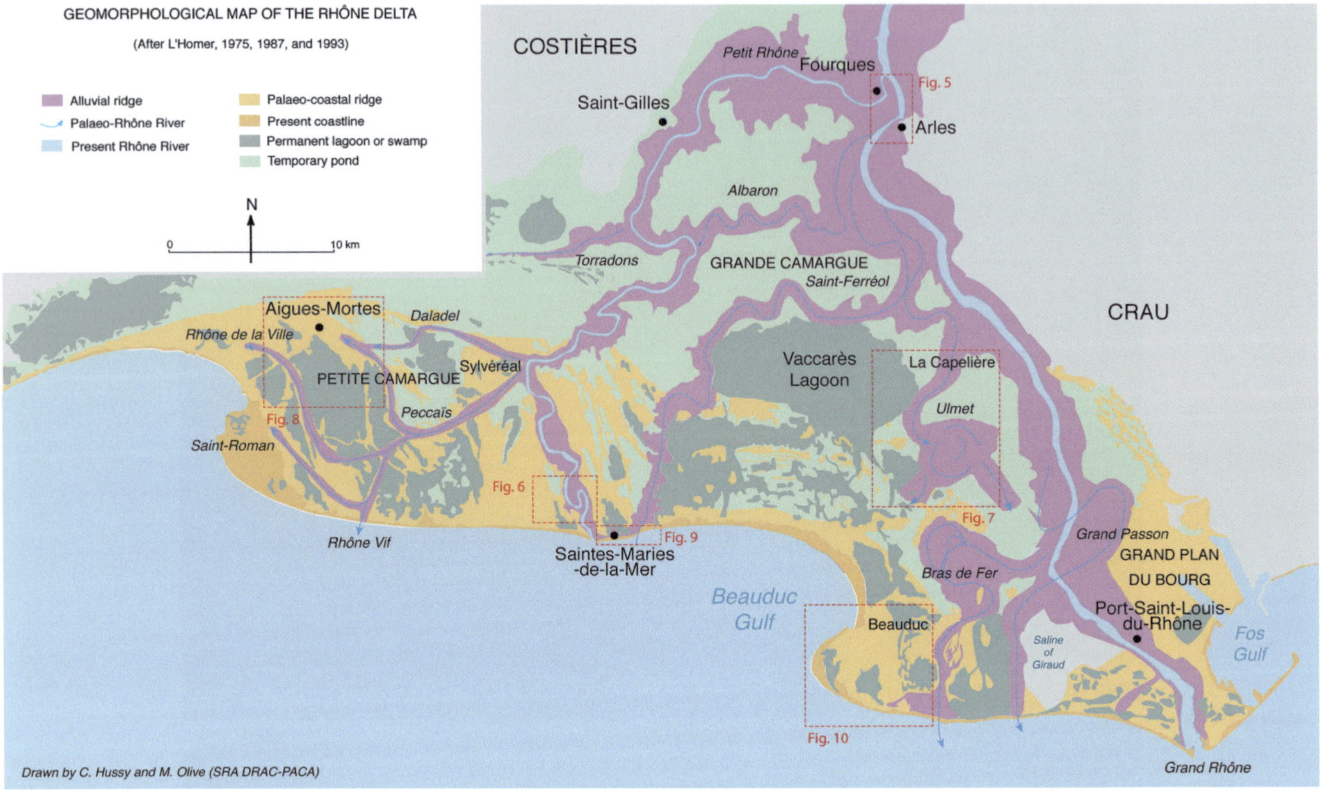

Fig. 20.3 Geomorphological map showing relationship between inherited and active landforms of the Rhône delta

and subsequently hydroelectric dams) induced channel adjustments resulting in a fluvial "metamorphosis" with a conversion of braided river pattern to a single, sinuous one (Arnaud-Fassetta 2003; Antonelli et al. 2004; Bravard 2010). Responses include simplification of channel morphology, erosion of the channel bed and infill of the alluvial margin, where the riparian forest is of great ecological richness (Provansal et al. 2010). The regulated channel is open for river navigation through a system of locks.

The lower Rhône River is located 70 km downstream of the Marcoule industrial radioactive releases: several studies give evidence for long-term retention of particle-reactive long-lived radionuclides in the fluvial margins and the "Camargue" delta (Provansal et al. 2012).

20.3.2 The Delta

The present plain corresponds to the top of a Holocene deltaic accumulation forming a sedimentary prism of more than 70-m thick to the south. Fluvial (alluvial ridges), brackish (salty lagoons) and coastal (sandbars) deposits unconformably overlie the Pleistocene alluvial gravel. River supplies lead to the advance of land surface since about 7,500 years ago, from a maximum shoreline situated to the north of the Vaccarès Lagoon (Fig. 20.3).

The Rhône River was divided, according to periods, in several branches, which patterns and mouths moved over time. Successive fluviatile channels, abandoned by avulsion, are still visible in the landscape: the oldest ones (6,500–2,000 BP) are located to the west, whereas the youngest ones are observed in the centre and east of the delta, including the two last branches: the Bras de Fer (1586–1712) and the Grand Rhône (since 1712).

The alluvial ridges of the former Rhône River (see site 4.4. La Capelière) and the coastal spits abandoned by shoreline advance form an irregular "grid" of sedimentary bodies higher (>2–3 m), sandy and better drained (the "*montilles*"), covered by crops and farms (the "*mas*"; Fig. 20.3). They delimit moist silty "lockers" which most of the time swampy. Fresh water dominates in the north, associated with *Phragmites australis*. Salt water characterises the southern part of the delta, associated with halophytic vegetation (the "*sansouire*"), where birds like to stay, especially flamingos, the emblem of the Camargue. The Vaccarès Lagoon (6,500 ha) is the largest and, like the other ones, protected by international labels (RAMSAR, UNESCO). Pesticides from agriculture (rice), however, disrupt the water quality. The rapid shoreline advance at times isolated marine lagoons, like that of Aigues-Mortes (see site 4.5).

The Grand Rhône River is slightly sinuous, while the Petit Rhône River develops several meanders, determined by Holocene geomorphic inheritance (see site 4.3. The Petit

Rhône River). Both branches were closely embanked in the late nineteenth century. In the Petit Rhône River, breaches in the riverbanks and dykes induce frequent inundation in the central or western part of the delta. The secular rise in sea level and storm surges, more frequent and higher than the surroundings during the twentieth century, considerably reduced the ability of the Rhône River to evacuate floodwaters to the sea, increasing the risk of flooding near river-mouths. The current coastline is formed by 70 km of sandy beaches backed by low dunes (<3 m) and unevenly degraded. Sectors of rapid erosion (see site 4.6. Les Saintes-Maries-de-la-Mer) alternate with well-developed sand spits (see site 4.7. Beauduc), but the overall sediment balance is negative on the coastline. Since many years, the sandy input from the two deltaic branches is insufficient to compensate erosion of the deltaic fringe.

20.4 Sites of Cultural Values and Touristic Promotion

20.4.1 The Site of Avignon

In Avignon, the Rhône River flows through a narrow rocky defile. Until the late nineteenth century, the river corresponded to the border between France (right bank) and the Holy Roman Empire and then the Papal lands (left bank). Two magnificent monuments face each other, the Chartreuse and Philippe Auguste Tower on the French side and the Palace of the Popes on the opposite side.

The Rhône River is now dammed by a hydroelectric dam. The slowdown of floodwaters upstream of the rocky defile and downstream of the city by the Durance confluence contributed over the centuries to the formation of many fluvial islands, mobile first and then stabilised. These islands were used as a substrate to build during the twelfth century, the famous Saint-Bénézet Bridge or *Pont d'Avignon*, sung by all the children of France (Fig. 20.4a).

From the fourteenth century, the terrible floods of the LIA, compounded by the effect of wars, eventually destroyed the bridge, despite several attempts to repair. Only a few arches have survived until today, siting on the limestone substrate on the left bank (Fig. 20.4b). Recent research has helped to locate the missing piles, to describe the "braided" pattern of the historical river, in which the main channel moved several times. At the entrance of the bridge, a beautiful museum gives the history and simulates the 3-D image of the original bridge.

20.4.2 Arles

The city of Arles is crossed by the Grand Rhône River, which meanders between a limestone outcrop and the beginning of deltaic land (Fig. 20.5). The river is a key element of the history and the development of the city since the Iron Age. Remarkable monuments were built during Roman times, such as the Antic theatre, the arena and the Alyscamps cemetery or the Roman circus. Because of its important heritage, Arles is classified by UNESCO since 1981. The *Musée départemental de l'Arles antique* was built in 1995 on the margin of the alluvial ridge where the Roman circus sited. Archaeological collections are particularly rich, and it will be possible to admire the oldest known bust of Caesar, who has been recently (2008) found at the bottom of the Grand Rhône River.

Considered as an important fluvial harbour since antiquity, the site of Arles has persisted despite the hydrological risk (Arnaud-Fassetta et al. 2005). The left bank in the upstream part of the city was affected by breaches and crevasse splays during protohistoric and Roman times; in the crossing of the city, it was also submerged several times (district of Sainte-Luce). The right bank at Trinquetaille recorded several flood events associated to crevasse splays and avulsions during Roman times, leading to the building of dykes and drainage structures. The plain extending south of the city (district of archaeological museum) was regularly flooded in the antiquity. Finally, the current urban sprawl spread in the floodplain, taking advantage of the low frequency of floods in the years 1960–1990.

20.4.3 The Petit Rhône Branch

The Petit Rhône is one of the two present fluvial branches of the Rhône delta, which marks the western edge of the Camargue. From its diffluence with the Grand Rhône in Fourques (north of Arles), the Petit Rhône is embanked all the way to the sea and joins the Mediterranean near Les Saintes-Maries-de-la-Mer after a sinuous/meandering course of 60 km through the deltaic plain. It is bordered by a riparian forest in its upstream section and by tamaris and "*sansouïre*" downstream where the channel is under influence of the salt wedge. It is possible (and recommended) to cross the river downstream by taking the scenic *Bac du Sauvage*.

The Petit Rhône branch is prone to the occurrence of crevasse splays during flood events (e.g. 1993–1994 and 2003; Arnaud-Fassetta 2013). Both human and material consequences of the December 2003 flood event were severe. However, this branch of the Rhône delta remains the most beautiful and most picturesque to make walking along its riverbanks, along which are installed fishermen's huts and large nets (the "*carrelets*"). Visiting the Petit Rhône by kayak or riverboats with a pair of paddle wheels allows one to explore its numerous meanders where there is an extraordinary freshwater avifauna (Fig. 20.6). The Petit Rhône is also a very popular river for sport fishing for giant catfish, a species introduced into the canal joining the rivers Rhône and Rhine in 1868.

Fig. 20.4 Hydraulic structure facing vagaries of time. (**a**) The broken bridge of Avignon (engraving, 1608). (**b**) The Avignon Bridge and the Palace of Popes (From Wikipedia 2004). It did not resist to flood events during the LIA

20.4.4 The Archaeological Sites on the Former Branches of the Rhône River, La Capelière

In the Camargue, the first known archaeological sites date back to the sixth century BC. The deltaic lands were an obvious strategic position for the city of Marseilles, founded by the Greeks who wanted to control the Rhône River and the hinterland. From the first century BC, habitats (grouped settlements; *villae*) multiplied, mainly in the north of the Camargue. The southern Camargue revealed the existence of a mosaic of large *villae* and agricultural settlements during the antic period. The archaeological sites essentially established along the old branches of the Rhône River (Arnaud-Fassetta and Landuré 2003). Societal vulnerability faced to flood hazard was lessened, by settling on the highest points of the alluvial ridges.

Some sites were temporarily abandoned or unoccupied during period of hydrographical disorder. On the site of La Capelière, a small rural settlement was built on the bumpy

Fig. 20.5 The site of Arles (Photograph by P.-L. Ferrandez). Note the Grand Rhône River in the background and development of the antic city of Arles on the *left* bank in the foreground

Fig. 20.6 The last tight meander of the Petit Rhône River (Photograph by Tiki III, *mini-croisière camarguaise*) before the river mouth

Fig. 20.7 The archaeological site of La Capelière (fifth century BC to fifth century AD) on the Ulmet alluvial palaeo-ridge bordering the eastern bank of the Vaccarès Lagoon (Photographs by C. Hussy) (**a**), G. Arnaud-Fassetta (**b**), and C. Landuré (**c**)

sandy-silt flood deposits right from the beginning of the occupation (fifth century BC). The site suffered from floods and crevasse splays. That is why human occupation levels of the first century BC were completely recovered by sandy-silt deposits of a crevasse splay, in which structures and amphoras were fossilised (Fig. 20.7). After these catastrophic

Fig. 20.8 The site of Aigues-Mortes (photograph by F. Hédelin). Constance Tower is in the foreground, salt pans in the background

floods, which probably appeared in the second half of the first century BC, the settlement was abandoned and reoccupied some decades later (~AD 30). Visiting the site, it is now possible, by joining the Camargue National Reserve, to observe traces of old channels of the Rhône River (the "*lônes*") and archaeological site of La Capelière now scattered in the salt marshes where salicornia dominates.

20.4.5 Aigues-Mortes

The landscapes around Aigues-Mortes are composed of wet zones (flood plains, ponds and lagoons) associated to sand dunes, which form the Petite Camargue (Fig. 20.8). Delta construction was relatively late in this area since no deltaic lobe was developed before Roman times. The main part of the deltaic infilling occurred during the historical period, thanks to the sedimentary supplies from several palaeochannels located to the south of Aigues-Mortes (Fig. 20.3).

It is in Aigues-Mortes that Saint Louis, King of France, sailed in 1248 for the Seventh Crusade. The harbour of Aigues-Mortes was more inland on the edge of the Marette Lagoon. At the beginning of the fourteenth century, Philippe le Bel used the fortified site of Aigues-Mortes to incarcerate the Templiers in the Tower of Constance (Fig. 20.8). This tower is also notorious because it was particularly a Protestant women's place of imprisonment after the revocation of the Edict of Nantes. The Desert Museum pays a warm tribute to these Cévenol prisoners who resisted to the repression of Louis XIV, while their husbands were dying on the galleys.

From the early fifteenth century, significant work was undertaken to facilitate access to the sea from Aigues-Mortes since the site is exposed to recurring silting. The year 1806 saw Aigues-Mortes become a fluvial harbour through the "*Canal du Rhône à Sète*".

20.4.6 Les Saintes-Maries-de-la-Mer

At the mouth of the Petit Rhône River, the site of Les Saintes-Maries-de-la-Mer has been inhabited since the Middle Ages, as evidenced by the fortified church built in the eleventh to twelfth centuries. Legend has it that the three *Saintes Maries* (Madeleine, Jacobé and Salomé), fleeing Egypt, landed here at the beginning of the Christian era. The Pilgrimage of the Gypsies from all over Europe commemorates this event

Fig. 20.9 (**a**) The beach of Les Saintes-Maries-de-la-Mer in 1888 (V. Van Gogh, Public domain). (**b**) Coastline between Les Saintes-Maries-de-la-Mer (background) and La Gacholle in the National Reserve of Camargue, south of the *Digue à la Mer* on March 16, 2010 (Photograph by M. Gauthier Clerc)

every spring. The poet Mistral made die also his heroine, Mireille, whose statue is established on the main place.

Les Saintes-Maries-de-la-Mer offers a beautiful synthesis of traditional symbols of the Camargue: salt marshes with typical houses of the "gardians", white horses and black bulls, arena devoted to the bull race (the "*Course à la cocarde*") and sandy beaches immortalised by Van Gogh (Fig. 20.9a). But the site, which was originally more than 1 km from the sea, is currently threatened by shoreline erosion (several metres per year; Fig. 20.9b). Indeed, the Petit Rhône River supplies no more enough sediment in order to resist to wave erosion, whose impact is exacerbated by the present sea-level rise (Sabatier et al. 2006). Many structures (dams, wave breakers, groynes) degrade the natural landscape and do not protect the coastal fringe in the medium term. The site is now part of an important challenge for choosing a sustainable coastal development.

Fig. 20.10 The Beauduc spit. Note the breach trough the coastal ridge, the wash-over channel and the wash-over fan *left* (west) to the Beauduc lighthouse (Photograph by M. Gauthier Clerc)

20.4.7 The Site of Beauduc

The beach and dunes of Beauduc are located in the central part of the deltaic fringe. We can reach this site by a dirt road 12-km long through the marshes where flamingos have now a sustainable habitat. Located at the bottom of a large gulf occupied by shallow water, the beach is very wide, further expands each year and is surrounded by a vast field of wild, white sand dunes (Fig. 20.10). This "World's End" is the end of a spit, still well supplied by sand derived from the erosion of the old mouth of the Bras de Fer (Sabatier et al. 2006; Fig. 20.3).

This is a very original site, between nature and culture, frequented by clam fishermen, surfers, loners and the international gentry. It is famous for its hamlet of eclectic "sheds", built after the Second World War, from recycled materials, old fishermen's huts, caravans and buses. In order to preserve this unique environment, the management of this space was disputed between the locals and State agencies. But the hamlet will be saved in order to preserve its original cultural heritage.

20.5 Conclusion

The lower valley and the delta of the Rhône River are attractive areas for their natural richness, the beauty of their landscapes and the importance of historical and cultural heritage, raising contemporary issues on the relationship between societies and their environment.

On the side of natural resources and landscapes, we highlight the large wetlands, whose ornithological richness is known worldwide; the vast sandy beaches of the Camargue; and the biodiversified riparian forest that runs along the river. Both historical and cultural heritage certainly deserve a visit, for the beauty of the monuments still in place in the old cities of Arles and Avignon, the deltaic sites and popular festivities.

Finally, this famous area is now the holder of an interesting debate concerning the limits of acceptable risk, whether induced by rigid hydraulic structures dating from the nineteenth to twentieth centuries or by the effects of climate change on sea-level rise.

References

Antonelli C, Provansal M, Vella C (2004) Recent morphological channel changes in a deltaic environment. The case of the Rhône River, France. Geomorphology 57:385–402

Antonelli C, Eyrolle F, Rolland B, Provansal M, Sabatier F (2008) Suspended sediment and ^{137}Cs fluxes during the exceptional December 2003 flood in the Rhone River, southeast France. Geomorphology 95:350–360

Arnaud-Fassetta G (2003) River channel changes in the Rhône Delta (France) since the end of the Little Ice Age: geomorphological adjustment to hydroclimatic change and natural resource management. Catena 51:141–172

Arnaud-Fassetta G (2013) Dyke breaching and crevasse-splay sedimentary sequences of the Rhône Delta, France, caused by extreme riverflood of December 2003. Geografia Fisica e Dinamica Quaternaria (in press)

Arnaud-Fassetta G, Landuré C (2003) Hydroclimatic hazards, vulnerability of societies and fluvial risk in the Rhône Delta (Mediterranean France) from the Greek period to the Early Middle Ages. In: Fouache E (ed) The Mediterranean world environment and history. Proceedings of the international conference "environmental dynamics and gistory in mediterranean areas" held in Paris, 24–26 April 2002. Elsevier, Paris, pp 51–76

Arnaud-Fassetta G, Bruneton H, Berger JF, Beaudouin C, Boès X, Provansal M (2005) A 8000-yr record of palaeohydrology and environmental change in fluviatile-influenced sediments from the urban Arles-Piton core, upper Rhône Delta, France. Z Geomorphol 49:455–484

Bravard JP (2010) Discontinuities in braided patterns: The River Rhône from Geneva to the Camargue delta before river training. Geomorphology 117:219–233

Bruneton H, Arnaud-Fassetta G, Provansal M, Sistach D (2001) Geomorphological evidences for fluvial change during the Roman period in the lower Rhône valley (southern France). Catena 45:287–312

Provansal M, Villiet J, Eyrolle F, Raccasi G, Gurriaran R, Antonelli C (2010) High-resolution evaluation of recent bank accretion rate of the managed Rhone: a case study by multi-proxy approach. Geomorphology 117:287–297

Provansal M, Ferrand E, Eyrolle F, Raccasi G, Monaco M, Gurriaran R (2012) Spatial variability in sedimentation rates and artificial radionuclide storage in alluvial banks of the lower Rhône River. Aquat Sci 74(4):735–750

Sabatier F, Maillet G, Fleury J, Provansal M, Antonelli C, Suanez S, Vella C (2006) Sediment budget of the Rhône delta shoreface since the middle of the 19th century. Mar Geol 234:143–158

Glacial Landscapes and Protohistoric Cultural Heritage of the Mount Bego Region, Southern French Alps

21

Jérôme Magail and Patrick Simon

Abstract

The region of Mount Bego, which is located in the Southern French Alps, is an interesting place to observe scenic inherited landforms as well as cultural remains left by the ancient human societies who lived there. Geomorphology, Archaeology and Anthropology are working jointly to reconstruct the history of landscapes and societies. In particular, the Bronze Age people used the perfect stony tables created by glacial erosion. Indeed, they engraved approximately 40,000 petroglyphs on rocks polished by the ice during the Last Glacial Maximum. At that time, the climate was very cold, corresponding to the full expansion of glaciers. Only the peaks of Mount Grand Capelet and Mount Bego were protruding from the ice. The glacial imprint is widespread in the landscapes of the Mount Bego region, which displays about 40 cirques, abundant moraines and erratic boulders and 30 glacial lakes. Periglacial conditions, which developed after glacier melting, favoured the formation of rock glaciers before the Postglacial warming which induced peat formation. The first traces of the landscape transformation by the human societies date back to the Neolithic, three millennia before the Protohistoric societies left the outstanding petroglyphs of the so-called *Vallée des Merveilles* (Valley of Wonders).

Keywords

Mount Bego • Southern Alps • Glacial geomorphology • Peat bog • Bronze Age petroglyph

21.1 Introduction

Mount Bego, which is surrounded by the Merveilles and Fontanable valleys, is located 30 km from the Mediterranean Sea. This region peaks at altitudes ranging from 2,500 to 3,000 m and displays Last Glacial landforms such as glacially carved valleys and cirques offering a rocky landscape dotted with kettle lakes (Fig. 21.1). This wonderful landscape shaped by the Quaternary glacial actions has been the focus of several geomorphological studies (Faure-Muret 1955; Orengo 1973; Julian 1980). The morphostructural frame inherited from the Alpine movements and Plio-Quaternary evolution has been modified by the installation and the melting of Pleistocene glaciers. The glacial imprint is variable due to the variety of rock types. After the deglaciation, periglacial and temperate conditions continued to reshape the landscapes before human societies added their touch by transforming the vegetation cover and inducing soil erosion. But in the Mount Bego area, around 2000 BC, the Bronze Age people not only left traces of their breeding and agriculture activities; they also left thousands of engravings which provide valuable information about their way of life. They focused most of their rock art on the two largest areas of ice-smoothed rocks (Begin-Ducornet 1991; de Lumley 1995), especially Permian rocks which resist to weathering and have remarkably preserved the petroglyphs (Fig. 21.2).

J. Magail (✉) • P. Simon
Museum of Prehistoric Anthropology of Monaco, 56 bis Bd du Jardin Exotique-MC, Monaco 98000, Monaco
e-mail: jerome.magail@map-mc.com; patrick.simon@map-mc.mc

M. Fort and M.-F. André (eds.), *Landscapes and Landforms of France*, World Geomorphological Landscapes,
DOI 10.1007/978-94-007-7022-5_21, © Springer Science+Business Media Dordrecht 2014

Fig. 21.1 The Merveilles Valley landscape with the pyramidal Mount Grand Capelet in the background. The Merveilles Peak in front of it has a large area of ice-smoothed rocks. The Long Inferieur Lake in the foreground is partly filled with a peat bog (Photo J. Magail)

Fig. 21.2 The famous Bronze Age engraving named "the wizard" has been done on an oxidised ice-smoothed outcrop displaying grooves and striations (Photo J. Magail)

21.2 Geological Context

The Merveilles Valley is situated at the transition zone between the middle and the high mountains (2,000–2,500 m), at the eastern end of the Argentera-Mercantour crystallophyllian massif or basement (Fig. 21.3). This NW-SE elongated massif belongs to the external crystalline massifs (like the Mont Blanc and Pelvoux massifs) according to the classic Alpine nomenclature. It is mainly made of metamorphic and granitic rocks above which sedimentary layers ranging from the Carboniferous (300 Ma) to the Pliocene (6 Ma) periods are resting unconformably. These sedimentary sequences present numerous stratigraphic lacunae due to the complexity of the regional tectonic history.

The Argentera-Mercantour crystalline rocks are mainly represented by the Argentera leucocratic intrusive granite (batholith type) and the anatectic granite of Valmasque. They are associated with metamorphic rocks, mainly inherited from the Variscan (Hercynian) period. They include several metamorphic complexes (Faure-Muret 1955) such as the migmatitic Chastillon-Valmasque complex found in the Mesches Lake and the Minière Valley areas (Fig. 21.4).

Sporadic but outstanding detrital Permian series (about 250 Ma) overlie these basement rocks. Classic sedimentary facies have been deposited in basins during the dismantling of the Hercynian chain. They consist of

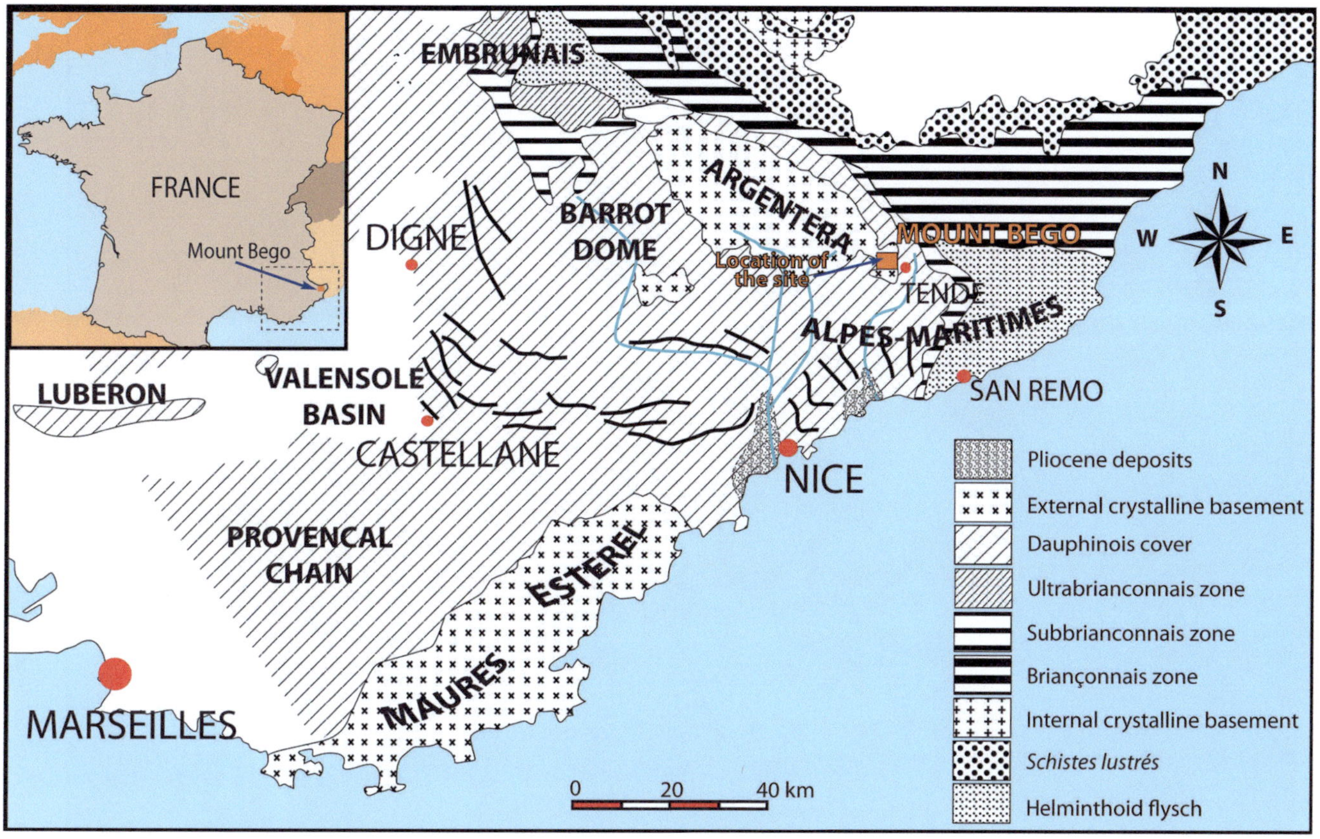

Fig. 21.3 General geologic map of the SE France (After Campredon and Boucarut 1975 – CAD Burle) CAD: Computer Assisted Drawing

conglomerates, sandstones, arkoses and mudstones. Faure-Muret (1955) identified four sets in the area: the Inferno, Meraviglie, Bego and Capeirotto series. These are mainly detrital, either coarse-grained and heterogeneous (Inferno and Bego) or pelitic and mostly homogeneous (Meraviglie and Capeirotto). Volcanism evidences can be found locally such as an interbedded dacite flow and the presence of rhyolite and rhyodacite pebbles. The total thickness of the Permian detrital series reaches about 4,000 m in the Tende province.

The post-Permian sedimentary cover restricted to the southeastern end of the Argentera-Mercantour massif dates back to the Triassic and Jurassic periods. Southwards, in the Nice and Castellane subalpine arcs, sedimentary series display impressive Jurassic limestones and Cretaceous marly limestones. Northwards and eastwards, autochthonous terranes are hidden by the first thrust sheet of the Subbriançonnais and Briançonnais zones (Eocene Helminthoid flysch). The tectonic uplift inherited from the Cenozoic Alpine evolution has been modified by the Pleistocene glaciers, with special reference to the glacial imprint left by the Last Glacial Maximum (OIS 2–20,000 BP).

21.3 The Glacial Imprint on Landscapes

During the Last Glacial Maximum (20,000 BP), the trimline (upper limit of the ice) reached the base of Mount Bego, as indicated by glacial landforms including microscale features (Orengo 1973; Finsinger and Ribolini 2001). The presence of Permian blocks on the Plan Tendasque at an altitude above 2,000 m indicates that the Minière Valley glacier spilled over its left bank and deposited these erratic blocks. The degree of preservation of such glacial imprints depends on the rock hardness and its weathering rate since deglaciation.

21.3.1 The Glacially Carved Valleys and Cirques

The Minière Valley, which extends from the Minière Lake (1,500 m) to the Pas de Colle Rousse (2,200 m), displays a series of three rock basins (*ombilics*) separated by three rock bars and the hilly passage of the Vallon de l'Enfer closed by a cirque (Fig. 21.5). The upstream lacustrine area is located at the confluence of three mountain valleys: the Merveilles, Huile Lake and Muta Lake valleys, characterised by a trough

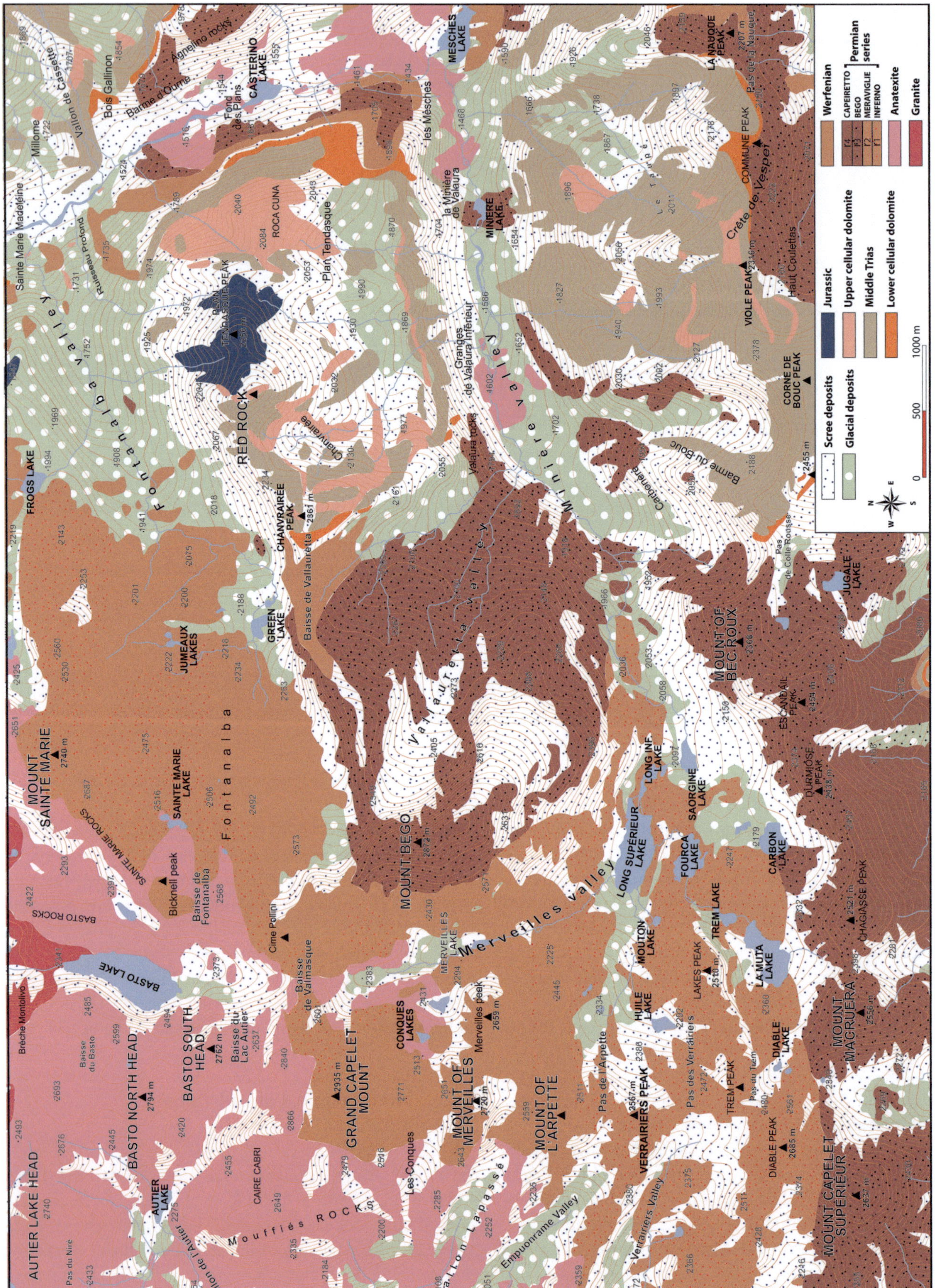

Fig. 21.4 Geologic map of the region of Mount Bego (After Goguel 1967 – CAD Burle) CAD: Computer Assisted Drawing

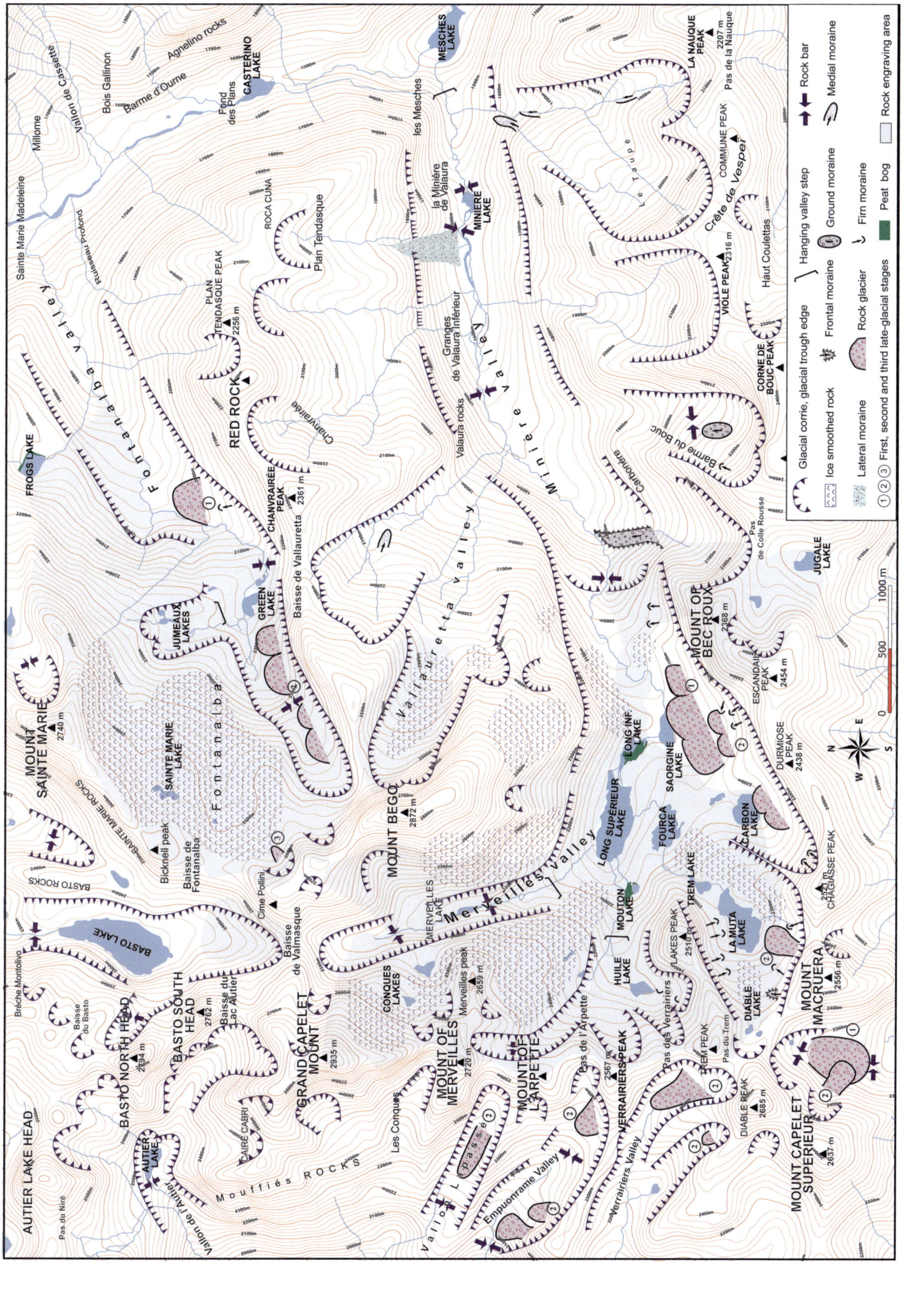

Fig. 21.5 Geomorphological map of the region of Mount Bego (After Orengo 1973 – CAD Burle) CAD: Computer Assisted Drawing

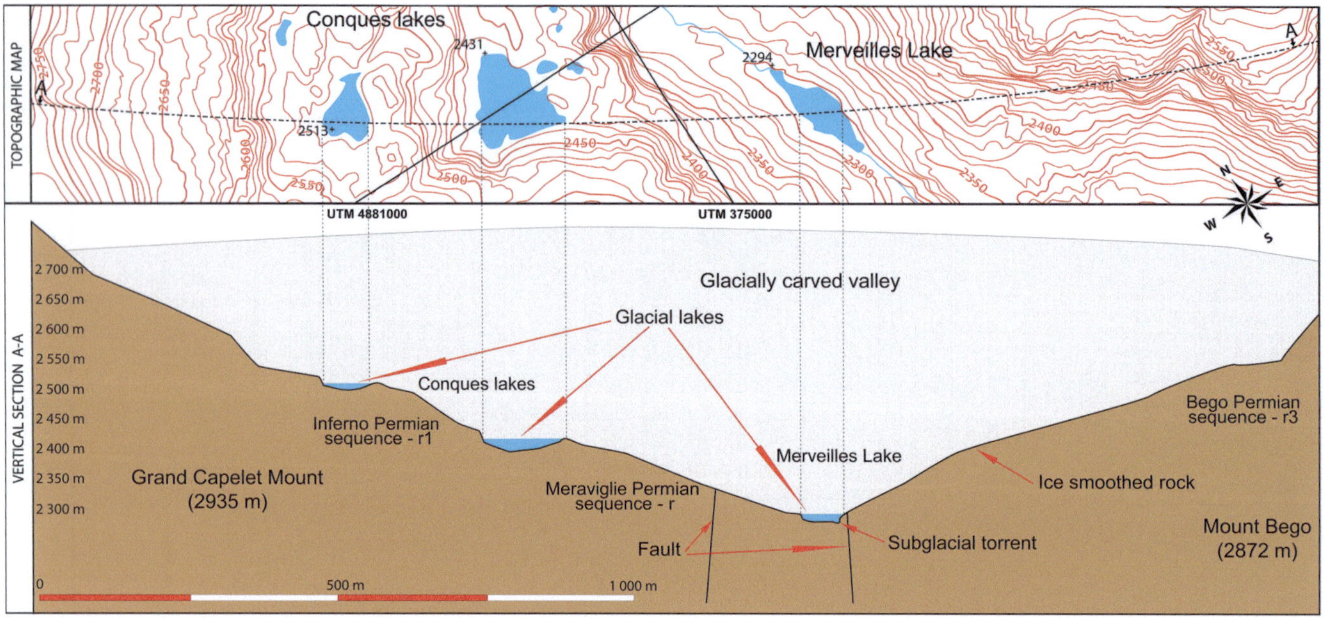

Fig. 21.6 Geomorphological cross-section in the Conques Lake area (Credit Magail – CAD Burle) CAD: Computer Assisted Drawing

profile. The Merveilles Valley is the longest and the narrowest and is dominated by the two highest mountains of the area: Grand Capelet (2,935 m) and Mount Bego (2,872 m).

Over 40 cirques are present in the region of Mount Bego. Mainly located on north-facing shady slopes, the largest are located on the right bank of the Minière Valley and are 2,000 m high. On the left bank, the cirque of Valauretta Valley is carved into the Mount Bego northeastern face. The Plan Tendasque peak area displays much smaller cirques carved in Triassic rocks; this area is characterised both by its glacial and karstic imprints. From the Pas de Colle Rousse to the Nauque peak as well as the Plan Tendasque, some outcrops of Jurassic limestone and mid-Triassic dolomite are prone to karst formation. However, the volumes are too low for large sinkholes to develop. The only karstic evidence is provided by large rock shelters of shallow depth situated in cliffs of mid-Trias in particular near the Pas de Colle Rousse and above the Granges de Valaura. Only one 50 m deep sinkhole was found in the region, namely, the Aven Primard, situated 5 km south of the Minière Valley, in the mid-Trias of Plateau de la Céva.

Other smaller cirques are found in the upper side of the N-S ridge passes across the Head of Basto and Mount Capelet Superieur, some belonging to tributary valleys of the Gordolasque Valley and others to the Minière Valley. The base of the highest cirques is at 2,600 m, as shown by the Conques Lakes (Fig. 21.6).

21.3.2 Scouring and Polishing Effects

The ice-smoothed rocky areas offer numerous examples of glacial scouring and polishing. Those present in the Pic des Merveilles area have particularly striking grooves and ridges, carved into a remarkable Permian eponymous series. The famous engraving known as the *Sorcier* (wizard) has been done on a scoured outcrop displaying large grooves (Fig. 21.4). In other places, grooves are observed, caused by the friction of ice-covered rocks in motion, which are accompanied by real scars left on larger blocks. At places, the ice-smoothed rock surface has been delicately polished. This is the case with pelite outcrops that display large curved or flat surfaces, perfectly smooth.

Current observations of scraping and polishing are possible on the Permian rocks thanks to the quality of this material. By contrast, the glacial polish and striations on Triassic outcrops have been erased by 13,000 years of limestone solution by rain waters. In the Merveilles Valley, there is an outstanding case of glacial polish on a so-called *paroi vitrifiée* (glazed wall). This massive wall in Permian rocks, 10 m high and 25 m long, is a N-S-oriented slickenside (28 °W) linked to a probable strike-slip fault which was active during the late Alpine orogenic phase. Then, a silica film developed on this plane upon pelitic rocks. Its vitrified and shining appearance is probably due to the friction of the silica film by the ice although a listric origin (tectonic polish) cannot be excluded (Fig. 21.7).

Following deglaciation, ice-scoured walls and horizontal surfaces were rapidly in contact with the atmosphere and submitted to iron oxidation inducing a reddish coloration of the surface layer. By hitting the rocks, Bronze Age engravers blew this oxidised film and made the original pelite colour appear at the bottom of their engravings. The bottom of the oldest engravings already displays an oxidised patina (Fig. 21.8).

Fig. 21.7 Remarkable glacial and/or listric polish on the *paroi vitrifiée* (glazed wall) of the "Valley of Wonders" (Photo J. Magail)

Fig. 21.8 Bronze Age engravings on oxidised Permian rocks (Photo J. Magail)

21.3.3 Late-Glacial Moraines and Drift Boulders

During the Late-Glacial period, the glacier that occupied the cirque beneath the Pas de Colle Rousse produced the largest frontal moraine as displayed across the right bank of the Minière Valley. It was deposited while ice was retreating from this valley. This 400 m long moraine is located on a steep slope, starting from 2,000 m altitude and ending near the stream at 1,770 m altitude. It is crossed by the access road to the Merveilles Valley. Located at higher elevations (2,400 m) under the Pas du Diable, another frontal moraine has a much smaller size. Ground moraine occupies the centre of the Barme du Bouc cirque. The Mount Bego also displays traces of glacial transportation during the Late-Glacial period, with drift boulders supplied by the slopes of Mount Bego deposited near the Long Inferieur Lake and Superieur Lake while ice was melting (Fig. 21.9).

21.4 Rock Glaciers and Peat bogs

21.4.1 Rock Glaciers

Periglacial conditions have allowed the formation of rock glaciers which belong to three generations (Orengo 1973; Finsinger and Ribolini 2001). Those of Saorgine Lake, sheepfold of Fontanalba and probably the one of Pas du Diable, were formed during the first phase. Other rock glaciers have been formed after the retreat of the lower part of glaciers. This is during this second phase that a second series of rock glaciers of Saorgine Lake develops itself, like those of Carbon Lake, the ones which lead to the Gordolasque Valley and those located above the Vert Lake of Fontanalba. The third phase of rock glaciers evolution is more discrete. The one located more than 2,500 m above the sea level, north of the summit of Pollini, is an example. Two factors among others (Evin 1985) were favourable to the formation of these rock glaciers: firstly their geographical situation on

Fig. 21.9 Erratic boulder deposited by the glacier at the entrance of the "Valley of Wonders" (Photo J. Magail)

north-facing hillsides where climate is the coldest and secondly the rubble material supplied from the broken Permian walls, above the rim of the cirques.

21.4.2 Peat Bogs

The end of the Late-Glacial period had been already marked by lake filling, peat formation and vegetal colonisation. The cores drilled in the Long Inferieur Lake peat bog of the Merveilles Valley provided enough data to reconstruct the landscape history since the Older Dryas (de Beaulieu 1976; de Beaulieu and Jorda 1977). Pollen analyses show the installation of a first steppe vegetation with a predominance of *Artemisia* pollen, followed by the progression of the tree cover as indicated by *Juniper* and *Pinus* taxa. A marked cold recurrence occurs during the Younger Dryas with a regression of tree pollens. The last phase of the Postglacial period is marked by pastoral activities near Long Inferieur Lake between 3,500 and 3,000 BP. The Subboreal period is characterised by an increased frequency of grasses and plantains (*Plantago* sp.) indicative of agricultural activities. Analyses carried out in the vicinity of Mouton Lake (2,150 m) reveal a larch forest destruction by men at about the end of Subboreal. The Mount Bego population cleared the larch forest in order to cultivate or develop areas of pastures. At the end of the third millennium, pollen analyses detect intense larch deforestation in the valley, followed centuries later by agricultural activity, with cereal and plantain pollens appearing from 1500 BC near Long Inferieur Lake. It is highly probable that groups of people have settled down on the slopes of Mount Bego during the summer season. Rock engravings of ploughs attest their agricultural practices (Fig. 21.10). Recent studies in the Millefont Valley area showed that traces of human activity appeared as early as 4,750 cal. BP, i.e. during the Neolithic period (Brisset et al. 2012).

21.5 Human Activities and Popular Beliefs

Even if the first traces of landscape transformations by human activities date back to the Neolithic, the Mount Bego region seems to have been a place of passage since the disappearance of the glaciers. The Alpine chain is separating several major valleys and men who intend to move from one to another have to cross the passes. At the north lies the plain of Po and at the south, valleys flow into the Mediterranean Sea. People used to reach back valleys and crossed ridges since the end of the Palaeolithic. With the development of pastoral activities at the end of the Neolithic, the region has become the summer meeting place of transhumant shepherds from all neighbouring valleys to the pastures of Mount Bego (Magail and Giaume 2005; Barral and Simone 1990). People have not lived in the winter at high altitudes on the snowy hillsides of these mountains, for we know that thousands of Bronze Age petroglyphs were made between the summer solstice and the mid-September (Magail 2006). The limited iconographic themes (bovines, weapons, fields, etc.) suggest that every year the Bronze Age engravers repeated the same petroglyphs as a rite (Magail 2007). During the last two centuries, the

Fig. 21.10 Detail of an engraving showing two oxen hitched to a swing plough (Photo J. Magail)

shepherds have given a fantastic meaning to this strange landscape shaped by glaciers displaying numerous Protohistoric signs. The name *Vallée des Merveilles* (Valley of Wonders) or *Meraviglia*, which refers to an enchanted or "fairy-like" place, illustrates the special place occupied by the Mount Bego region in the vernacular culture. Similarly, the *Cime du Diable* (Devil's Peak) and the Valmasque Valley names also evoke the supernatural and the evil. These names are associated with popular legends and fairy tales.

21.6 Conclusions

The numerous cirques and lakes scattered around the Mount Bego are indicative of the strong pressure exerted by the glaciers during the Last Glacial optimum period. The huge quantities of water released during deglaciation have caused significant erosion by reshaping lateral moraines and creating alluvial fans. Above 2,000 m, sparse vegetation gave way to a landscape marked by ice-smoothed and polished outcrops. The freshness of small-scale glacial features is amazing. Following deglaciation, this rocky landscape has also changed in colour. Some surfaces are oxidised and this iron oxidation gave a reddish colour to large areas. The entire plateau of the Merveilles Valley and the base of Fontanalba peak are characterised with this bright colour. Although the Mount Bego has been shaped by Quaternary glaciers, its landscape continues to evolve according to the seasons, with alternations of freezing and thawing inducing frost shattering and rainfall episodes responsible for various hydrological phenomena including flooding of the Minière Valley and the formation of potholes in its gorges.

Human activity is also responsible for landscape transformation. People cleared the forests and contributed to soil erosion since the Late Neolithic. The Mount Bego specificity remains the 40,000 engravings inscribed on the rocks around 2000 BC. Between 2,000 and 2,600 m, Bronze Age engravers selected rocks which had been polished by glaciers, especially the Permian rocks including the fine-grained schistose and pelitic rocks which were easy to engrave. The attractiveness of the Mount Bego region lies in a unique combination of scenic glacial landscapes and traces of ancient high-altitude human occupation.

References

Barral L, Simone S (1990) Who's who in Merveilles? Bulletin du Musée d'Anthropologie préhistorique de Monaco 33:71–80

Begin-Ducornet J (1991) Etude du site à gravures protohistoriques de la région du Mont Bego dans son contexte géologique et géomorphologique. Contribution à la compréhension du choix et de la répartition des pétroglyphes. Colloque International, Le mont Bego, Tende, pp 95–105

Brisset E, Guiter F, Miramont C, Delhon C, Arnaud F, Disnar JR, Poulenard J, Anthony E, Meunier JD, Wilhelm B, Pailles C (2012) Approche multidisciplinaire d'une séquence lacustre holocène dans les Alpes du Sud au Lac Petit (Mercantour, Alt. 2200m, France): histoire d'un géosystème dégradé. Quaternaire 23(4):309–319

Campredon R, Boucarut M (1975) Guides géologiques régionaux Alpes-Maritimes, Maures, Estérel. Masson & Cie, Paris

de Beaulieu JL (1976) Le paysage végétal depuis les glaciers. In: IX Congrès de l'Union Internationale des Sciences Préhistoriques et Protohistoriques, Vallée des Merveilles. Nice, pp 28–35

de Beaulieu JL, Jorda M (1977) Tardiglaciaire et postglaciaire des Alpes de Haute-Provence. Le glaciaire de la Blanche, Trois évêchés. Bulletin de l'Association française pour l'étude du quaternaire V14(3):3–15

de Lumley H (1995) Le grandiose et le sacré. Edisud, Aix en Provence

Evin M (1985) Contribution à l'étude de la macrofissuration dans les parois situées à l'amont des glaciers rocheux des Alpes du Sud. Revue de géomorphologie dynamique 34(1):17–30

Faure-Muret A (1955) Etudes géologiques sur le massif de l'Argentera-Mercantour et ses enveloppes sédimentaires. Imprimerie Nationale, Paris

Finsinger W, Ribolini A (2001) Late Glacial to Holocene deglaciation of the Colle del Vei del Bouc-Colle del Sabbione area (Argentera Massif, Maritime Alps, Italy-France). Geogr Fis Din Quat 24:141–156

Goguel J (1967) Notice de la carte géologique détaillée de la France, Feuille de St Martin-Vésubie – Le Boréon, BRGM, 1/50000, 947, Paris

Julian M (1980) Les Alpes Maritimes Franco-Italiennes, Etude Géomorphologique. Thèse de Doct. d'Etat, 2 vols. Honoré Champion, Paris

Magail J (2006) Les gravures rupestres du Mont Bego des activités et des rituels en leur temps. Bulletin du Musée d'Anthropologie préhistorique de Monaco 46:96–107

Magail J (2007) Mont Bego, un cadran saisonnier vieux de 4000 ans. Archéologia Ed Faton 446:10–19

Magail J, Giaume JM (2005) Le site du Mont Bego, Actes du colloque des15 & 16 mars 2001 à l'Université de Nice – Sophia Antipolis. Editions Serre, Nice

Orengo C (1973) Glaciaire et tardiglaciaire des vallées de la minière, des Merveilles et de Fontanalbe. Revue de géographie alpine 61(4):583–599

Karst Memories Above and Beneath the Sea: Marseilles and Its Continental Shelf During the Cosquer Cave Occupation

22

Jacques Collina-Girard

Abstract

In the south of France, the Cosquer Cave with its famous prehistoric paintings is located in a karstic area located between Marseilles and Cassis. This emerged and submerged karst is typical of karstic coasts submerged after the Late-Glacial Maximum. All the forms observed in the hinterland can be observed directly by scuba divers and indirectly on bathymetric charts: lapiaz, karstic archs, sinkholes, uvala and polje. The emerged and submerged landscapes are mainly the heritage of specific lithological conditions (Urgonian limestones) and tectonic conditions (vertical faulting network leading to coastal collapse in the Mediterranean Sea). Other elements of this submerged landscape are given by the traces of the last sea level rise (palaeo-shorelines and erosion platforms and notches). All the area between Marseilles and La Ciotat is now established as the Calanques National Park, including the Cosquer Cave with its upper Palaeolithic rock art paintings, which adds an international archaeological interest to this exceptional natural area.

Keywords

Continental shelf • Karst • Prehistory • Cosquer Cave • Marseilles

22.1 Introduction

Along the French Mediterranean coasts, the area between Marseilles and Cassis offers spectacular landscapes, well known over the world to tourists, scuba divers, rock climbers and geologists (Collina-Girard 2012). Under a blue sky swept by the northwest wind (*mistral*), the white of the rocks contrasts with the deep blue of the Mediterranean Sea penetrating in the hinterland in deep *calanques* (Fig. 22.1). These landscapes, which are evocative of the coasts of Croatia, Turkey or Greece are located in the immediate neighbourhood of the city of Marseilles. This wild area, known as the *Massif des Calanques*, will be soon protected as a National Park. The attractiveness of this emerged and submarine landscape can be mainly connected to three natural factors: specific lithology, specific structural pattern and typical Mediterranean climate. The final touch is yielded by the postglacial sea level rise submersion of this coastal karst. The discovery of prehistoric paintings in the famous Cosquer Cave (Clottes and Courtin 1996; Clottes et al. 2005), submerged by the sea and with an opening at −37 m, has added an exceptional piece of submarine archaeology to the celebrity of the *Massif des Calanques*, also named *Massif de Marseilleveyre*.

J. Collina-Girard (✉)
Assistant Professor Aix-Marseille 1, Géologie quaternaire et préhistoire LAMPEA, MMSH, UMR7269 du CNRS, 5, Rue du Château de l'Horloge, BP647, 13094-Aix-en-Provence, Cedex 2
e-mail: Jacques.Collina-Girard@univ-amu.fr; collina@mmsh.univ-aix.fr

22.2 Lithology and Coastal Landscapes Between Marseilles and Cassis

The *Massif de Marseilleveyre* belongs geographically to the southern calcareous Provence extending from the Rhône Valley (west) to the Mediterranean Sea (south) and to the Durance valley and the Verdon Canyon (north). Its geological and geomorphological story is marked by the pyreneo-provencal and alpine orogenic phases and the long-term

M. Fort and M.-F. André (eds.), *Landscapes and Landforms of France*, World Geomorphological Landscapes,
DOI 10.1007/978-94-007-7022-5_22, © Springer Science+Business Media Dordrecht 2014

Fig. 22.1 The peak of the vertically faulted *Grande Candelle* overlooking the *Calanque de Sugiton* (Photo J. Collina-Girard)

evolution of karst relief (Nicod 1967; Julian and Nicod 1989; Cavalera et al. 2010), mostly linked to the Messinian period (Mocochain and Clauzon 2010; see Chap. 19 in this book).

Navigating from the west to the east along the shore from Marseilles to La Ciotat allows to observe the beginning of the lithological sequence represented in the *Massif de Marseilleveyre* (Fig. 22.2). The Jurassic unit of the *Calanque de Montredon* is represented by dolomite strata with ripple marks, mud cracks and stromatoliths deposited in shallow marine waters. These formations are overlain by 400 m of calcareous marls and limestones (Lower Cretaceous formations). Above, a main lithological unit (around 250 m) of compact bioclastic limestones, with fossils of rudists, corresponds to the Barremian and Bedoulian (Urgonian) formations.

The karst of the *Massif des Calanques* is mainly established on Urgonian limestones. These white rocks, compact and homogeneous, were extracted as building stone in quarries around the city of Cassis during the modernisation of Marseilles at the end of the nineteenth century and were traditionally used in Provence to make kitchen sinks, hollowed in polished Urgonian slabs. These rocks were also exported. As an example, a part of the basement of the Statue of Liberty in New York was imported from Cassis. This homogeneous and compact formation is affected by a well-developed tectonic network dominated by vertical faults and joint pattern.

22.3 Structural Framework of the Karst Between Marseilles and Cassis

Between Marseilles and Cassis, the tectonic framework is characterised by vertical faults along which the coastal and submarine areas have subsided to the south and the Mediterranean Sea (Fig. 22.1). Their structural lineaments are mainly NW–SE and NE–SW and the fault throw could exceed 300 m. These two directions, NW–SE and NE–SW, can be observed, in fractal form, from metric outcrop in the field to the lineaments observed on aerial photographs and

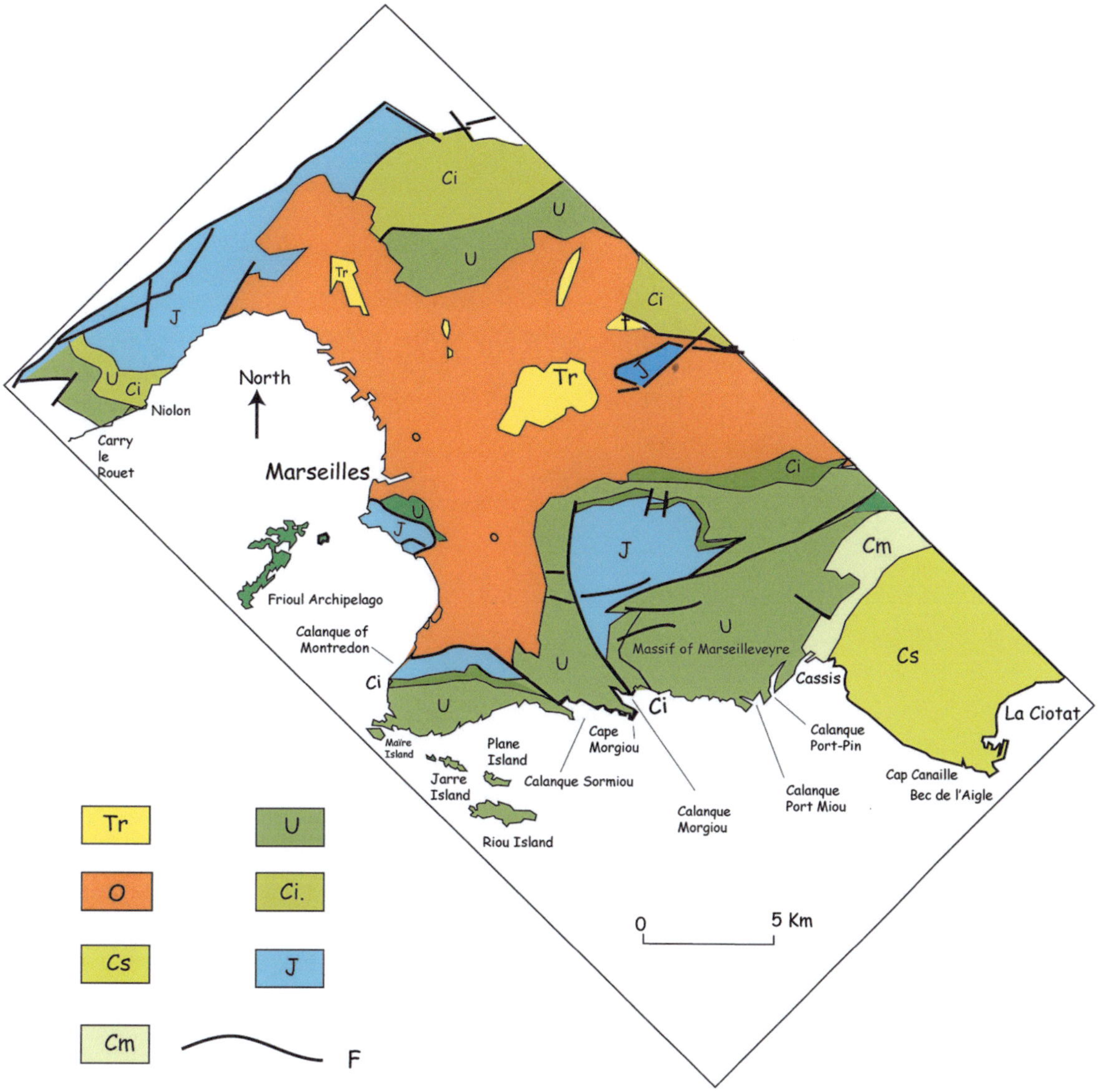

Fig. 22.2 Geological sketch of Marseilles and its surroundings. Legend: *Tr* Quaternary travertines; *O* clays, sands and pudding stones (Oligocene); *Cs* Upper Cretaceous limestones and conglomerates; *Cm* Middle Cretaceous clays; *U* Lower Cretaceous limestones (Urgonian); *Ci* Lower Cretaceous; *J* Jurassic dolomites; *F* main faults (After J. Collina-Girard 2012)

satellite images. There is also a clear correlation between the orientations of the coasts between Marseilles and Cassis and this structural framework, which constituted the preliminary lines for marine erosion during the last sea level rise (Fig. 22.3). These structural directions constitute also the implicit framework explaining the location and the orientation of the karstic thalwegs observed both on the continent and on the continental shelf. These directions determine also the location and morphology of caves and avens explored by speleologists. As an example, the planimetry of the Cosquer Cave, with its two main chambers oriented NE–SW, is linked to the same morphostructural framework. In the same way, the sinkholes submerged on the continental shelf are generally aligned, following the same structural trends.

22.4 Balancing Between Lithological and Structural Causalities: Some Frequent Karstic Forms

Karst formation is mainly related to the Tertiary period (Eocene and Miocene), but the Late-Glacial and postglacial periods have contributed to give its present aspect to the karstic landscapes of the Massif of Marseilleveyre.

22.4.1 The *Calanques*

Calanques is the local name, in Marseilles, of the coastal karstic valleys penetrating more or less deeply inland. These valleys

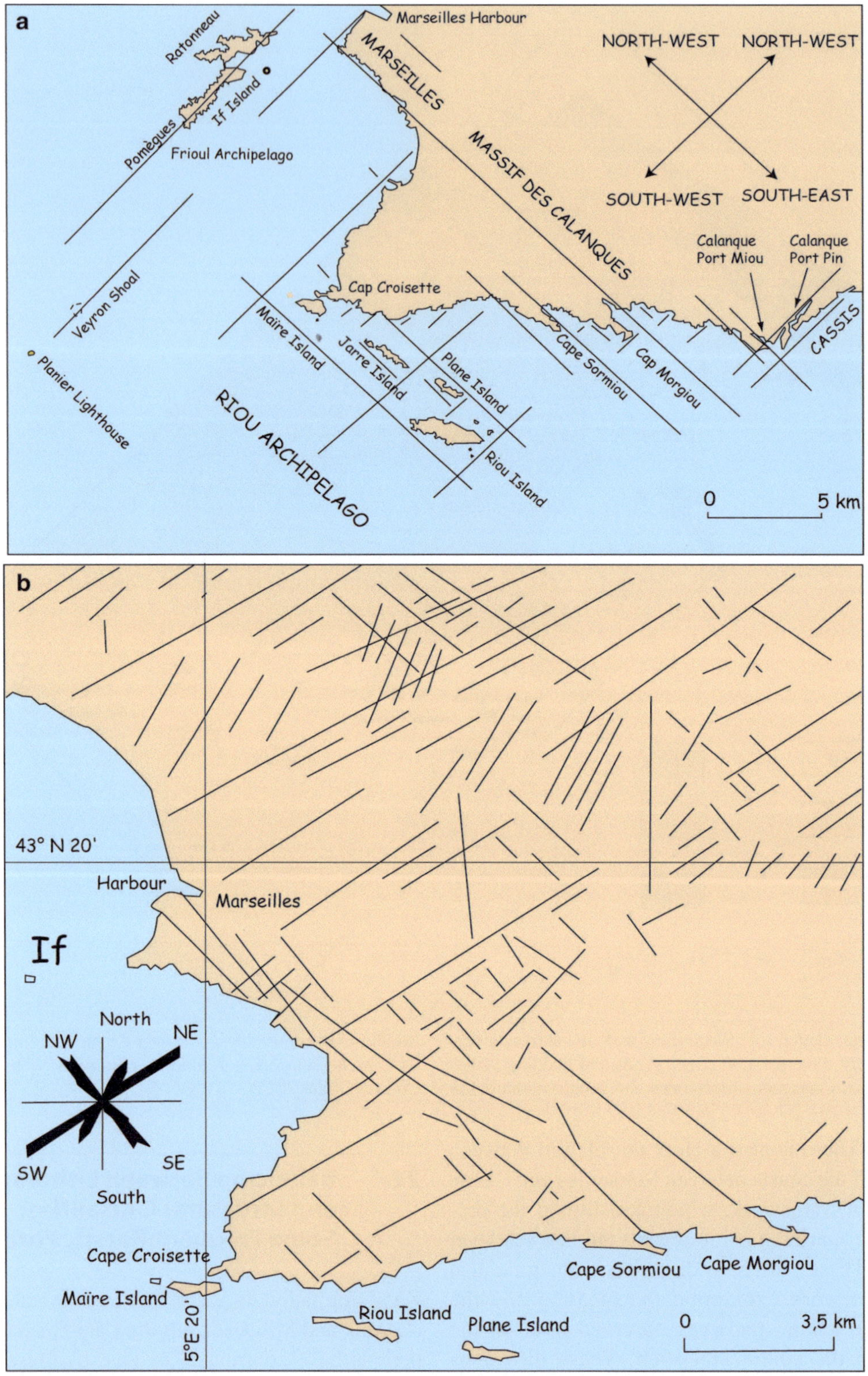

Fig. 22.3 Comparison between the orientation of the coast between Marseilles and Cassis (**a**) and the structural lineaments observed on aerial photographs and presented in a rose diagram (**b**) (Modified after Collina-Girard 2012 and S. Dupire 1985)

are generally located on faulted areas corresponding to the structural directions NW and SE. As an example, near Cassis, the *calanque* of En Vau is NW–SE, and the nearby *calanques* of Port Pin and Port Miou have a NE–SW direction. This case is particularly demonstrative of the widespread structural control imprinted on the coastline. These coastal valleys, like the fjords of Norway and the *rias* of Portugal, were partially submerged during the sea level rise following the Last Glacial (Fig. 22.4).

Fig. 22.4 The Cape Morgiou and the karstic Triperie Creek between the *Calanque* of Sormiou (on the *left*) and the *Calanque* of Morgiou (on the *right*). Cosquer Cave opens at a depth of –37 m (Photo J. Collina-Girard)

22.4.2 Submerged Karstic Arches and *Lapiaz*

Scuba divers appreciate the spectacular landscape of submarine karstic archs (Fig. 22.5a, b), well represented at the southern part of Maïre Island between –5 and –20 m but also on Plane Island at –35 m and near the diving spot of the Pierre à la Bague at –45 m. This lighthouse announcing the southern entrance of Marseilles Bay is built on the emerged part of a large shoal oriented SW–NE from Cap Croisette. The flat top of this submarine plateau is an extended karstified area displaying a spectacular submerged karren or *lapiaz* (Fig. 22.5c). On its shallower part, it is possible to explore a karstic gallery network extending between –15 and –25 m (*Plateau du Veyron*). Some geological sampling in these galleries yielded continental Quaternary sediments from the Late-Glacial period.

22.4.3 Submerged Sinkholes

In the hills surrounding Cassis, a few dolines can be identified from aerial photographs. Along the shore, it is possible to observe partially submerged sinkholes forming sheltered creeks and rounded *calanques* (*Calanque de Pouars* on Plane Island). The historical quarantine harbour of Port Pomègues (Frioul Islands) is a partially submerged uvala. Some of these rounded creeks correspond to the inactive parts of submerged karstic valleys or aven apertures (Triperie Creek on Cape Morgiou, *Calanque de Marlet* on Frioul Islands). Many of these forms can also be identified from bathymetric documents (Fig. 22.6) of the continental shelf off Marseilles (Collina-Girard 1996). This continental shelf terminates at the upper limit of the submarine canyons of Messinian origin (Planier Canyon, Cassidaigne Canyon). This slope change occurs approximatively between –150 and –200 m and increases up to –1,500 m towards the abyssal plain. The continental shelf comprises, from the coast, around 6–20 km of submerged karstic terrain, including territories of the prehistoric artists of the Cosquer Cave during the Late-Glacial Maximum (19,000 years BP) when the sea level was about –135 m.

This large submarine plateau is dotted with numerous small basins (1–5 m depth), generally aligned, following the main fracture directions. We have interpreted these forms, which are widespread on the continental shelf, as sinkholes submerged and preserved by the rapid post-LGM sea level rise (Collina-Girard 2004). Most of these dolines (Fig. 22.6) can be identified on the flat areas, between –100 and –150 m.

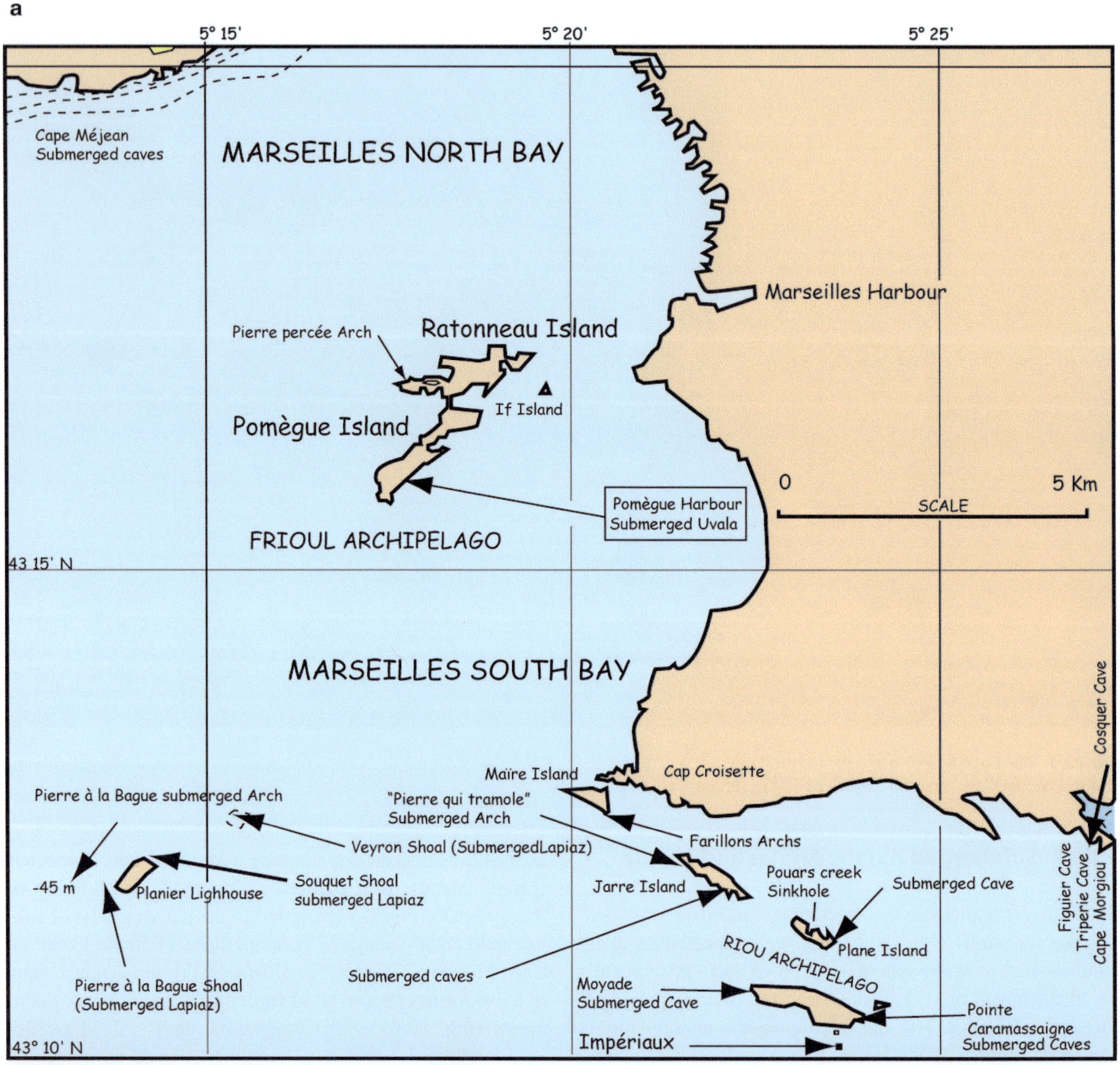

Fig. 22.5 Location of some scuba diving spots mentioned in the text (**a**); the submerged arch (–5 to –20 m) below Farillon Islet (**b**); the submerged karren of the *Pierre à la Bague*, near the Planier Lighthouse (**c**) (Photos J. Collina-Girard)

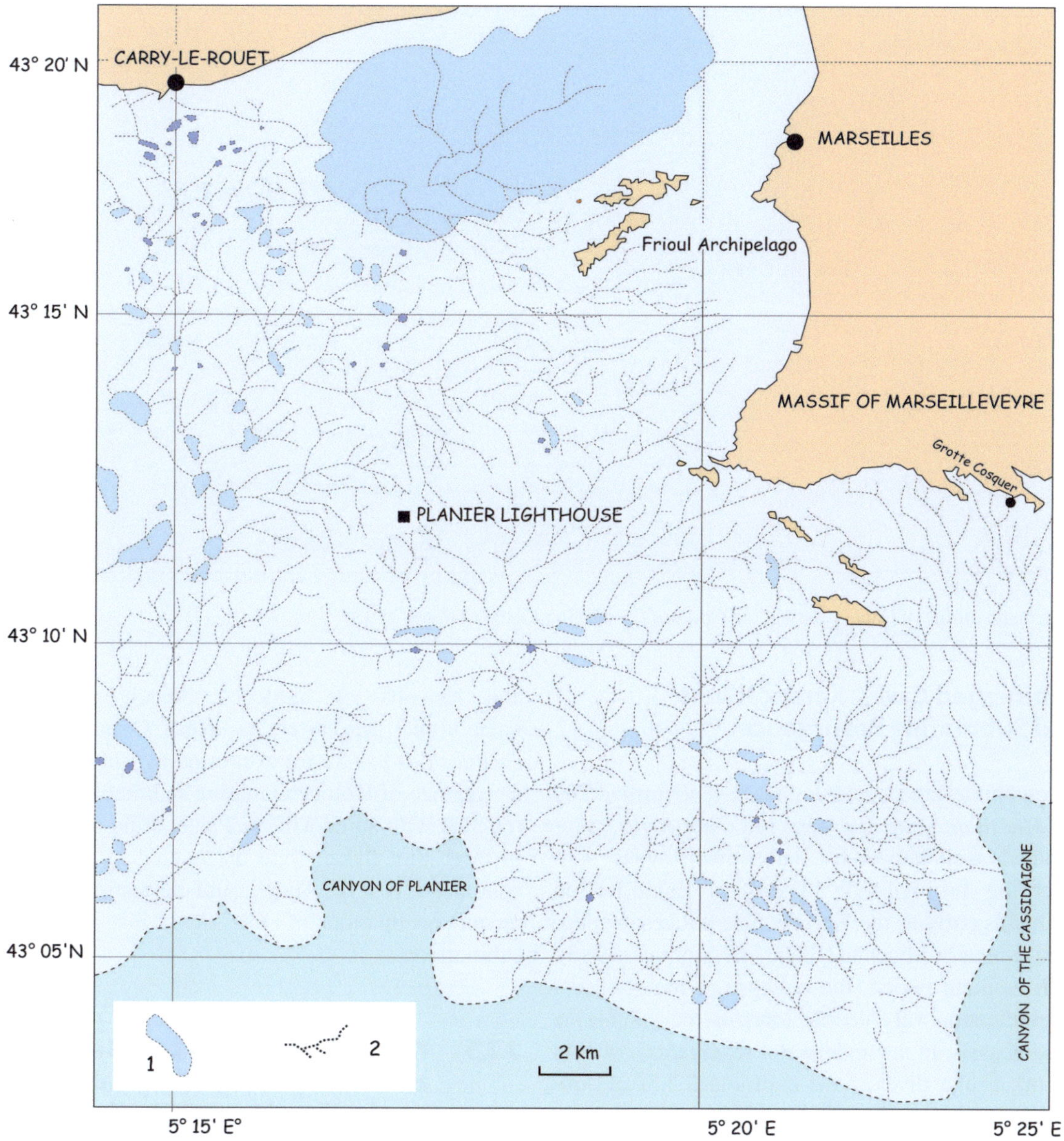

Fig. 22.6 The continental shelf off Marseilles. Legend: *1* submerged sinkholes and polje, *2* submarine valley network on the continental shelf (After Collina-Girard 2012)

We visited, using scuba diving, the shallower of these dolines located immediately at the southeast of the *Château d'If*, between −25 and −35 m. They appear as conical depressions filled by sands, but in other cases, seismic surveys had shown that these depressions affect, below the sand and mud cover, the limestone substratum.

22.4.4 Submerged Polje

An important submerged offshore sub-rectangular basin (10 km long, 5 km wide), oriented NE–SW, lies between Frioul Islands and Carry-le-Rouet, west of Marseilles (Fig. 22.6). The maximal depth reaches −78 m. This submarine graben constitutes an offshore extension of the Marseilles Oligocene Basin, where the city is established (Collina-Girard 2004). In the western part, some small depressions (sinkholes) underline a karstified limestone outcrop oriented NW–SE, probably in contact, along a faulted area with the Oligocene formations. This endoreic basin with its flat bottom is probably a karstic plain superposed on a collapsed basin, similar to the others forms, known in the emerged karstic area (Julian and Nicod 1989), northeast of Cassis (polje of Cuges-les-Pins and Sainte-Baume).

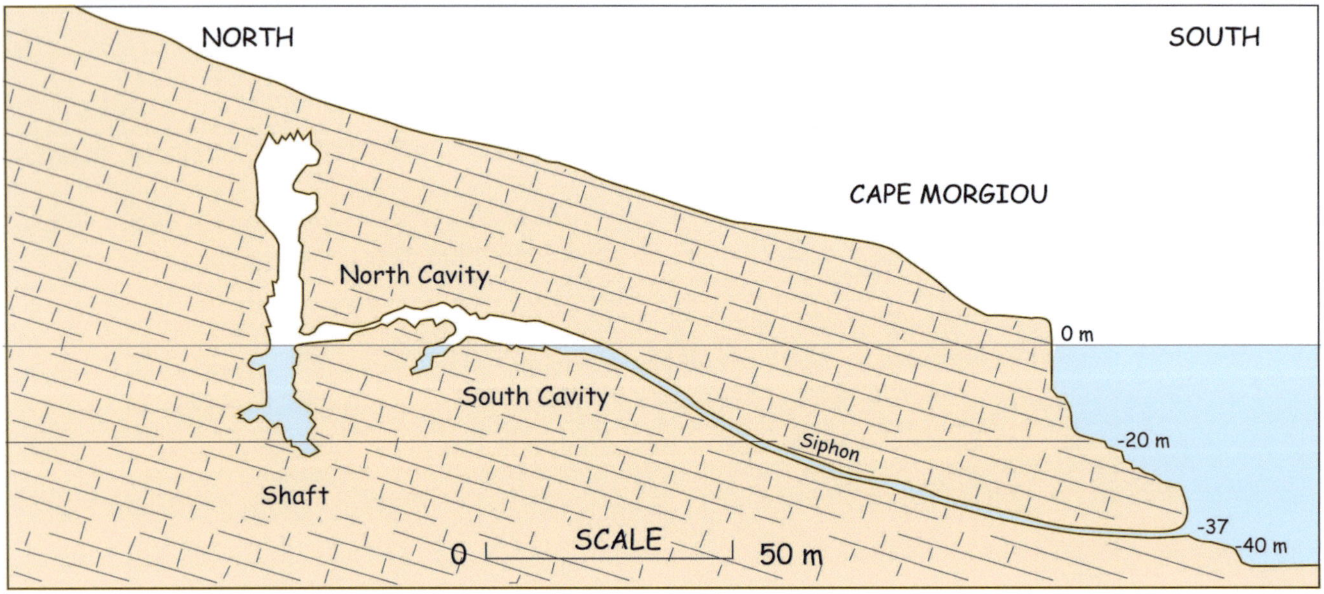

Fig. 22.7 Schematic profile of Cape Morgiou and Cosquer Cave (After Collina-Girard 2012)

22.4.5 Submerged Caves, Karstic Rivers and Submarine Resurgence

On the emerged hinterland, speleologists have identified 200 caves in the *Massif de Marseilleveyre*, but other caves, completely drowned, well known by scuba divers, were submerged after the Last Glacial Maximum. Karstification began probably as early as the Eocene. The thickest speleothems are Oligocene or Miocene (Blanc and Monteau 1988). During the Messinian period, the collapse of the sea level extended karstification very deeply towards the submarine canyons. Near Cassis, in the underwater resurgences of Port Miou (−20 m), scuba divers have explored 2 km of subhorizontal galleries that end in a vertical shaft explored down to −170 m but are continuing deeper. The hypothesis of a yet unknown submerged Messinian gallery, reaching, around −200 m, the *Canyon de la Cassidaigne* has been evoked by hydrogeologists (Cavalera et al. 2010). This seems to be the only explanation for the abnormal residual salinity of the freshwater after the building of a barrage blocking off marine saltwater in the resurgence.

The most famous cave is the Cosquer Cave (Clottes and Courtin 1996), discovered in the summer of 1991 at −37 m by Henri Cosquer below Cape Morgiou (Fig. 22.7). Prehistoric paintings and engravings largely covered the emerged part of the cavity. This prehistoric sanctuary is part of a network of emerged or submerged galleries, with several levels of karstification mainly associated with the Messinian period. Nearby the *Grotte du Figuier* is a very large cave, which might contain archaeological sediments, between −5 and −20 m. The *Grotte de la Triperie*, immediately neighbouring the Cosquer Cave, shows immerged speleothems and remains of hardened breccia containing Lower Palaeolithic flint artefacts. Near Cassis, the *Grotte des Trémies* (−9 to −20 m) excavated by Eugène Bonifay shows a sequence of Holocene marine sediments overlying breccia remains with some Middle Palaeolithic artefacts. A number of large flint flakes were discovered by divers in a second chamber at −5 m. They could be testimonies of the most recent occupation of the cave, just prior to its final submersion.

22.5 Karst and Sea: The Last Sea Level Rise and Its Geomorphological Testimonies on the Continental Shelf

The bathymetric map established for the continental shelf (Collina-Girard 1992, 2004) shows escarpments, submarine cleaves and abrasion platforms. The foot of these escarpments follows depth curves that are independent of the structural directions. These escarpments were interpreted as Late-Glacial and postglacial shorelines as also observed in Portugal and Brazil, where coastal cleaves were perfectly conserved by the post-LGM sea level rise, which was sufficiently rapid to avoid the erosion of these fragile coastal forms (Collina-Girard 2002).

Statistical analysis of the depth of the foot of these escarpment measured by scuba diving observation (Collina-Girard 2002) revealed well-defined modal values (−11, −16/−20, −25, −35/36, −55 m) in the entire area between Carry-le-Rouet and Cassis but also in other lithological contexts of metamorphic rocks from the Palaeozoic basement in the Western Mediterranean (Port-Cros, Porquerolles, Corsica,

Elba Island). A sequence of marine notches was discovered near Toulon by Jacques Laborel (Port d'Alon) at the same depth (−11, 16/−20, −25, −35/36 m). The −55 m abrasion platform, south of the Frioul Islands, can also be observed by scuba divers at Riou Islands in a rocky cleaves area. It is also possible to observe at the same depth a well-preserved cliff with overhanging surfaces and caves evoking the present coastal caves. In accordance with the radiometric dating of fossil coastal algae rims in contact with the bedrock, the age of this stillstand of sea level, at −55 m, is estimated at about 8,500–9,500 BP, which also dates the beginning of the biological colonisation of the rock by marine organisms in a period probably a little earlier than the arrival of sea level at this depth (Sartoretto et al. 1996).

Deeper down, an escarpment can be noticed on the bathymetric maps at about −90 m (south of Riou and Planier Island). A third level was observed at −100/−105 m, south of Planier Island. Coastal sediments (pebbles, gravel and sands) were directly cored in a shingle bar located at −100 m. These coarse sediments associated with *Mytilus* banks were ^{14}C dated at 13,850 ± 200 years BP (Collina-Girard 2004) and could correspond to Meltwater Pulse 1A. Finally a slope breaking area around −130/−140 m is probably related to the Late-Glacial Maximum shoreline. These escarpment levels and abrasion platforms, observed in Marseilles and in the whole Western Mediterranean Sea, are widespread worldwide suggesting a stepped sea level rise during the last deglaciation from 14,000 BP to present (Collina-Girard 2002).

22.6 The Karst as an Archaeological Trap: Cosquer Cave

Cape Morgiou has been carved out by the sea in Urgonian limestones dipping gently towards the southeast. The entrance of Cosquer Cave opens at −37 m, at the foot of a rock cliff, and is connected to an ascending siphon, along 116 m, following the general dip (30° towards the southeast) before joining the main cavity (Fig. 22.7). The cave includes two main chambers oriented NE–SW and linked by a narrow passage. At the northern extremity, a large shaft, partially submerged, extends vertically from −25 m to about +30 m above present sea level. All these cavities are connected in their submarine parts as shown by submarine surveys. Cosquer Cave appears presently as a main prehistoric sanctuary partially preserved by the sea level rise. It is the only case, worldwide, of a prehistoric cave painting, found in such a spectacular geological context. The sea level rise preserved from destruction this "Lost World" for over 20,000 years, evoking the British novels of the Victorian period written by A. Conan Doyle or H.G. Wells.

The first occupation took place at 26,000–29,000 years BP. After a gap of 10,000 years, the second period of occupation occurred between 17,000 and 20,000 years BP (Clottes and Courtin 1996).

Engravings and paintings are widespread on all the walls but with greater density in certain spots. In most of the drawings and paintings, charcoal was used as colouring material and was therefore ^{14}C dated: 177 naturalistic animal figurations, an unrealistic human-seal figuration, 65 negative handprints, 20 undetermined figurations, 216 geometric signs and erotic art (1 phallus and 6 engraved vulvae).

22.6.1 The Older Period: Gravettian (Around 27,000 Years BP)

During this first period, the walls of the cave weathered by moonmilk are covered by erratic finger marks. These marks are found throughout the cave, even in areas difficult of access. In Cosquer Cave negative handprints were generally drawn with red clay, but in the eastern part of the "shaft" (north of the cave), charcoal was used and therefore directly dated. These handprints were made blowing colouring material on the hand placed on the walls. Some animals (horses) and geometrical signs directly finger-drawn could be contemporaneous of this early period. The representations of the older phase (with the typical negative handprints) fit perfectly with the Gravettian rock art standard, well known in the Pyrenees.

22.6.2 The Younger Period: Tardigravettian (19,200–18,500 BP)

Most of the paintings and engravings were realised during this period. The horse is the most represented animal (35.6 %; cf. Fig. 22.8), followed by Caprinae, ibex and chamois (18.6 %), bovines (bison and auroch) and Cervidae (9.6 %). Drawings of some more rarely represented animals in prehistoric rock art, such as saïga, Megaceros, elk and panther are also present. Marine animals, penguins and seals, are present and contribute to the originality of this prehistoric site. An engraving of seal could be interpreted as a shamanistic mixed representation of a human and a seal (Clottes and Courtin 1996). This early artistic phase could be compared to that of others contemporaneous sites known in the Rhone Valley with similar stylistic qualities (caves of Ardèche).

Rock art is well preserved in the upper part of the cave, but the lowest submerged galleries certainly are also ornamented. Sometimes, engravings can still be distinguished few centimetres below the present sea level. The paintings and engraving are well preserved immediately above the present sea level, which obviously proves that the Holocene sea level never exceeded the present one in Cosquer Cave (Fig. 22.8). Following the model prediction for sea level

Fig. 22.8 The panel of the little horses (18,000 years BP) partly effaced by the sea in Cosquer Cave (Photo J. Collina-Girard and French Ministry of Culture, survey 1992)

change in Mediterranean Sea, the cave entrance was completely flooded by 9,000 ± 200 radiocarbon years (between about 9,800 and 10,300 calibrated years BP according to Lambeck and Bard 2000). Moreover, two samples of littoral algal concretions, collected at the entrance of the siphon at −37 m, provided two calibrated dates: 7740–7734 years BC and 5739–551 years 4 BC, probably a little earlier for the submersion of the entrance as supposed by the model prediction (Sartoretto et al. 1995). This submersion resulting in the trapping in karst of the "hole of memory" was represented by Cosquer Cave until the first explorations of 1991. The present sea level was reached during the Late Antiquity (Morhange et al. 2001).

22.7 Conclusion

The landscape between Marseilles and Cassis is entirely carved from Urgonian limestones and controlled the faulting network of the *Massif des Calanques*. These two geological factors favoured the development of a typical karst area that was partially submerged by the postglacial marine transgression. The final touch of this landscape is the luminosity of the Mediterranean climate associated with an unstable wind field. The growth of tourism has increased pressure on the area between Marseilles and La Ciotat, which has been established as National Park in 2012 and is now protected from uncontrolled degradation.

In the future, the submerged karst off Marseilles could be the ideal place for testing adaptations to deep-sea technologies developed by the COMEX Society and space technologies necessary for the colonisation of the Moon and Mars (Weiss et al. 2012).

References

Blanc JJ, Monteau R (1988) Le karst du Massif des Calanques (Marseille-Cassis). Karstologia 11–12:17–24

Cavalera TH, Gilli E, Mamindy-Pajany Y, Marmier N (2010) Mechanism of salt contamination of karstic springs related to the Messinian Deep Stage. The speleological model of Port Miou (France). Geodinamica Acta 23(1–3):15–28

Clottes J, Courtin J (1996) The cave beneath the sea: paleolithic images at Cosquer. Harry N. Abrams, Inc., New York (English translation by Marilyn Garner)

Clottes J, Courtin J, Vanrel L (2005) Cosquer redécouvert. Le Seuil, Paris

Collina-Girard J (1992) Présentation d'une carte bathymétrique au 1/25 000 du Précontinent marseillais (au large de la zone limitée par la grotte Cosquer et l'habitat préhistorique de Carry-le-Rouet). Géol Méditerr XIX(1):77–87

Collina-Girard J (1996) Bassin de Marseille, topographie du plateau continental. ANDRA-UNESCO, IUGS, EPGC (eds). I.G.N., Aix-en-Provence/Paris

Collina-Girard J (2002) Underwater mapping of Late Quaternary submerged shorelines in the Western Mediterranean Sea and the Caribbean Sea. Quat Int 92:63–72

Collina-Girard J (2004) Prehistory and coastal karstic area: Cosquer Cave and the "calanques" of Marseilles. Speleogenesis, The Virtual

Sci J, pp 1–13. http://www.speleogenesis.info/archive/publication.php?Type=publication&PubID=3265

Collina-Girard J (2012) La Provence immergée, Plongées à Marseille et ses abords

Dupire S (1985) Etude cartographique au 1/25 000° de la zone sud du Bassin de Marseille: les travertins de Marseille. Aperçu géomorphologique et néotectonique, Thèse Université de Provence (Aix-Marseille 1), 104 p

Julian M, Nicod J (1989) Les karsts des Alpes du Sud et de Provence. Z Geomorph NF Suppl Bd 75:1–48

Lambeck K, Bard E (2000) Sea-level change along the French Mediterranean coast for the past 30 000 years. Earth Planet Sci Lett 175:203–222

Mocochain L, Clauzon, G (2010) Spéléogenèse et crise de salinité messinienne de Mediterranée. In: Audra PH (ed) Grottes et karsts de France, vol 19, Karstologia Mémoires. Association française de karstologie, Nimes, pp 52–53. ISBN: 2-9504222-55

Morhange C, Laborel J, Hesnard A (2001) Changes of relative sea level during the past 5000 years in the ancient harbor of Marseilles, Southern France. Palaeogeogr Palaeoclimatol Palaeoecol 166: 319–329

Nicod J (1967) Recherches morphologiques en Basse-Provence calcaire. Editions Ophrys, Gap, 557 p

Sartoretto S, Collina-Girard J, Laborel J, Mohrange C (1995) Quand la grotte Cosquer a-t-elle été fermée par la montée des eaux? Méditerranée 3–4:21–24

Sartoretto S, Verlaque M, Laborel J (1996) Age of settlement and accumulation rate of submarine coralligene (−10m to −60m) of the northwestern Mediterranean sea, relation to Holocene rise in sea level. Mar Geol 10:317–331

Weiss P, Gardette B, Chirié B, Collina-Girard J, Delauze H-G (2012) Simulation and preparation of surface EVA in reduced gravity at the Marseilles Bay subsea analogue sites. Planet Space Sci t. 74(1): 121–134

Scenic Granitic Landscapes of Corsica

23

Charles Le Coeur

Abstract

Corsica appears as "a mountain into the sea". The high ridge is nowhere far from the coast and is shaped into steep slopes resulting from Tertiary deep incision along fault lines. Selective denudation has dug into granitic Corsican basement rocks, intruded by large plutonic bodies. Granite landforms offer a large set of mesoscale features and rock slopes carved by differential surface weathering. High mountain landscapes display Quaternary glacial and periglacial features, whereas lower hills and basins show evidence of long-term weathering. Scenic landforms are seen in the ring complex areas where selective erosion could either dig out easily weathered rocks (Porto basin) or leave salient high towers (Bavella Needles). At a local scale, contrasted slope patterns appear such as impressive granite slabs dominating Mediterranean scrubland and extensive saprolite-mantled basins with large boulders and castellated tors (Tavaro basin). Surface weathering processes have carved numerous microforms such as grooves, flutes, *gnammas* and tafoni either on large slabs or on boulders. Early human settlements are nested into granitic lowlands (Filitosa) where water is abundant and deep soils allow cultivation, whereas hills were partly deforested and are currently covered with extensive scrubland.

Keywords

Ring structure • Mesoscale features • Selective erosion • Surface weathering • Tafoni

23.1 Introduction

Corsica offers a large set of granitic landscapes. Large slope features and microforms carved into granitic boulders are located within the mesoscale features which result from differential denudation. These landforms offer spectacular landscapes combining bare summits, large hills and deep gorges. Rock slabs dominate irregular slopes covered by pine forests or large bush areas (Fig. 23.1).

Corsica was defined by F. Ratzel as "a mountain into the sea". The mountain ridge culminates between 2,700 and 2,000 m and is located not further than 45 km from the sea.

It is why coastal landscapes are quite close to glacierised high hill slope sceneries in a Mediterranean environment. The western coast is shaped into huge mountainous promontories framing large bays.

The island is shaped along a high granitic ridge 120 km long stretched from NNW to SSE between Monte Cinto (3,007 m) and Monte Includine (2,134 m) and the high towers of Bavella (1,855 m). This backbone is cut by a set of NE-SE trending deep valleys that leads towards several high passes.

C. Le Coeur (✉)
Laboratory of Physical Geography (CNRS-Meudon),
University of Paris 1, Paris, France
e-mail: charles.lecoeur@cnrs-bellevue.fr

M. Fort and M.-F. André (eds.), *Landscapes and Landforms of France*, World Geomorphological Landscapes,
DOI 10.1007/978-94-007-7022-5_23, © Springer Science+Business Media Dordrecht 2014

Fig. 23.1 The granitic crests of Bavella Needles dominating the Laricio pine forest (Photo V. Lucchi)

23.2 Geological Setting and Geomorphological Evolution

Corsica corresponds to a large granitic Hercynian basement over which Alpine metamorphic series were thrusted on westwards during Alpine orogeny. This pattern results from Oligo-Miocene anticlockwise rotation movement on the Corso-Sardinian block and Apennines compressions. Contrasting with the western granitic Corsica, the eastern part of the island granite is shaped into metamorphic Alpine series culminating at 1,307 m as in the narrow Cap Corse peninsula and at 1,285 m in the Castagniccia massif. A narrow sedimentary aggradation plain (less than 12 km wide) spreads along the eastern coast (Gauthier 2006) (Fig. 23.2).

Granitic basement is Hercynian in age (350–300 My). It is made of granodioritic series including large NE-SW elongated large subalkaline granitic bodies. Into this framework, Permian volcanism (250 My) produced a huge lava pile that still outcrops in Scandola peninsula. Permian hypovolcanic complexes were intruded at different levels into the granitic basement (Bonin 1982):

1. Monte Cinto granitic pluton framed by rhyolitic ring complex corresponds to the higher summit, on the NW part of the Corsican backbone.
2. Porto-Otta complex is excavated into a narrow basin on the western coast.

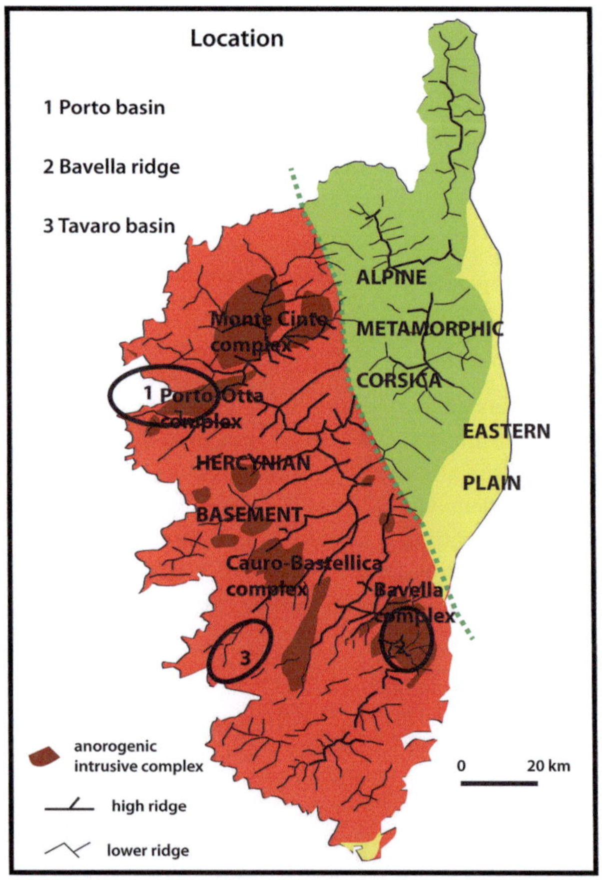

Fig. 23.2 Geological schematic map and location of study areas

3. Cauro-Bastellica ring complex offers a large set of selective denudation landforms dissected by Prunelli river.
4. Bavella ring is shaped into high pinnacles at the southern end of the Corsican main ridge.

Smaller intrusions provide more diversity in the plateau landscapes of the granitic basement.

These alkaline plutons are either shaped into large mountainous ridges that are cut by fault-controlled deep valleys or carved into arcuate or elongated basins. Differential erosion landforms are controlled both by the structural level of the intrusive bodies and by granite lithology (Grelou-Orsini 1977).

Major landforms in western Corsica offer gently southwestward sloping high ridges (1,200–700 m) cut by deep valleys. It appears as a remnant of an uplifted and tilted erosion plateau. These large landscape features are dominated by steep culminating hills as residual mountains shaped into more resistant granite (Rondeau 1962). New data from apatite fission tracks (Danišík et al. 2007) and 10Be on granite surfaces let relate the upper extensive planation level to the Early Miocene period (Kuhlemann et al. 2005). Late Miocene faulting and tilting due to Alpine compressive tectonic is responsible for major uplift. A lower planation surface was developed later. It results from extensive chemical weathering (Kuhlemann et al. 2007). These large plateau features are dissected by large basins and deep, mainly fault-controlled river valleys, in response to Late Tertiary uplift of the island and to the Messinian sea level lowering.

Three examples of granitic landforms illustrate the diversity of Corsican landscapes seen at different scales: Porto-Otta granitic basin and the Calanches of Piana, Bavella Needles and Tavaro basin with conspicuous remnants of early human settlements in granitic boulders.

23.3 Porto-Otta Valley and the Calanches of Piana

Porto-Otta valley belongs to the Scandola-Porto natural reserve which is inscribed since 1983 on the UNESCO World Heritage List. The valley itself is deeply carved into an elongated Permian intrusive body (Fig. 23.3). It is surrounded by a couple of high arcuate mountains (1,200 m). The southern hills extend westwards into a 500–600 m high ridge hanging directly above the sea through steep rocky slopes, the Calanches of Piana. This area displays spectacular sceneries both on mesoscale denudation features and on microforms chiselled into red granite.

Major landscape features result from deep erosion affecting a ring complex intruded south of the high level Cinto complex. The Porto-Otta pluton was intruded into the regional calco-alkaline rock (Evisa granite). It offers evidence of emplacement with repeated pulses: an outer body (pink granite) shows resistant microgranite chilled margins and encloses a large screen of country rock. The inner body is composed of biotite granite and includes larger gabbroic bodies in the lower part (Platevoet 1992). Basic rock outcrops are visible in the valley bottom and in the low benches of the eastern part of the basin.

The surrounding hills appear as a large granitic cupola excavated by selective erosion: high slopes are shaped into huge convex slabs above steep slopes dominating screes with scattered scrubs. These steep rock walls correspond to more resistant lines associated to intrusive discontinuities and to hardened country rock screens incorporated into the pluton. These high rock walls display long flutes carved by water running down the convex slabs. Lower slopes are gentler and covered by pine forest. This concave part leads to a ledge (120–200 m), remnant of the basin floor carved into the easily weathered inner granite. Several steam sections and some pits show sandy weathered granite embedding round boulders

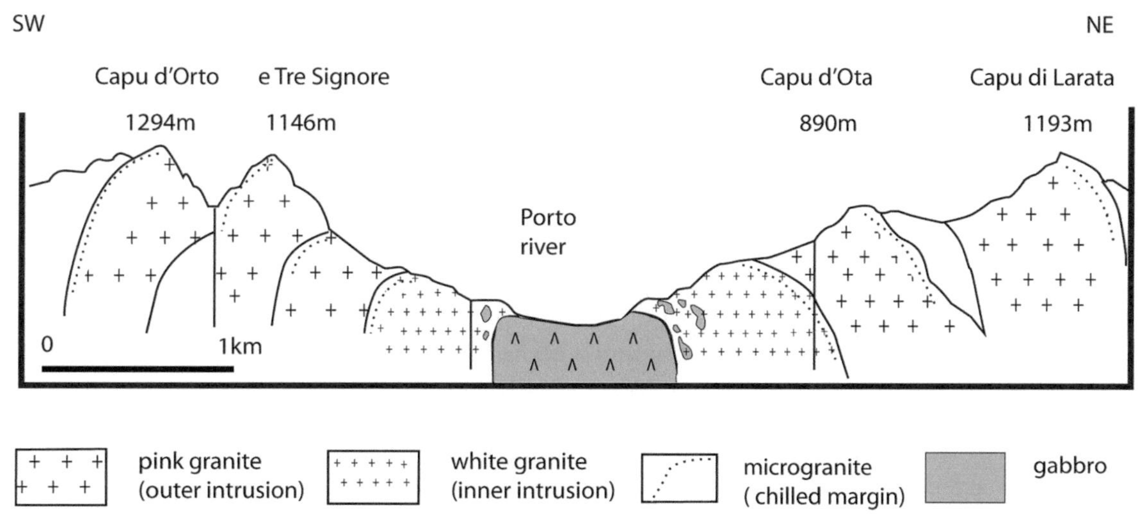

Fig. 23.3 Section across the Porto-Otta basin (Photo C. Le Coeur)

Fig. 23.4 Calanches of Piana: granite tower over the Gulf of Porto chiselled with tafoni resulting from surface weathering and intense salt disintegration (Photo C. Legac, with permission)

between large gabbroic inclusions. This bench, where the Otta village is located, is cut by several downslope gullies, and it can be seen on both sides of the valley, gently sloping westwards from 300 to 120 m. Steep slopes overlook the Porto river gorge. The river itself flows out from a 400 m deep canyon (Spelunca gorge) cut into the granitic Evisa plateau (800–900 m). A nice walk from this perched village offers dramatic sceneries of Mediterranean forest growing on slabs and pinnacles overhanging the gorge, down to the river. Fluvial erosion has cut across some rock bars and dug numerous potholes in the river bed. Downstream the valley still flows into a narrow channel, flowing across large boulders and gravel banks. The lower part of the valley widens towards the coast into an alluvial fan 1 km long. Coarse material carried by the stream is accumulated into the large pebble beach of Marine de Porto.

The southern mountainous ridge enclosing the Porto valley extends westwards into the magnificent Calanches of Piana. The elevations decrease from 900 to 600 m with impressive slopes of red granite dominating directly the bay of Porto. The crest splits into two asymmetrical north-facing scarps, controlled both by intrusive granite cupolas and by fault lines. Steep rocky slopes and high cliffs shaped into red granite offer the most stunning landscapes on the island. Tors, boulders and pinnacles of coarse-grained red granite and rhyolite are hanging above the sea (Fig. 23.4).

This castellated landscape is dissected by narrow gullies sheltering scattered pine trees among bushes and rounded rocks. A scenic road leading uphill from Porto to Piana village swings across towers and rock buttress in a fantastic landscape, and the granite towers offering strange shapes have received various names as the castle, the dog's head, the bishop hat, the lovers, etc. Small trails lead to spectacular block chaos or to shredded huge rocks surrounded by thick bush. The nineteenth-century French writer Guy de Maupassant described them this way: "I stopped stunned in front of these amazing rocks of pink granite, 400 m high, strange, tortured, eaten into by time, bloody under the final fires of twilight and taking every form – like fantastic people out of a fairy tale, petrified by a supernatural power." These spectacular landforms are carved into well-jointed resistant red granite divided by large sets of fractures. Intense runoff erosion has removed weathered materials from fractures and transported them down to the sea. It has dissected the ridge into large rock towers, crumbling boulders. In detail, rocks are rounded or dug by large cavities called tafoni. These features result from granular disintegration operating on microfissured rock surfaces. Such surface weathering, which is controlled by water penetration into rock minerals, is more efficient on shaded rock faces where mistiness keeps humidity. It is why cavernous holes were produced by continuous wetting of bare rock surfaces mainly on the shadow face of large blocks, as overhanging or north-facing surfaces. Tafoni range from 0.5 to 3 m in diameter; their formation is increased by salt erosion on rocks exposed to strong sea winds in coastal areas.

23.4 Bavella Needles

Bavella offers spectacular landforms of high granitic needles that develop in a continuous crest as the southern stronghold of the mountainous Corsican backbone. These hills offer an impressive landscape of high rocky towers overhanging the deep valley of Solenzara River. The rocky crest is chiselled by deep gullies running straight down the valley 1,000 m below. Huge vertical granitic walls rise above large Laricio pine forest covering the lower slopes around the Bavella pass. The rocky crest corresponds to the major water divide, opposing long westward slopes and steep eastern side. This asymmetrical relief arrangement induces strong contrasts in scenery.

From west, this castellated ridge can be seen above the Zonza plateau (700–800 m) dissected into long backbone hills by deep valleys flowing down into the 30 km long Rizzanese watershed. On the east side, Solenzara River plunges within 15 km 1,200 m down into the Tyrrhenian Sea. In the northern part of the crest, the Needles emerge from the high smooth ridge of Includine mountain (2,134 m). The high crest is arranged as an east-facing crescent cut by the deep NE-SW fault-line valley (Fig. 23.5a).

These spectacular hills are shaped into a granitic ring complex intruded into the Hercynian Corsican basement. The ring structure consists of alkaline granites that are very

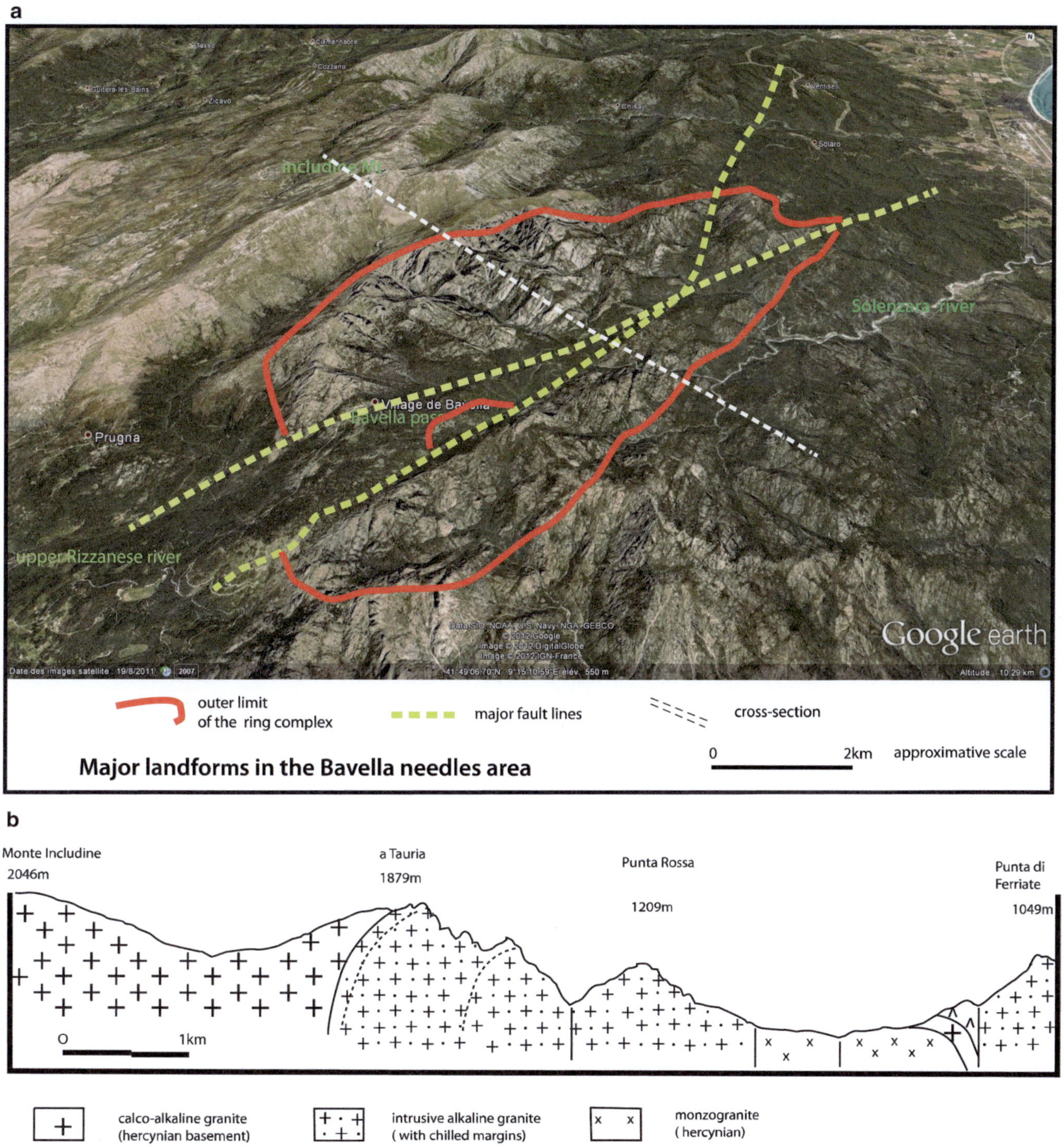

Fig. 23.5 The Bavella ring complex: general geomorphological arrangement (**a**) and cross-section across Bavella Needles massif (**b**)

resistant among calco-alkaline and monzogranite country rocks. Selective erosion dug deeply into faulted and less resistant material corresponding to the upper Solenzara valley. This basin dug into monzogranite is closed by a large pink granite rock bar (Ferriate ridge, Punta Ornucciu and Monte Rossu) which corresponds to the eastern part of the Bavella ring structure. Two streams flow through 250 m deep gorges into Solenzara River.

The Bavella Needles appear as 400 m high rocky towers which are carved in poorly jointed pink microgranite that corresponds to the chilled margin of the pluton (Fig. 23.5b). Fine-grained rocks with drusy cavities are shaped into huge pinnacles on the surface of which surface weathering processes could carve numerous small cavities. Main slopes on the inner side of the ridge offer large sets of convex slabs (50–100 m high) showing long erosion grooves. The different summits are separated by

Fig. 23.6 Culminant periglacial landscape of Bavella Needles: high granite towers are cut by gullies partly filled with scattered blocks resulting from Quaternary frost activity (Photo C. Le Coeur)

deep fracture-controlled gullies partly covered by screes. Narrow pine forest spots are nested on benches separating slabs and on the lower talus slopes. In the basin the slopes are moderate and mainly forested. Downstream of the valley, some stream sections are cut into thick regolith, and numerous core stones supply important sediment to the river.

This spectacular landscape results from two successive styles of evolution.

Late Tertiary differential denudation left the outer ring prominent and dug out a large weathering basin in the inner part of the complex. Asymmetrical uplift of the island induced important valley deepening in the eastern flank.

The Quaternary cold periods induced strong periglacial slope processes (Fig. 23.6). The Bavella ridge was not high enough to feed any glacier: only one small glacier was nested on the eastern side of Includine mountain, from 2,000 to 1,700 m (Conchon 1986), but was covered by large snow accumulations feeding avalanche gullies. It suffered from intense frost action along the fracture lines cutting the ring structure. The massive rock towers remained isolated among a set of E-W and SW-NE gullies. Gaps in the high ridge are covered by remnant coarse sand and scattered blocks sloping down to steep scree strips. Therefore the crescent ridge has been dissected by intense mechanical weathering and efficient debris removal on the eastern steep slopes down to the Solenzara valley.

23.5 Filitosa and the Lower Tavaro Basin

The Tavaro basin is located in the southern part of the large peninsula separating the Gulf of Ajaccio from Propriano bay, on the SW side of the island (Fig. 23.7). Tavaro river flows from the hills of central Corsica into a deep valley cut into the high plateau (1,200 m), and it merges into a low lying basin 12 km long and 4 km wide before reaching the sea. This lowland is framed by two NE-SW low ridges (400–700 m), remnant of a Tertiary erosion surface uplifted and tilted to the SW. On the western side, a long backbone (500–700 m) is capped by large tors and rock slabs corresponding to resistant leucogranitic bedrock, reinforced by numerous aplitic dykes; on the east side round hills (500–700 m) are shaped into monzogranites and gabbro-dioritic bodies alongside leucogranitic mountains (Monte San Petru 1,400 m).

The Tavaro basin has been excavated into easily weathered coarse-grained granodiorites of the Hercynian basement during the Late Tertiary (Marre and Rossi 1981). Two lines of crushed granite frame the low area above which a few small round hills correspond to minor gabbro and diorite intrusions. The landscape offers gentle slopes, a large alluvial plain and marshes along the shore. Convex mounds offer evidence of deep weathering mantle with numerous large granite boulders packed into sandy regolith, and castellated blocks carved with tafoni protrude from dense vegetation at places (Gauthier 2006) (Fig. 23.8).

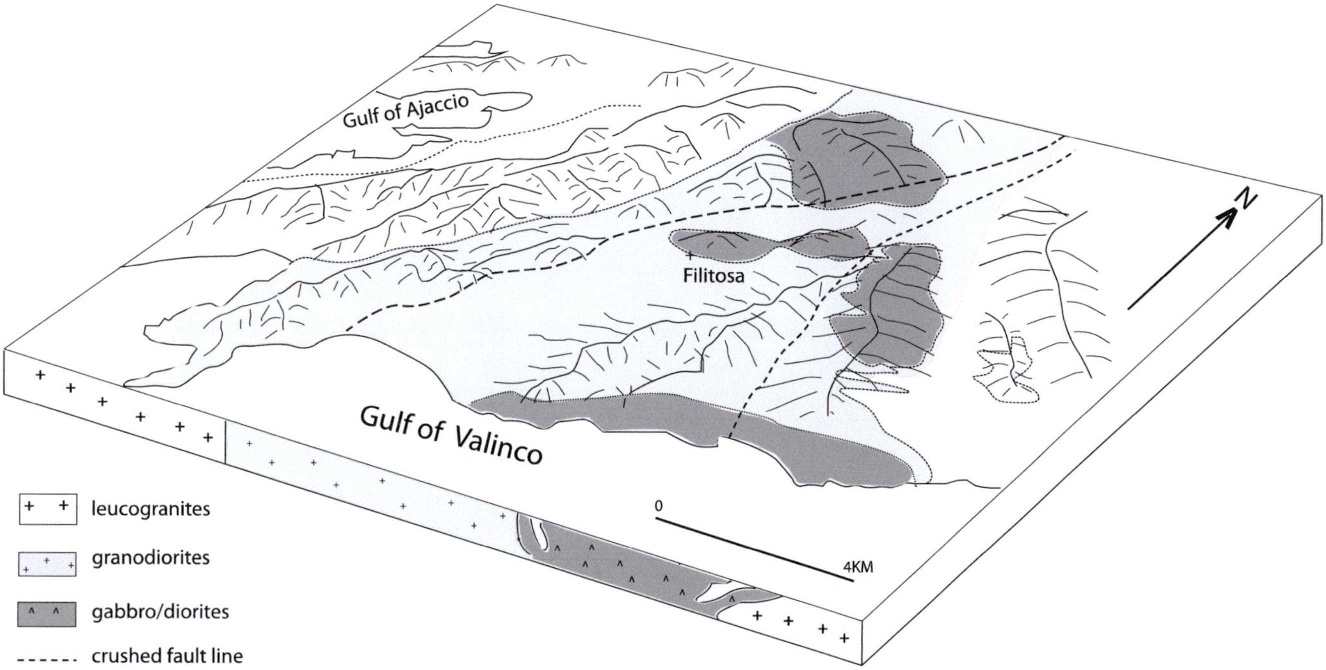

Fig. 23.7 The Tavaro basin (© C. Le Coeur)

Fig. 23.8 Granodioritic castellated blocks carved with tafoni are protruding from dense scrubland dominating the Tavaro lower valley (Photo V. Lucchi)

Fig. 23.9 The Filitosa megalithic site: Late Bronze Age settlement displaying anthropomorphic megaliths carved in weatherable granodiorite (Photos V. Lucchi)

This geomorphological landscape with deep soils and abundant water could provide favourable conditions for early human settlements in Corsica. Several megalith sites offer evidence of Mesolithic (2000–1500 BC) to Bronze Age (1500–1000 BC) dwellings nested among large blocks either on small hilltops or downslope. Filitosa, which is the most famous site in this area, offers a large collection of standing stones showing anthropomorphic sculptures (Fig. 23.9).

The megalithic sculptors used easy to carve core stones of crumbly granodiorite. It is why some sculptured stones suffered from present-day surface disintegration. The older megaliths,

2–3 m high, showing large human faces are related to the Early Bronze Age. Later standing stones were erected as a circle around a small mound and sculptured with distinct heads and arms-bearing bodies. It can be associated to the Late Bronze Age when *Torrean* people built defensive forts. Traces left by of stone masons and remains of a supposed sanctuary can be seen among large boulders in the scrubland. Some large tafoni were used as dwellings or sheepfold in a large open landscape with scattered old trees. Thus early populations seem to have had some knowledge of the granitic landform diversity and natural resource in Mediterranean environments.

23.6 Conclusion

These different examples of Corsican granitic landforms offer similar feature collections: high hill slopes shaped into rock slabs caped with massive tor in more resistant granites, deep valleys passing from narrow gorges to wider basins carved into less resistant rocks and microform-rich slope feature offering various tors and boulders carved by grooves, *gnammas* and tafoni scattered in extensive scrubland. Corsican ring complexes offer mesoscale landforms alike circular hill arrangements seen in the Permian intrusion in the Oslo Graben or in the Tertiary ring complex in western Scotland. Granite slopes show various scenic rock features such as tor and large slabs resulting from thick regolith removal during valley deepening sequences. Intense surface weathering under Mediterranean climatic conditions has produced a large set of microforms similar as the granitic rock features seen in Sardinian hills or in numerous granite hills in Aegean Sea.

References

Bonin B (1982) Ring complex granites and anorogenic magmatism. Studies in Geology. North Oxford Academic Publisher, Oxford

Conchon O (1986) Quaternary glaciations in Corsica. In: Sibrava V, Bowen DQ, Richmond GM (eds) Quaternary glaciations in the Northern Hemisphere. Quat Sci Rev 5:429–432

Danišík M, Kuhlemann J, Dunkl I, Székely B, Frisch W (2007) Burial and exhumation of Corsica (France) in the light of fission track data. Tectonics 26:1001–1029

Gauthier A (2006) Des roches, des paysages et des hommes. Géologie de la Corse, Éd. Albiana, Ajaccio

Grelou-Orsini C (1977) Données nouvelles sur les granites de Corse: géologie et érosion différentielle. Méditerranée 28(1): 35–44

Kuhlemann J, Székely B, Dunkl I, Molna G, Tima G, Frisch W (2005) DEM analysis of mountainous relief in a crystalline basement block, Cenozoic relief generations in Corsica (France). Z Geomorph NF 49:1–21

Kuhlemann J, van der Borg K, Bons P, Danišík M, Frisch W (2007) Erosion rates on subalpine paleosurfaces in the western Mediterranean by in-situ 10Be concentrations in granites: implications for surface processes and long-term landscape evolution in Corsica (France). Int J Earth Sci 97(3):549–564

Marre J, Rossi P (1981) Architecture d'un secteur du batholite corso-sarde: les régions d'Ajaccio et Sartène. Bull BRGM 2(1–2):153–167

Platevoet B (1992) Mélanges magmatiques dans le complexe annulaire de Porto, Corse. CR Ac Sc Paris 294:907–910

Rondeau A (1962) Recherches géomorphologiques en Corse. These d'Etat, University of Paris, A. Colin, Paris

Exceptional Coral Reef and Lagoon Geomorphology in French Polynesia

24

Samuel Etienne

Abstract

Tahiti and her islands were exemplified by C. Darwin to support his theory on coral reef evolution around oceanic islands. The Austral Isles, nearby Tahiti, offer an illustration of this theory – now explained by the plate tectonic theory – but also present singular deviation due to the presence of several hotspots leading to localized uplifting. Volcanic geomorphology and karst development on uplifted coral reef are then peculiar ingredients of the archipelago and deserve a close outlook. High-energy events, such as tropical cyclones, are an essential forcing agent of morphogenesis. Finally, reefal bioerosion adds its own flavour to offer a unique Polynesian landscape.

Keywords

Oceanic islands • Pacific Ocean • Volcano-karstic landforms • Coral reefs • Tropical cyclone • Biogeomorphology

24.1 Introduction

French Polynesia is a vast territory located in the South Pacific Ocean, comprising five archipelagos and 121 islands. Emerged lands represent less than 3,500 km^2 disseminated over 5.5×10^6 km^2 of deep ocean (exclusive economic zone). Tahiti, the largest and most populated island, covers 1,042 km^2 (Collectif 1993). Polynesian islands can be classified in two physiographic categories: high volcanic islands with steep slopes (Fig. 24.1) and low-lying islands (atolls), mainly represented in the Tuamotu Archipelago where land levels rarely exceed 10 m a.s.l. In some cases, uplifted atolls lead to the appearance of high limestone islands fringed with calcareous cliffs with altitudes reaching 100–150 m (e.g. Makatea, Rurutu). Both island categories and intermediate stages are genetically linked as they represent one evolutionary process from the slowly emerging submarine volcano (seamount stage) to the almost drowned island (atoll stage), illustrating the theory proposed by Charles Darwin after his voyage in the South Pacific (including a visit to Tahiti and her islands) on the *HMS Beagle*.

During this plate tectonic-driven evolution, coral reefs appear on the volcano shores and evolve with typical morphologies from fringing reef to barrier reef and ultimately atoll (Fig. 24.2). Within this scheme, the Austral archipelago (Fig. 24.3) is quite exceptional: at first sight, it illustrates properly the Darwin's theory, but in the details, it also presents unusual and interesting geomorphic singularities. The Austral Isles, the least visited archipelago of French Polynesia, are those which offer the greatest diversity in terms of geomorphological landscape, and they really deserve a focus (and a visit!).

24.2 Geographical and Geological Settings

The Austral Isles are the southernmost archipelago of French Polynesia, located between 22.4°S and 27.7°S and 144°W and 153°W. It is composed of five oceanic islands roughly aligned along a N110 direction with a broadly increasing age towards the north-west. Tubuai and Rurutu are the nearest

S. Etienne (✉)
Ecole Pratique des Hautes-Etudes, Labex CORAIL, Paris, France
e-mail: samuel.etienne@ephe.sorbonne.fr

M. Fort and M.-F. André (eds.), *Landscapes and Landforms of France*, World Geomorphological Landscapes,
DOI 10.1007/978-94-007-7022-5_24, © Springer Science+Business Media Dordrecht 2014

Fig. 24.1 Bora Bora Island and her blue lagoon (Photo S. Etienne). Mount Otemanu (727 m) is made of metre-thick lava flows with an overwhelming alkali basalt composition. Its particular molar shape is the result of intense erosion of the 3–4-Ma-old shield volcano

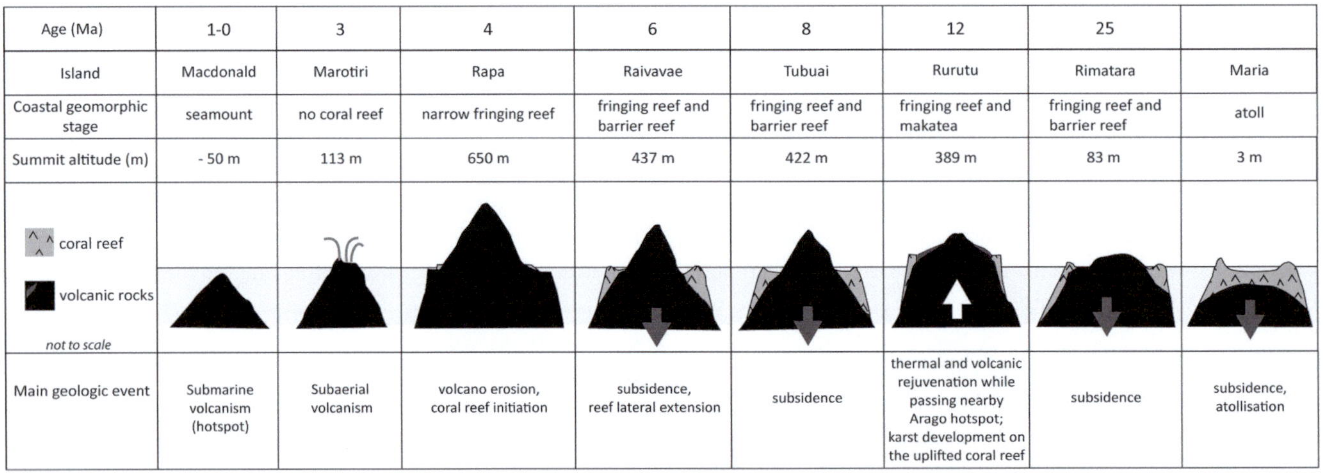

Fig. 24.2 Coral reef development on oceanic volcanoes in relation to hotspot activity and tectonic plate drifting. Example from the Austral archipelago (Modified from Mottay 1976)

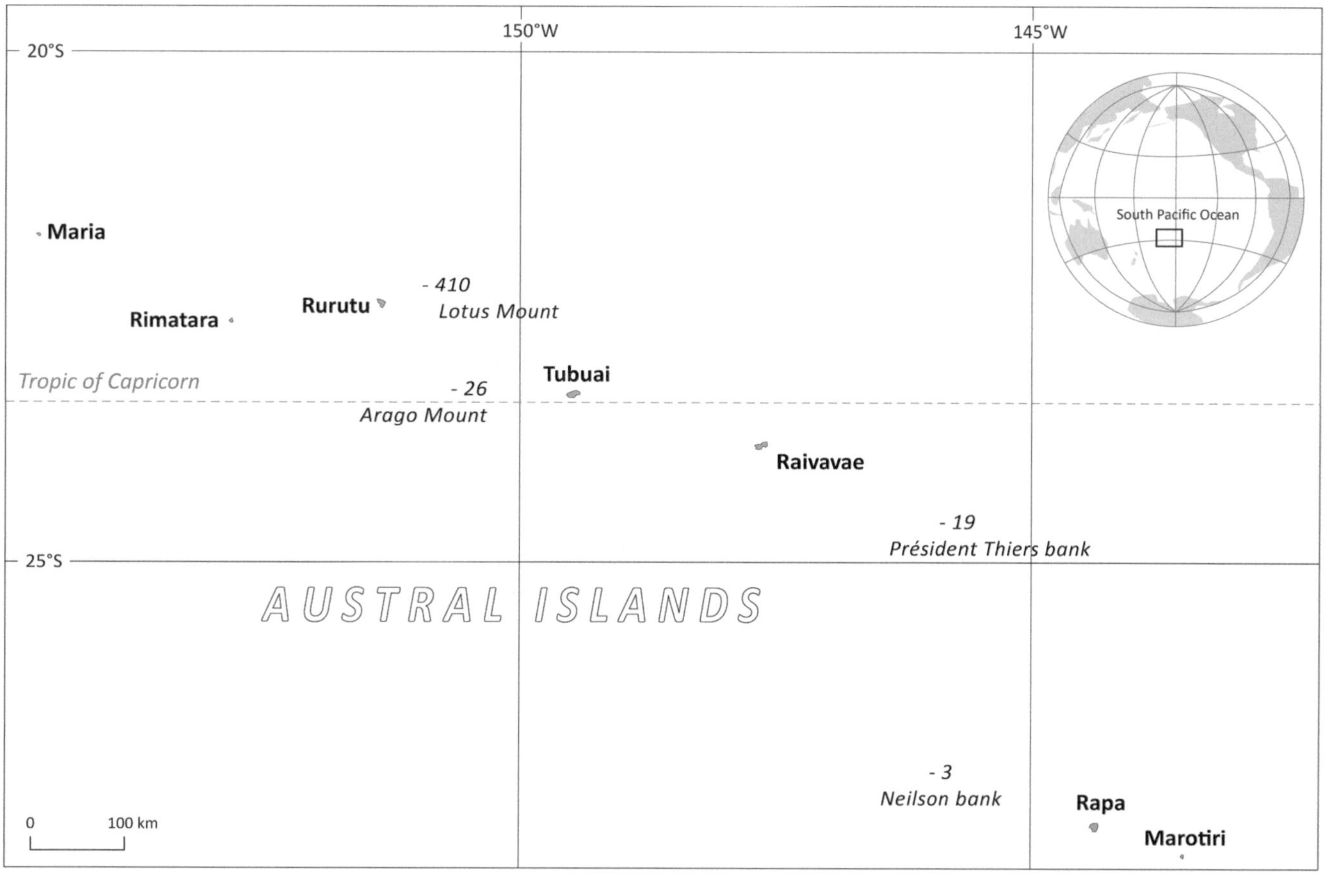

Fig. 24.3 Location map of the Austral Islands, French Polynesia

Austral Islands from Tahiti (c. 600 km). The Austral Isles have an oceanic subtropical climate with annual precipitation around 2,000 mm and mean monthly temperature ranging from 18 to 23 °C. These volcanic islands were active while fed by the MacDonald hotspot and underwent downward movements after disconnection from the hotspot and subsequent thermotectonic subsidence of the seafloor. The subsidence sequence, characterized by the erosion of the volcano and the coral reef building, was disrupted by the rejuvenation of some islands when passing a nearby second hotspot: this phenomenon heated the oceanic plate, led to localized uplift (and, sometimes, new volcanic activity) and the emergence of the whole coral reef (e.g. Rurutu and Rimatara). Tubuai Island is expected to undergo the same uplift – in a few million years – as it will come closer to the recently discovered Arago hotspot (Bonneville et al. 2002). The coral platforms, especially those inherited from Holocene sea-level highstand, support the coastal plain which is the main area of settlement in high islands. Elevation of coastal plains is very low, almost always below 10 m a.s.l.

Tubuai Island (45 km^2) is a shield volcano formed between 10 and 8.8 Ma by the accumulation of alkali basalt and basanitic lavas, followed by nephelinitic and tephritic flows and finally intrusion of phonolitic plugs. The highest summit reaches 422 m a.s.l. (Taitaa). In the south-west, the Mounts Tavaetu (327 m), Hanareho (324 m) and Mareura (327 m), three intrusive nephelinitic plugs, draw a small chain called by locals *La Femme couchée* (The Lying Woman) due to its anthropomorphic profile when seen from the lagoon (Fig. 24.4). The coastal plain is filled with Holocene bioclastic sands derived from the reef barrier and the lagoon.

Rurutu Island is also a small (38 km^2) volcanic island but presents a more complex geological history and offers a richer variety of geomorphic landscapes, including unusual karst landforms. The first phase of evolution (13–10.5 Ma) was very similar to Tubuai: a submarine volcano slowly emerged over the sea level, followed by the accumulation of subaerial alkali basalt flows building a shield volcano. During the next four million years, coral reef developed all around the coastline (10.5–6.5 Ma), building a massive coral limestone unit characterized by wackestone and packstone textures. Approaching the Arago hotspot, the whole island was uplifted (thermal rejuvenation) prior to the onset of a new volcanic phase with basanitic and hawaitic lava flows (1.1–0.6 Ma). This particular geological history, involving both uplift of the coral reef and volcanism, has intrigued scientists for centuries and adds a valuable cultural dimension to the site. Rurutu was discovered by Europeans in August

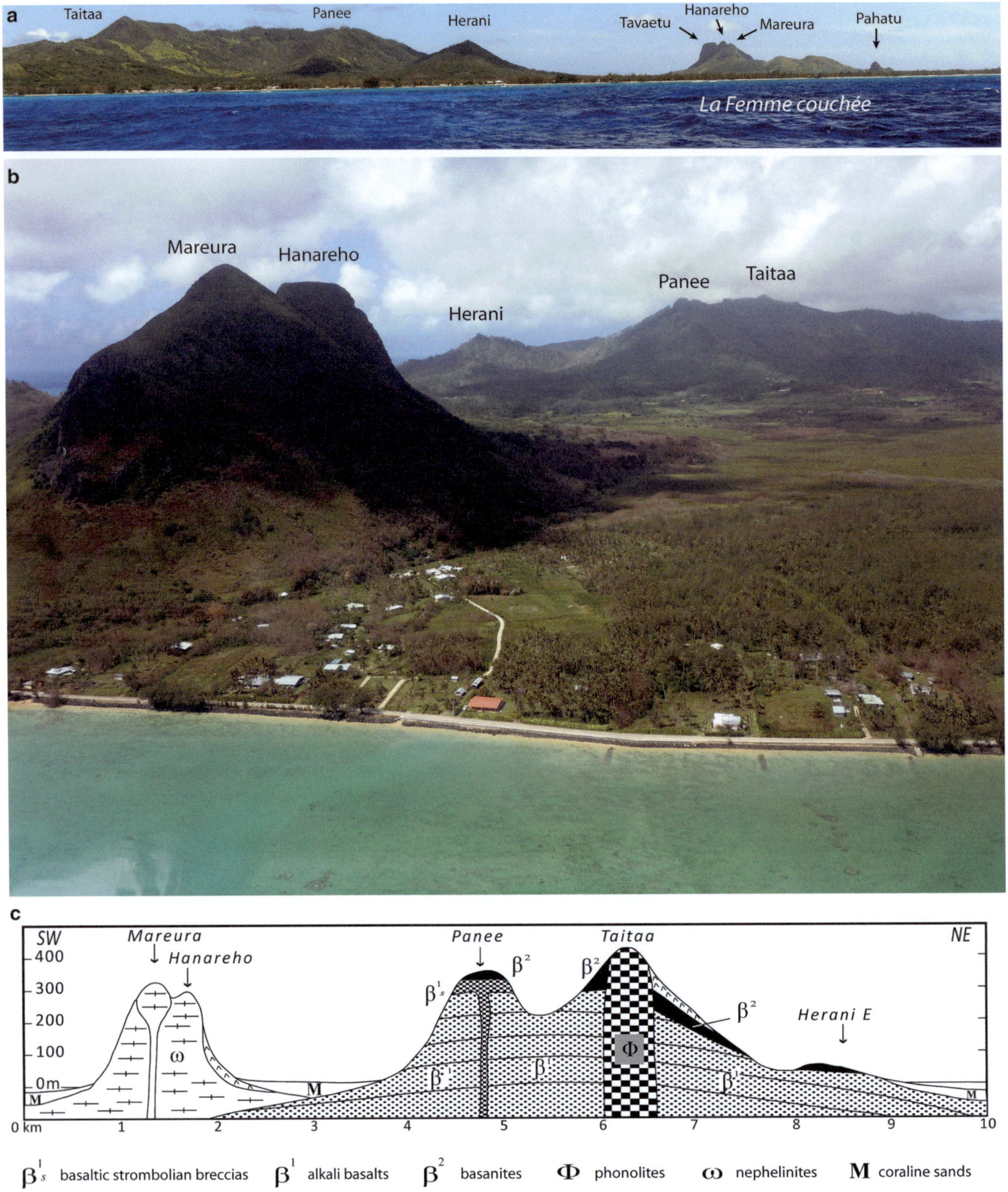

Fig. 24.4 Volcanic landforms of Tubuai Island. Eroded shield volcano and exposed intrusive plugs (Photos S. Etienne). (**a**) North-south profile of the island seen from the lagoon. (**b**) Aerial view towards the NE: coastal plain and necks. (**c**) Sketch of the volcanic structures (Redrawn from Maury et al. 2000)

1769 during James Cook's first voyage on the *HMS Endeavour*. It became well known by geoscientists for its uplifted reefs when D. Tyerman and G. Bennet (1840) related their surprise to '*perceive that it* [the plateau] *was a coral mass, rising to the perpendicular height of two hundreds feet above the beach.* [...] *The unanswerable question naturally arises. Was the level of the sea, at any remote period, above these formations, as the coral insects are never known to work upward beyond high-water mark? Or have these prodigious fragments of animal labours been heaved from their ocean foundations, by some convulsion of nature, which has left them in situations where they never could have been raised by their minute architects, in the ordinary course of providence?*' Rurutu Island was then mentioned by C. Darwin (1842) who had not visited it personally and later by the coral reef specialist J.D. Dana (1853) who recognized the uplifted nature of the island (p. 129). Later, W.M. Davis considered in his *Coral Reef Problem* (1928) the issue of elevated reef, referring to the work of Chubb (1927). Davis did not appreciate new insights into the erosional development of raised reefs which led to '*a vigorous renewal of controversy over the coral reef problem in which the conclusions reached differed substantially from those of Davis*', as stated by Stoddart and Spencer (1987), but he admitted that the elevated barrier reefs of '*Mangaia and Rurutu have been well enough studied to demonstrate that the conditions and processes of their origin confirm the postulates of Darwin's theory*' (Davis 1928, p. 410).

More recently, the Austral Isles became famous among geochemists when Vidal et al. (1984) showed that lavas from Tubuai have a very specific isotopic composition with strong radiogenic Pb and low radiogenic Sr and Nd. Then, studies on Rurutu and Mangaia (Cook Islands) revealed the same trend and allowed the definition of a new mantle pole known as HIMU (for high μ where $\mu = {}^{238}U/{}^{204}Pb$). HIMU is one of the main geochemical mantle components that give rise to oceanic island basalts. The HIMU component is thought to be recycled oceanic crust that lost part of its Pb through hydrothermal processes prior to and during subduction (Chauvel et al. 1992). For this reason, the Austral Isles are already a geosite, but their landform variety deserves to be a geomorphosite as well.

24.3 Evolution and Landforms

Tubuai and Rurutu offer the opportunity to observe a very interesting set of geomorphic features including unusual volcano-karstic landforms, Quaternary sea-level notches, extensive beach ridges and tropical cyclone coarse deposits.

24.1.1 Makateas

A major geomorphic feature of the Austral Isles is the presence of a series of volcano-karstic islands called *makatea* (Fig. 24.5). Makateas are tablelands of Neogene limestone, representing uplifted fringing reefs, and form annular girdles surrounding volcanic island cores (Dickinson 1998). This configuration is the result of a succession of submersion and emersion of the island related to isostatic movements rather than eustatic variations. During the submersion stage, coral reef grows around the volcanic island building a fringing reef, while during the next emersion stage, humid tropical climate induces a karstification of the coral body. A makatea is a quite rare landform also encountered at Rimatara (Austral), Mitiaro, Atiu, Mauke (Cook Islands) and Niue in the Pacific Ocean.

Morhange (1990) has identified four types of makateas in the Austral-Cook with a conspicuous evolution towards the NW (Fig. 24.6): in Rurutu where the uplift movement is the most important (at least 150 m), the volcanic core dominates an irregular and discontinuous coral belt. North-westwards along the alignment of volcanic edifices, the relative importance of the coral body increases, and the height of the volcanic terrain diminishes (Mangaia, Atiu, Mauke). Ultimately, the karstic landscape dominates a central depression (Mitiaro Island). The makatea's topography fed a 50-year controversy, summarized by Stoddart and Spencer (1987), about the importance of subaerial erosion. These authors presented a convincing scenario explaining the formation of the extensive depression between volcanic and limestone bodies for the Cook Islands' makateas (Stoddart et al. 1985), but Rurutu itself appears again as another deviation from the model: the drainage pattern in Rurutu is generally away from the makatea, whereas in Mangaia, streams are directed towards the makatea, concentrate in a swampy depression and then promote solutional undercutting of the limestone bodies.

The existence of makatea series is the result of the tectonic plate drifting over two close hotspots: the first one creates the series of aligned volcanic oceanic islands, whereas the second one induces a thermal rejuvenation of the plate (and, hence, of the volcanoes and its coral crown). Volcanic reactivation might be concomitant to the uplift, as it can be observed in Rurutu where uplifted Miocene coral series are partially fossilized by Quaternary volcanic lavas. When observed more closely, the limestone cliffs are characterized by seaward dipping layers of coral (15–24°) suggesting progressive series typical of an outer reef position. Karst processes have allowed the development of an underground network of caves with speleothems which are valuable palaeoclimate proxies.

Fig. 24.5 Makatea south of Avera, Rurutu Island (Photo S. Etienne). The uplifted coral limestone body is anchored to the volcano. After the mid-Holocene highstand, the sea level declined to its present position, and a modern narrow fringing reef has developed. Cave exsurgences underline the karstification of the makatea; notches on the cliff face are remnants of Quaternary sea-level variations

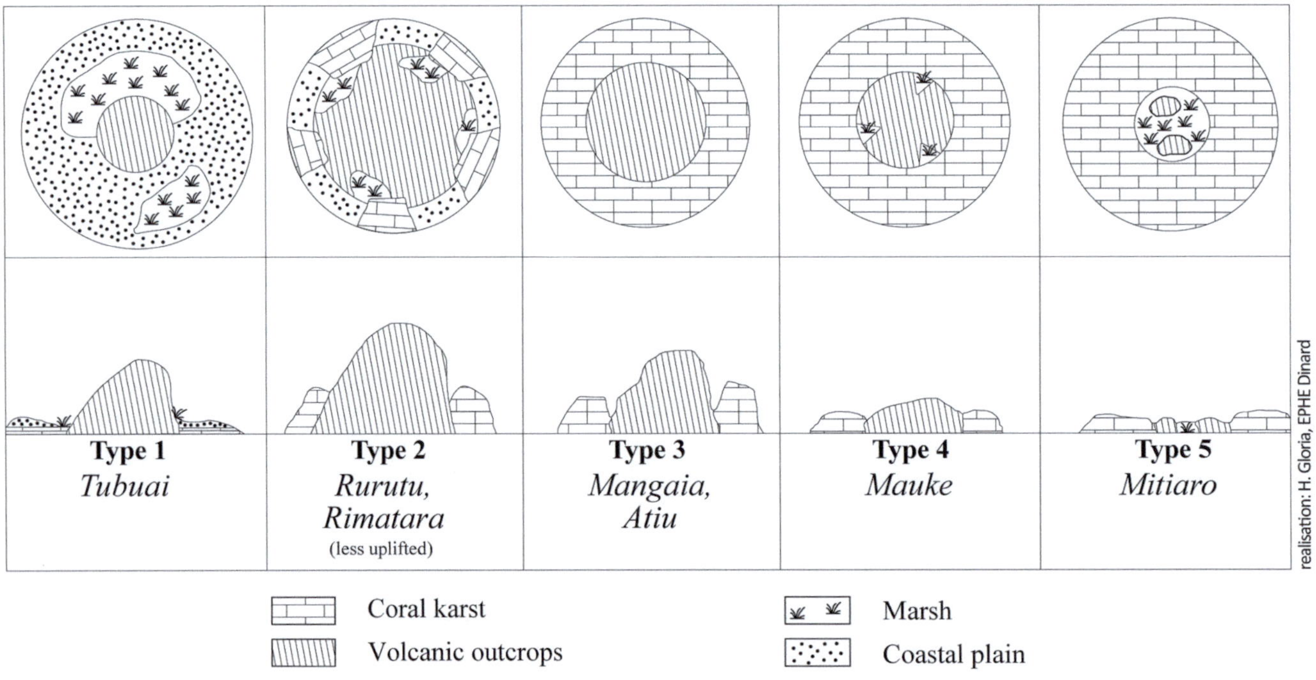

Fig. 24.6 Conceptual model of makatea's evolution (Redrawn from Morhange 1990)

24.3.2 Palaeo-Sea Levels

On the sea face, different levels of deep horizontal notches intersect the limestone cliffs. Some of them are remnants of Quaternary palaeo-sea-level notches. The most recent ones are 2–3 m high and 3–6 m deep and directly connected to reef flat on the same topographic surface. Their existence is being threatened by modern sea-level rise which contributes to the destruction of these heritages (Evelpidou et al. 2012). Tropical Pacific islands have experienced a mid-Holocene highstand (MHH) in sea level, produced by early Holocene eustatic rise followed by late Holocene hydro-isostatic decline in regional sea level (Dickinson 2003). MHH timing and duration vary over the Pacific Ocean; in the Cook-Austral Islands, it stands between 2000 and 0 BC (*ibid.*). This MHH appears as a second notch level standing between +1.2 and +1.7 m. Between 8 and 10 m a.s.l., a third level of notches appears. Coral colonies in growth position have been sampled on some notch floor and dated by Pirazzoli and Salvat (1992) between 118 and 126 ka. Thus, they represent a last interglacial (marine isotope stage 5) sea-level heritage. Notch floors are sometimes covered by cemented coral fragments, and speleothems are present inside the notch (Fig. 24.7).

Fig. 24.7 Palaeo-sea-level heritage: marine notches on a makatea cliff, north of Avera, Rurutu (Photo S. Etienne). Foreground: Slightly emerged Holocene notch (+1.2 m above present mean sea-level). Background: MIS 5e notch 8–10 m a.s.l. Karstic cave resurgence on the cliff top face (Picture taken at high tide, 20 October 2011)

24.3.3 Impacts of Tropical Cyclones on Coral Reef Geomorphology

Cyclones represent the second type of major coastal hazard in French Polynesia, especially in the Austral Isles where cyclone frequency is one event every 6–7 years. In February 2010, tropical cyclone Oli struck French Polynesia and allowed the observation of geomorphic impacts of such high-magnitude, low-frequency meteorological event (Etienne 2012). It appears that cyclones are necessary for the long-term sediment replenishment on shorelines (Fig. 24.8), but in the short term, such storms will seem to have mainly destructive effects. Observed reef damages are widespread but unevenly distributed: submarine geomorphic impacts include, in the most exposed areas, uprooting of massive (diameter >1 m) coral colonies down to 25 m b.s.l. and total destruction of branching corals down to 16 m (Tubuai); reefs not affected directly by cyclonic waves show little damage below a depth of 10 m, but sediment (sand, gravel) transfer appears important there, with clear evidence of sediment extraction from grooves and landward transport. Outer reef slope angle appears as a major control factor on coral destruction with vertical submarine cliffs relatively shielded compared to gentle slopes. On the reef flat, sandy banks (cays) and coarse sediment islets (motus) have suffered drastic changes during the cyclone passage: a cay lost nearly 50 % of its spatial extent during the event but recovered almost fully within a year. Motus appear more resistant: their geographical limits are almost unchanged on a vertical view, but changes are more important in the vertical plan where layers of coral rubble have accumulated at the surface of the loose island. Large coral boulders (c. 1 m^3) inherited from cyclone passage can be observed on the outer reef flat surface. High-energy event might have a long permanence in the landscape: in Rurutu, near Moerai, a coral boulder field lying on the reef flat appeared after the passage of TC Wasa in 1991 and is still a prominent feature of the coastal landscape.

24.3.4 Reef Biogeomorphology

Tahiti and her islands are among the best places in the world for scuba diving. Although the reef biodiversity is not as rich as in the Coral Triangle, close encounters with schools of sharks and dolphins and humpback whales have become world-famous attractions. However, smaller animals are of greater interest for geoscientists and bioscientists interested in biogeomorphology (Fig. 24.9). Carbonate sediment

Fig. 24.8 The lagoon and the coastal plain of Tubuai (Photo S. Etienne). The western plain built during the mid-Holocene highstand when bioclastic sands provided by the reef have accumulated at the base of the volcanic slopes. A succession of arcuate beach ridges, probably resulting from high-energy event, alternates with swampy areas. On the *left*, beach sand overwash can be observed along the airstrip. This is the result of the passage of Tropical Cyclone Oli, 5 days earlier (Picture taken 11 February 2010 towards east)

production due to parrotfish grazing over coral colonies ranges from 0.7 to 3.3 kg/m^2/year in Moorea lagoon, and sea urchin bioerosion is also of primary importance for sediment production. Some triggerfish build intriguing stone circles during the spawning, while hawksbill turtles overturn large quantities of coral fragment seeking for sponge to eat. All contribute greatly to the production and transfer of sediment inside the coral reef system.

24.4 Conclusions

In French Polynesia, the Austral Isles concentrate almost every type of landforms resulting from the interaction of oceanic volcanism and coral reef development. Following Darwin's theory on coral reef associated with oceanic island, reefs should evolve from the fringing stage (Rapa) to the atoll (Maria) accompanying the subsidence of the volcanic islands. In the Austral Isles, the existence of a second hotspot induces a thermal rejuvenation of the tectonic plate and the uplift of the volcano/reef bodies (Rurutu, Rimatara). The spectacular volcano-karstic landforms of Rurutu are the result of this perturbed coral reef evolution. Unusual karst landforms (makatea) have developed on the uplifted reefs sometimes associated with new volcanic activity. Sea-level variations during the Quaternary add some notch lines on the limestone cliffs, the existence of the lower ones being threatened by modern sea-level rise. Modern and ancient high-energy events contribute to shape the reef, the lagoon and the coastal plain, whereas coral bioerosion remains a key process for sediment production. The Austral archipelago is the home of important geomorphic sites to understand the evolution of oceanic islands which deserve further investigations.

Fig. 24.9 Diversity of biogeomorphic processes in coral reef systems of French Polynesia (Photos S. Etienne) (**a–d**) and M. Augeyre (**e**). Large marine animals like manta ray (**a**) attract scuba divers from all over the world, but geomorphologists might better enjoy observing coral-boring bivalve (*Tridacna crocea*) (**b**); grazing by parrotfishes using their beak-like mouth armed with jaw teeth (**c**); ocean triggerfish *Canthidermis maculata* digging coral rubble during spawning season (**d**); or hawksbill turtles *Eretmochelys imbricata* (**e**) overturning coral fragments when looking for benthic sponges

References

Bonneville A, Le Suavé R, Audin L, Clouard V, Dosso L, Gillot PY, Janney P, Jordahl K, Maamaatuaiahutapu K (2002) Arago seamount: the missing hotspot found in the Austral Islands. Geology 30:1023–1026

Chauvel C, Hofmann AW, Vidal P (1992) HIMU-EM: the French Polynesian connection. Earth Planet Sci Lett 110:99–119

Chubb LJ (1927) Mangaia and Rurutu: a comparison between two Pacific islands. Geol Mag 64:518–522

Collectif (1993) Atlas de la Polynésie française. Éditions ORSTOM, Paris

Dana JD (1853) On coral reefs and islands. Putman, New York

Darwin C (1842) The structure and distribution of coral reefs. University of Arizona Press, Tucson [1984]

Davis WM (1928) The coral reef problem. American Geographical Society, New York

Dickinson WR (1998) Geomorphology and geodynamics of the Cook-Austral island-seamount chain in the South Pacific Ocean: implications for hotspots and plumes. Int Geol Rev 40:1039–1075

Dickinson WR (2003) Impact of mid-Holocene hydro-isostatic highstand in regional sea level on habitability of islands in Pacific Oceania. J Coast Res 19:489–502

Etienne S (2012) Marine inundation hazards in French Polynesia: geomorphologic impacts of tropical cyclone Oli in February 2010. In: Terry J, Goff J (eds) Natural hazards in the Asia-Pacific region: recent advances and emerging concepts, vol 361, The geol soc London spec publ., pp 21–39

Evelpidou N, Kampolis I, Pirazzoli PA, Vassilopoulos A (2012) Global sea-level rise and the disappearance of tidal notches. Glob Planet Change 92–93:248–256

Maury RC, Guille G, Guillou H, Blais S, Chauvel C, Brousse R (2000) Notice explicative, carte géologique de la France (1/50000), feuille Rurutu et Tubuai – Polynésie française (1154). BRGM, Orléans

Morhange C (1990) Mise en évidence d'une déformation de la lithosphère à partir de l'étude quantitative de l'évolution géomorphologique des îles Australes et Cook du Sud en Polynésie. Rev Géomorph Dyn 1:1–13

Mottay G (1976) Contribution à l'étude géologique de la Polynésie française, archipel des Australes, Mehetia. Unpublished PhD thesis, Paris XI

Pirazzoli PA, Salvat B (1992) Ancient shorelines and Quaternary vertical movements on Rurutu and Tubuai (Austral Isles, French Polynesia). Z Geom NF 36:431–451

Stoddart DR, Spencer T (1987) Rurutu reconsidered: the development of makatea topography in the Austral Islands. Atoll Res Bull 297:1–19

Stoddart DR, Spencer T, Scoffin TP (1985) Reef growth and karst erosion on Mangaia, Cook Islands: a reinterpretation. Z Geom Suppl-Bd 57:121–140

Tyerman D, Bennet G (1840) Journal of voyages and travels. London Missionary Society, New York

Vidal P, Chauvel C, Brousse R (1984) Large mantle heterogeneity beneath French Polynesia. Nature 307:536–538

La Réunion Island: A Typical Example of a Basaltic Shield Volcano with Rapid Evolution

25

Nicolas Villeneuve, Patrick Bachèlery, and Jacobus Kemp

Abstract

La Réunion Island is a volcanic area located in a tropical climate context. Volcanism has built up a 7 km high cone over the last five million years. For the last two or three million years, magmatic, tectonic and weathering processes have been engaged in a gigantic race to shape the island's morphologies, with dramatic consequences in terms of landforms such as volcanic peaks (the dormant Piton des Neiges and active Piton de la Fournaise), cliffs or "ramparts", deep gorges and waterfalls which, together with endemic flora and fauna, helped the central part of La Réunion to be recently recognised by UNESCO as a World Heritage Site. Yet, induced natural hazards such as volcanic eruptions and landslides remain a threat for the population.

Keywords

Basaltic shield volcanoes • Volcanic geomorphology • Evolution • Natural hazard

25.1 Geographical and Geological Setting

La Réunion Island (Fig. 25.1a, b), located in the western Indian Ocean (21° south and 55° east), has a subaerial extent of around 2,500 km^2, composing an elliptic island shape (60 km × 70 km) elongated along a N120° axis. The island is a volcanic complex composed of at least three edifices. Two of them are currently visible in the landscape (the Piton des Neiges, 3,070 m, and the Piton de la Fournaise, 2,632 m), and the third one, called Les Alizés volcano, considered to have grown before about 600 Ka, is now completely dismantled (Lénat et al. 2012) by large mass-wasting events (Le Friant et al. 2011; Oehler et al. 2008) and erosion (Saint-Ange et al. 2011; Salvany et al. 2012).

The Piton des Neiges volcano, located on the northwestern part of the island, is inactive since 12 Ka (Deniel et al. 1992). It is currently accepted that the Piton des Neiges built up in four main stages. The Piton de la Fournaise, on the southeastern part, is considered one of the most active volcanoes in the world with an average of 0.8 eruptions per year since 1998.

La Réunion is situated in the tropics, just north of the Tropic of Capricorn. This intertropical, oceanic setting gives the island a humid tropical climate, with an annual average humidity of between 70 and 80 %. Two main seasons can be distinguished: the southern winter and the southern summer, with winter bringing polar air closer to the island resulting in lowered temperatures.

During the summer months, specifically January to March, La Réunion is strongly affected by tropical depressions and cyclones. These systems sweep over La Réunion with

N. Villeneuve (✉)
Laboratoire Géosciences Réunion, Université de La Réunion, Institut de Physique du Globe de Paris, Sorbonne Paris-Cité, UMR 7154 CNRS, 97715 Saint-Denis, France
e-mail: nicolas.villeneuve@univ-reunion.fr

P. Bachèlery
Laboratoire Magmas et Volcans, UMR 6524, CNRS-IRD-Université Blaise Pascal, Observatoire de Physique du Globe de Clermont-Ferrand, 5 rue Kessler, 63038 Clermont-Ferrand, France
e-mail: P.Bachelery@opgc.fr

J. Kemp
Department of Geography and Environmental Studies, Stellenbosch University, Private Bag X1, Matieland 7602, South Africa
e-mail: jkemp@sun.ac.za

M. Fort and M.-F. André (eds.), *Landscapes and Landforms of France*, World Geomorphological Landscapes,
DOI 10.1007/978-94-007-7022-5_25, © Springer Science+Business Media Dordrecht 2014

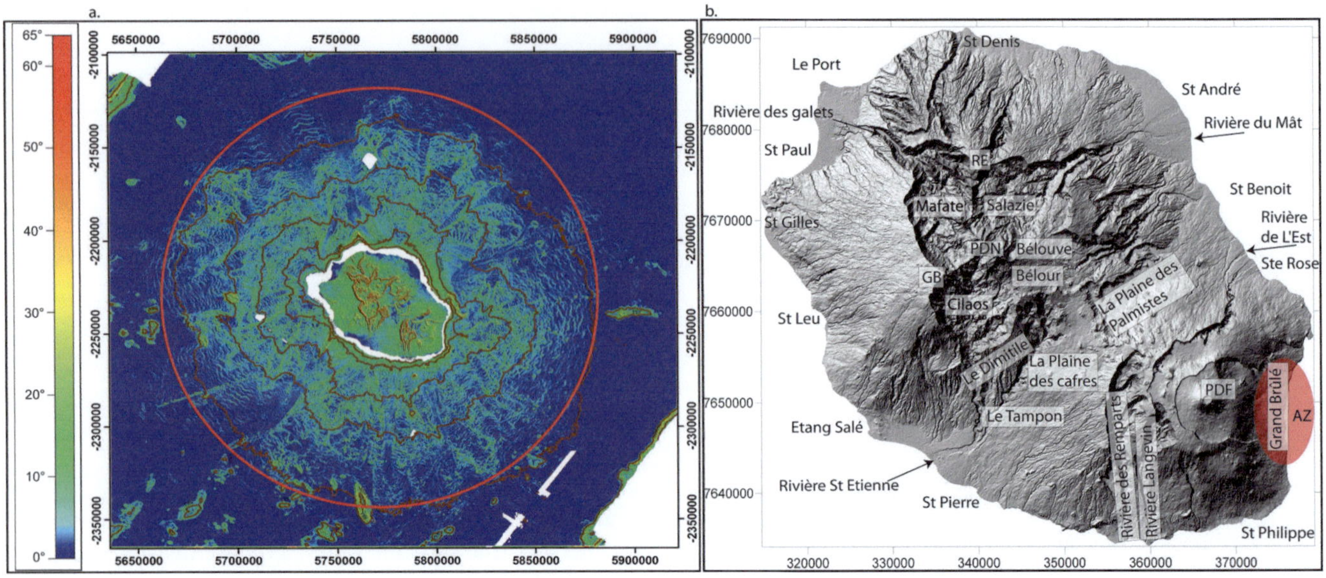

Fig. 25.1 (**a**) Slope gradient computed from FOREVER and ERODER1 swath bathymetry and IGN DEM (1997) for the aerial part. *Brown lines* corresponded to –4,000, –3,000, –2,000 –1,000 and –500 m isobath. The *red circle* shows the approximate boundary between La Réunion and the abyssal plain. (**b**) Shaded relief map of Réunion Island showing the location of Piton des Neiges (PDN), Piton de la Fournaise (PDF) volcanoes and place names. The *red oval* (AZ) corresponds to the hypovolcanic complex Les Alizés volcano as defined by Salvany et al. (2012). Read Roche Ecrite and Grand Bénare for RE and GB. Coordinates in metres (WGS84, UTM40S)

wind speeds that can reach 250 km/h and release enormous amounts of precipitation onto the island. In fact, La Réunion holds several world records for precipitation, being the place where the highest rainfalls for periods between 1 h (261.5 mm on 29 February 1964, Cyclone Giselle) and 15 days (6,083 mm between 14 and 28 January 1980, Cyclone Hyacinthe) have been recorded. Globally, cyclones are very well studied, followed and known. Nevertheless, in terms of risk on La Réunion, they are the most significant factor responsible for deaths and injuries. They are also responsible for much damage to the infrastructure on which the community and local economy relies.

Outside the cyclone period, precipitation is highly variable in both time and space. The rainfall distribution on La Réunion is most strongly influenced by its topography. In winter, the southeasterly trade winds (known locally as *Alizés*) carry humid air to higher elevations, causing adiabatic cooling and precipitation on the eastern mountain slopes. Mean annual precipitation for La Réunion ranges from around 750 mm on the sheltered west coast to more than 10,000 mm on the slopes of Piton de la Fournaise in the east (Fig. 25.2).

These significant quantities of rainfall had dramatic implications for soil erosion and sediment transport by the island's rivers. With La Réunion's high rainfall, steep slopes and young reliefs, rates of erosion and sediment transport are huge and often considered some of the highest in the world (Garcin et al. 2005).

25.2 Gross Morphology of La Réunion Complex

Around La Réunion Island, large amounts of debris avalanche deposits and numerous secondary submarine slope instabilities, resulting from the sliding and spreading of the submarine volcano flanks, have been recognised (Oehler et al. 2008; Le Friant et al. 2011).

Several authors have proposed different interpretations of morphologies of La Réunion's bathymetry linked directly to onshore geological processes. Recently, high-resolution bathymetric acquisition during the FOREVER and ERODER cruises allowed very fine scale analyses and models that complement those of Oehler et al. (2008).

La Réunion, from the abyssal plain to the summit of the Piton des Neiges volcano, is a large 7 km high cone (Fig. 25.1a) with a total volume of about $5.01 \cdot 10^4$ km^3. Considering that the subaerial part has a volume of $2.1 \cdot 10^3$ km^3, only 4 % of the volcanic complex is above sea level. The contact between La Réunion Island's volcanic products and the abyssal plain can be considered to lie at an average depth of 4,200 m.

The La Réunion Island volcanic complex is relatively well contained within a circle (radius 80–100 km) centred on the Piton des Neiges present summit (red circle, Fig. 25.1a). This mainly corresponds to the slope transition between the abyssal plain and the cone of the volcanic complex, and Le Friant et al. (2011) explained the stretching towards the southeast as a consequence of a series of massive debris

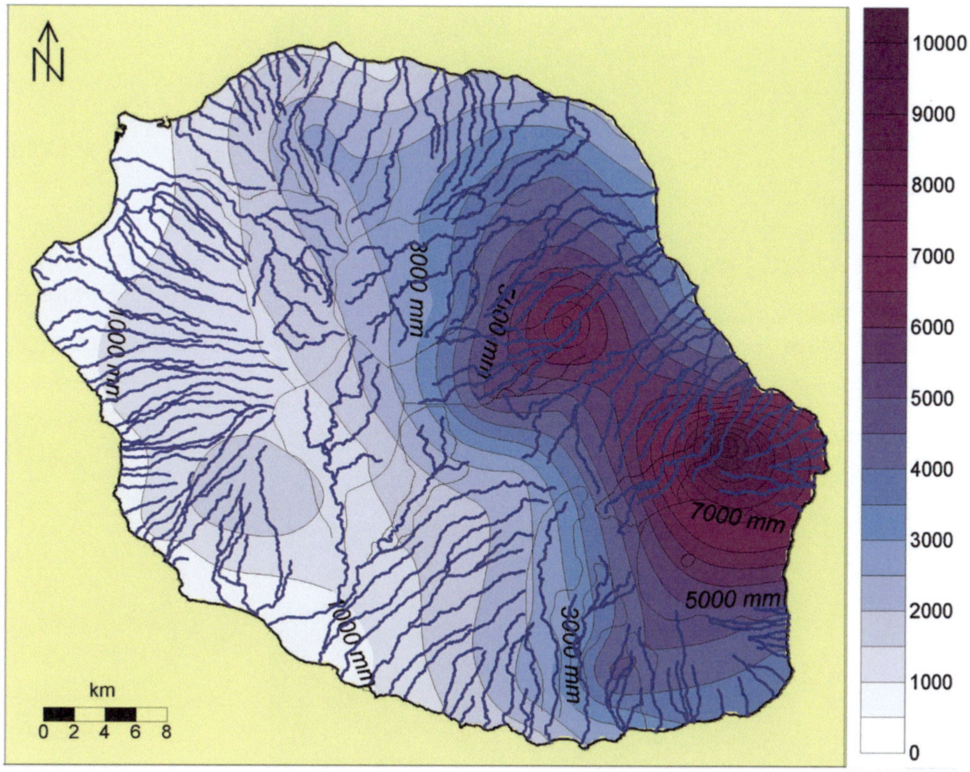

Fig. 25.2 Mean annual precipitation (between 1971 and 2000) overlain by the hydrologic network for La Réunion. Isohyets have a contours spacing of 500 mm (Data from Meteo-France)

avalanches that occurred at Piton de la Fournaise (red oval, Fig. 25.1b). These authors mainly proposed that flank destabilisation, onshore and offshore, would be less massive on Piton des Neiges than on Piton de la Fournaise.

25.3 Geomorphic Landscape of the Volcanic Island

25.3.1 A Very Rapid Evolution

La Réunion Island is characterised by a rapid landscape evolution resulting from a favourable geology combined with a tropical climate. Relationships between dramatic climatic events, small-scale and large-scale landslides, torrential transport of sediments and active volcanism lead to rapid evolution in terms of building, dismantling and/or refilling of recently formed geomorphic features.

The most remarkable example occurred during the large 2007 eruption of the Piton de la Fournaise when the centre of Cratère Dolomieu dropped 350 m in height in only a few hours (Fig. 25.3a, b) without significantly modifying the crater contour (Fig. 25.3c). Conversely, paleo-valleys on La Réunion have been infilled by new lava flows, as can be observed in the upper reaches of the Rivière des Remparts (Fig. 25.4).

The coexistence of contrasted (building *vs* dismantling) and rapid evolution of the landforms in the same area, as highlighted by these two examples, certainly is one of the most interesting features of geomorphology concerning very active volcanoes.

Unstable cliffs and slopes are often sites of mass wasting such as collapses and debris avalanches, responsible for large to massive scale landforms like Cirques. The most massive landslide is the Saint-Gilles slide on the western flank of Piton des Neiges (Bachèlery et al. 2003). The various breccia units cover an area of 16 km^2. Bathymetric surveys show the offshore prolongation of "Saint-Gilles breccias" up to a depth of 2,500 m, allowing us to estimate the total area covered by the debris avalanche deposits at more than 500 km^2.

The main river systems on La Réunion also drain the edifices of Piton des Neiges and Piton de la Fournaise in a generally radial pattern. Applying the Ollier (1988) model to La Réunion, Piton de la Fournaise would be in the intact stage, while Piton des Neiges would be in the planeze stage.

25.3.2 Main Geomorphological Features of the Piton des Neiges Volcano

In terms of emerged surface extent, the Piton des Neiges volcano represents two-thirds of the total island area. At the

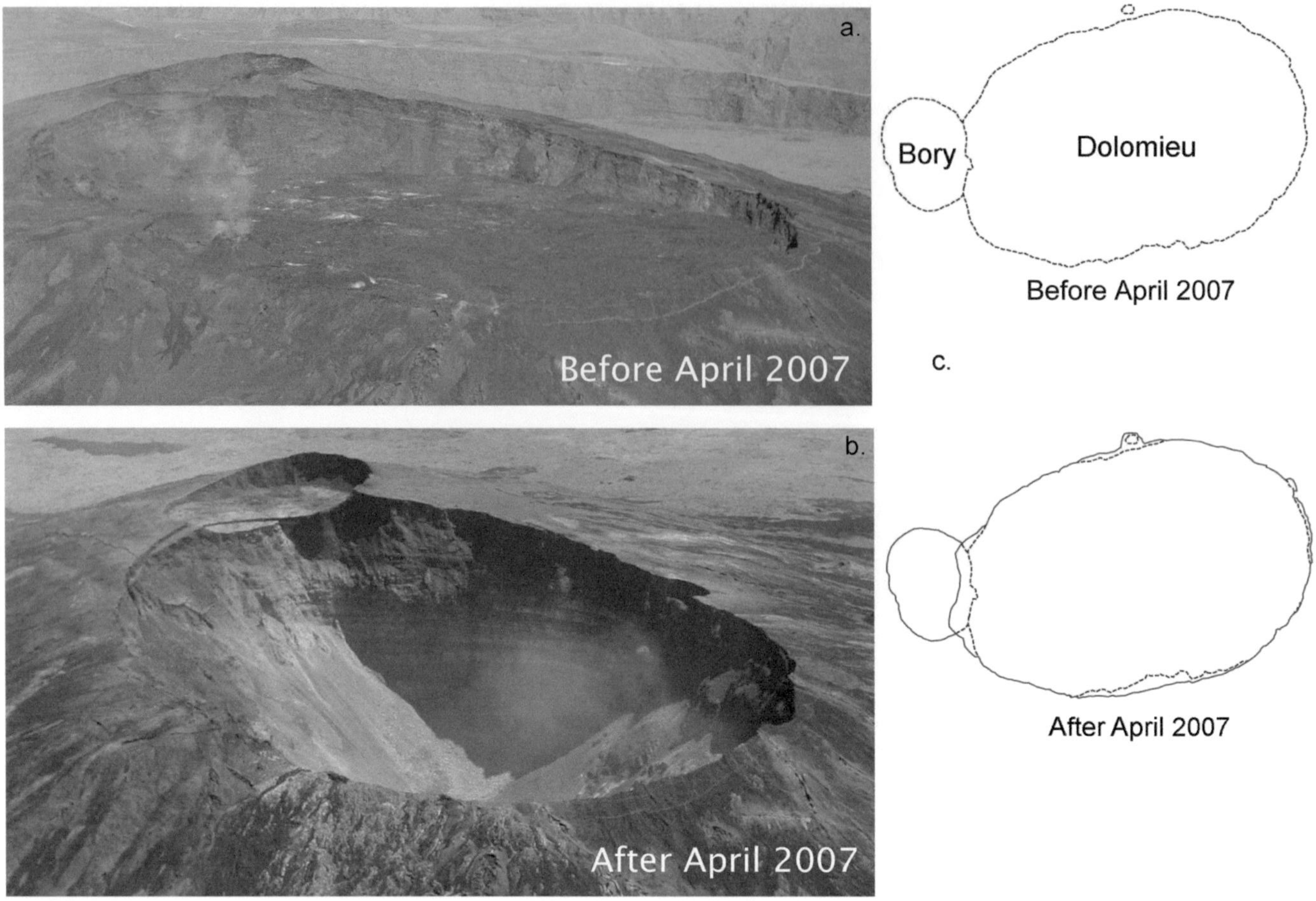

Fig. 25.3 Southeastern aerial photos of the Piton de la Fournaise summit before (**a**) and after (**b**) the collapse in April 2007. The mapping of the Dolomieu (foreground) and Bory (background) crater outlines from orthophotos acquired in 2003 and 2007 illustrates the relatively small change to Piton de la Fournaise in plan view compared to the extent of the vertical collapse (**c**)

core of Piton des Neiges, three large depressions (locally known as "Cirques") exist, namely, the Cirque de Salazie drained by Rivière du Mât to the northeast, the Cirque de Cilaos drained by Rivière St-Etienne to the south and the Cirque de Mafate drained by Rivière des Galets to the northwest (Figs. 25.1b and 25.2). They are the main and the most spectacular features of the Piton des Neiges edifice.

The origin of these major depressions has been widely discussed. Several authors have assumed either a volcano-tectonic or erosional origin, sometimes associated with massive debris avalanches. Oehler et al. (2008) proposed a multi-process origin, combining erosion, structural events and massive landslides. Except for the high cliffs surrounding the Cirques and some other rare reliefs inside, breccia deposits largely blanket the bottom of the three Cirques. It is quite difficult to recognise the source of the breccias and to distinguish between an erosional or tectonic origin for the landscape. The two processes were certainly acting together for millennia to build this spectacular relief. Currently, weathering can efficiently proceed on breccia materials that compose the cirque floor, and the debris inside is very easily mobilised, supplying a strong transport of sediment off the island.

As already said, Piton des Neiges built up in four main stages. Salvany et al. (2012) assumed that the Cirques began to form with the lowering of the lava output rate of Piton des Neiges, around 180 Ka ago (Phase I). Around 12,000 years ago (Deniel et al. 1992), all magma production on Piton des Neiges stopped, and only erosive mechanisms remained, which widened and deepened the Cirques to their current state, accounting for negative volumes of more than 101 ± 44 and 105 ± 41 km^3 for the Mafate and Cilaos Cirques. For Salazie cirque, Salvany et al. (2012) considered that it had been filled by lava 70 Ka ago. The same erosional volume has been obtained in only 70 Ka.

The external slopes of Piton des Neiges volcano are less dismantled. With an almost triangular shape dissected by three rivers, they form large Planezes, stretching from Le Grand Bénare to the western coastline for the Western Planeze, from la Roche Ecrite to the northern coastline for the Northern Planeze and from the Piton des Neiges to the eastern coastline and to the Plaines area for the Eastern Planeze.

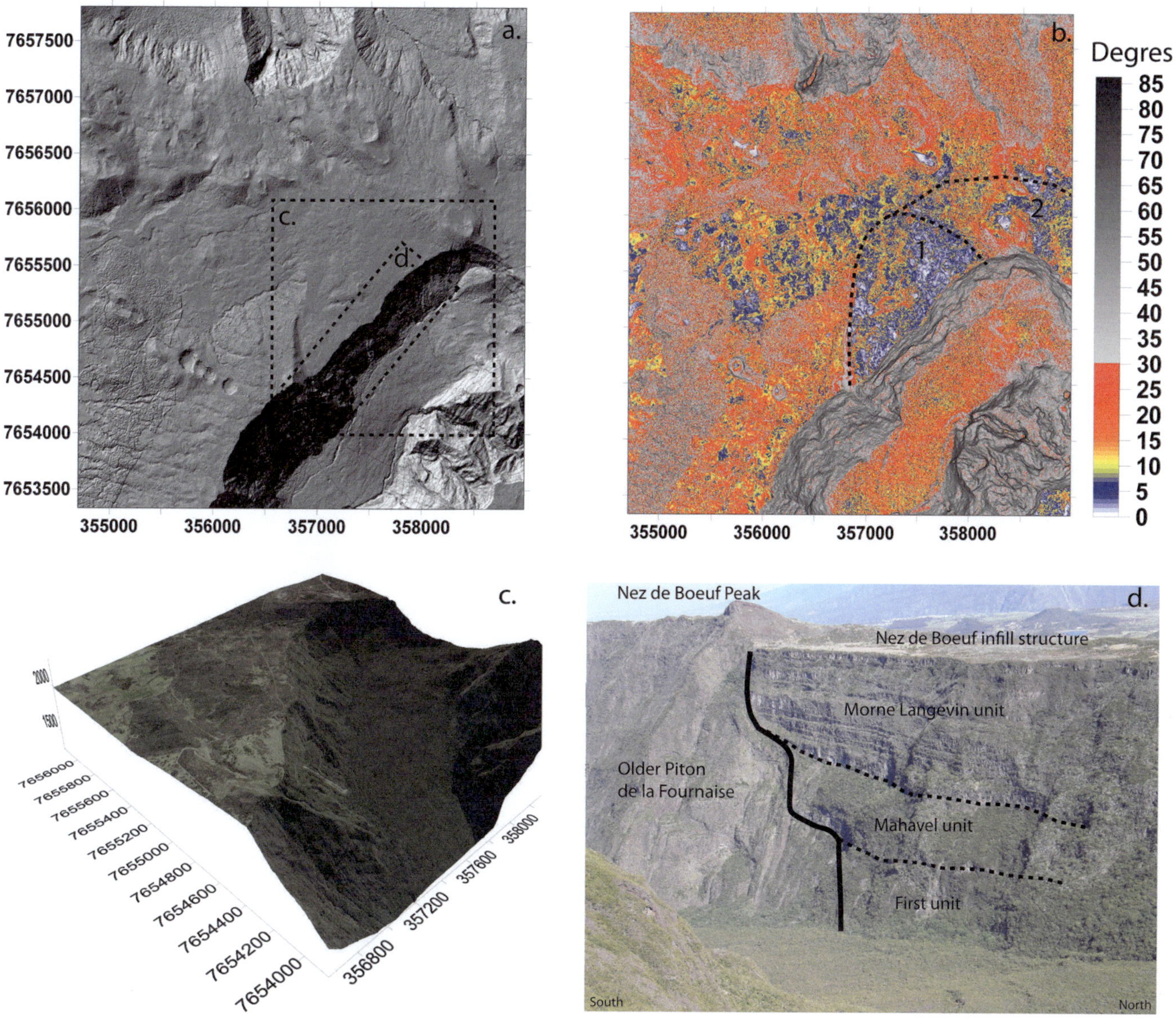

Fig. 25.4 Infilling of a paleo-Rivière des Remparts on the northeastern side of Rivière des Remparts. (**a**) Shaded relief of 5 m DEM (IGN); (**b**) slope gradient computed from the 5 m DEM, 1 and 2 on the north Rivière des Remparts head correspond to the possible extension of the river before infilling; (**c**) a 3D surface (2012 orthophoto from IGN); (**d**) outcrop view from east (From Merle et al. 2010)

The Western Planeze, largely covered by recent lavas (Phase IV), is particularly smooth (Figs. 25.1 and 25.5). This is the result of both recent refilling of erosion morphologies by lava flows and of its relative protection against weathering. For the Northern Planeze, the La Montagne edifice is much more eroded. This area has not been overflowed by Phase III (partially) and Phase IV of the Piton des Neiges lava flows (Fig. 25.5). This signifies that soils have been eroded during a long period estimated at least to 450 Ka.

The Le Dimitile edifice and patchwork on the Eastern Planeze have the same configuration. This Planeze is very complex. First, its southeastern extent is limited by the Piton de la Fournaise formations. Secondly, the Bebour-Belouve area is a deep depression that is considered a paleo-cirque infilled with recent lava flows.

25.4 Main Geomorphological Features of the Piton de la Fournaise Volcano

Piton de la Fournaise is composed of at least four different landscapes (Figs. 25.6 and 25.7): the Enclos Fouqué and Grand Brûlé structures, the northeast and southeast rift zones, the main rivers scars and the Saint Pierre-Les Plaines-Saint Benoit corridor.

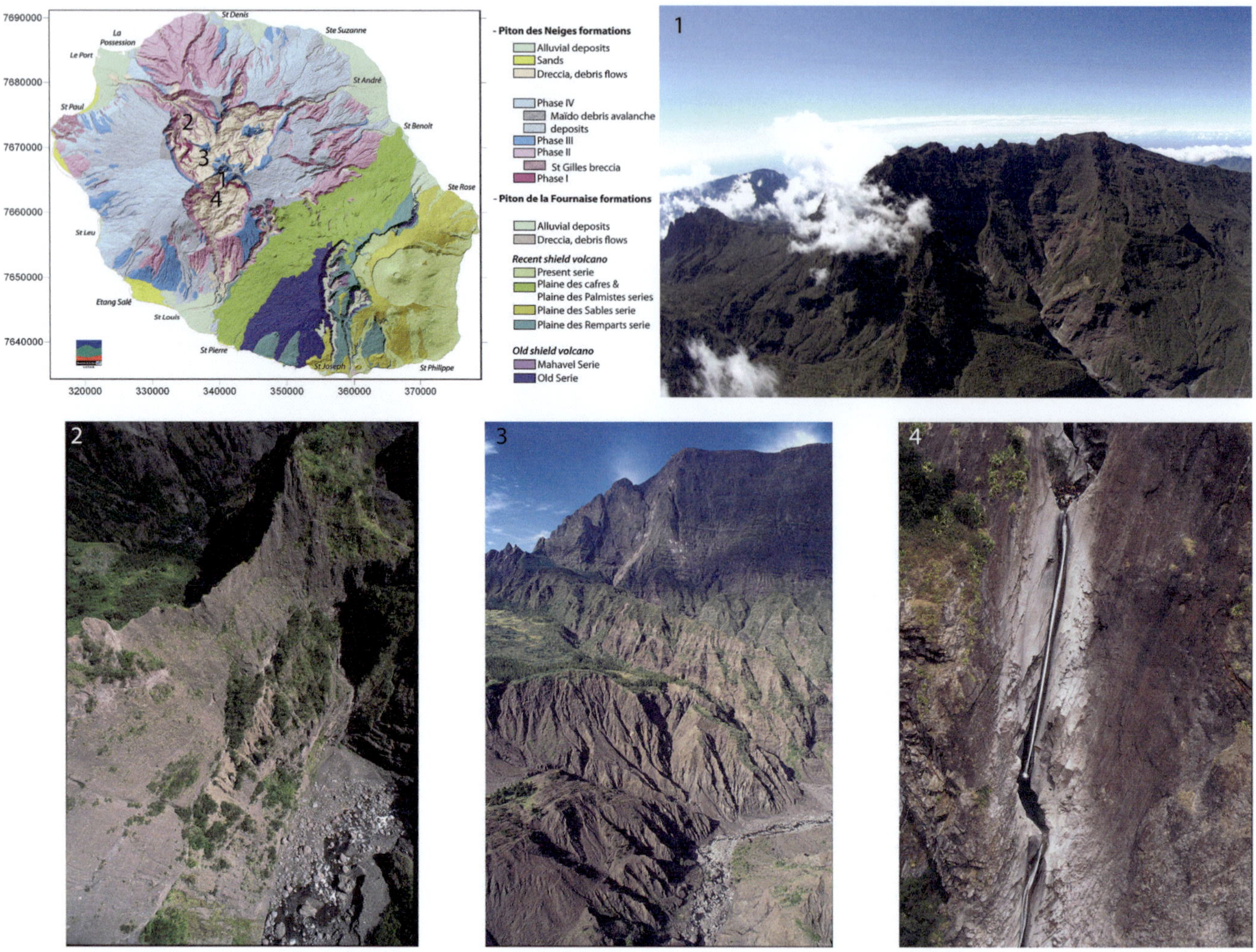

Fig. 25.5 Geological map of La Réunion (From Fèvre 2005). Photos: Eroded landscapes in the Cirques. *1* Old, eroded Piton des Neiges, aerial view from the west; *2* very narrow crest formed by collapse of unstable materials in Mafate; *3* Badlands developed in unstable materials in Mafate; *4* the huge "Fleur Jaune" waterfall dropping from massive materials in Cilaos

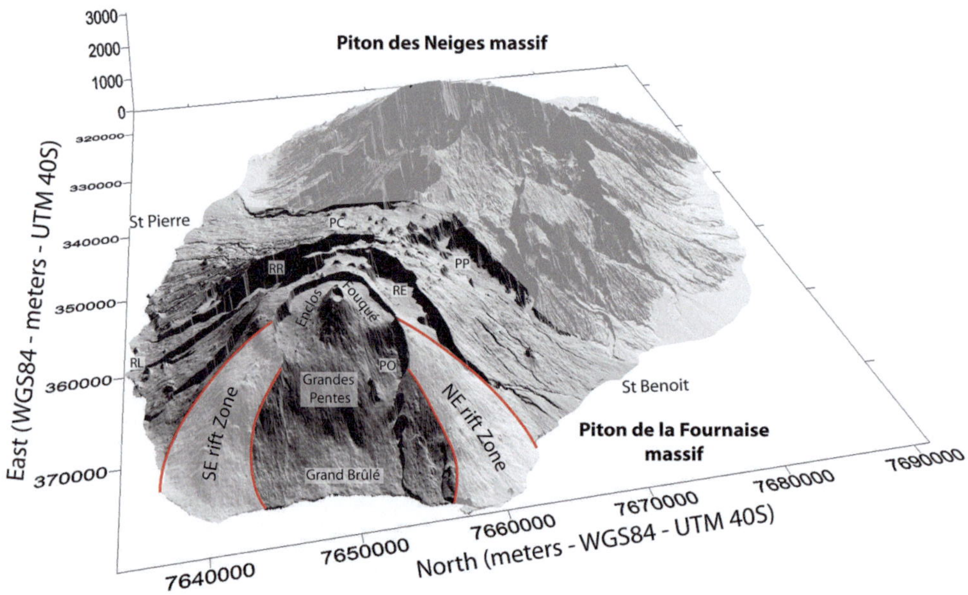

Fig. 25.6 Eastern view of Piton de la Fournaise volcano (25 m DEM from IGN). Read Rivière de l'Est for RE; Rivière des Remparts for RR; Rivière Langevin for RL; Plaines des Cafres for PC; Plaine des Palmistes for PP; Plaine des Osmondes for PO

Fig. 25.7 Synthesis of different morphologies observable on the Piton de la Fournaise edifice. (**a**) Summit of the Piton de la Fournaise with the Cratère Dolomieu and the Cratère Bory; (**b**) Château Fort crater vent: a very typical scoriaceous cone shape; (**c**) pahoehoe lava flow overlapping an "aa" lava during the August 2006 event; (**d**) Rivière des Remparts viewed from north; (**e**) Plaine des Cafres and Rivière des Remparts (*left*); (**f**) lava flows from the Chisny eruption which had entered Rivière Langevin 1 Ka; (**g**) upper catchment of Rivière de l'Est called Savane Cimetière, viewed from north; (**h**) a lava lake enclosed within a scoriaceous cone during the August 2006 eruption

25.4.1 The Enclos Fouqué and Grand Brûlé

This is a large uninhabited horseshoe structure of ~100 km^2, open to sea on the eastern part of the island. Four main morphological units can be observed: (1) the summit area, (2) the "Remparts", (3) the Grandes Pentes and the Plaine des Osmondes and (4) the Grand Brûlé. The main geomorphological features of the Enclos Fouqué are related to the volcanic and tectonic histories of the area associated with recent events. Presumably, the formations have been shaped by a series of events whose initial process is a combination of structural instabilities and local collapses linked with phreatomagmatic and phreatic explosions that could be paroxysmal. Morphologies are also linked to lava infilling and smoothing.

Bory (250 m × 350 m) and Dolomieu (750 m × 1,100 m) craters constitute the summit part of Piton de la Fournaise, as viewed from the west, where they top a 400 m high central cone. These landforms owe their origin to repeated collapses; though relatively rare, these collapse events usually occur in relation with high lava effusion rates. Michon et al. (in press) describe 14 events between 1791 and April 2007. Ten of these events are responsible for phreatic and/or phreatomagmatic activities. These events generate a real risk for summit hikers (~120,000/year) and Pas de Bellecombe visitors (~270,000/year). In April 2007, a ~$90 \cdot 10^6$ m^3 caldera completely modified the geometry inside the Cratère Dolomieu (Fig. 25.3a, b).

The two summit craters are surrounded by high slopes with an average gradient of between 15° and 25°. They are relatively constant in all directions for 1.6 km from the centre of the Cratère Dolomieu. Several small eruptive vents are visible on the slope of the terminal cone. Between 30 and 50 cones per square kilometre can be counted in this area.

The western part of the Enclos Fouqué is relatively flat. Lénat et al. (2001) named this area the "Champ de Lave de l'Enclos Fouqué". Pahoehoe lava mainly composes this large lava field and is probably related to the late activity of the Cratère Bory during the seventeenth and eighteenth centuries.

The Enclos Fouqué/Grand Brûlé area is edged by ~35 km of cliffs, namely, the "Remparts". These are 100–200 m high subvertical cliffs. They are natural barriers that protected inhabited areas against lava flows coming from more frequent eruptions.

The Grandes Pentes correspond to ~25 % of the total area of the horseshoe depression. Here, the slope gradients range between 25° and 30°. Their formation is probably linked to landslides and/or spreading of the volcanic edifice.

The Plaine des Osmondes, localised northeast of the central cone, is a depressed area (2.5 × 4 km), edged on the north, the west and the south by the "Remparts", the Grandes Pentes and Le Piton de Crac, respectively. The eastern part opens this area towards the Grand Brûlé. Courteaud (1996), from geophysical data, has interpreted the Plaine des Osmondes as the head of paleo-valley that could have been built during a long period of quiescence of the volcano, at least in this region.

The Grand Brûlé is a relatively flat (4° to 8°) corridor between the Grandes Pentes and the coast. Its morphology is probably related to lava accumulation. Below the sea level, the Grand Brûlé stretches into the Chenal Vincent unit (Oehler et al. 2008). From Michon et al. (submitted), lava flows of about 28 eruptions have crossed the Grand Brûlé and reached the ocean since the beginning of the sixteenth century. The emplacement of lava flows crossing the coastal road (used daily by about 3,200 vehicles) and reaching the seashore occurs relatively often. Lava flows entering the sea are very impressive. Nevertheless, the difference in temperature generates explosions on the unstable platform built up by the progression of the lava flow, and acid plumes generated by seawater vaporisation can cause respiratory problems and/or environmental damage.

25.4.2 The Northeast and Southeast Rift Zones

Both rift zones are convex surfaces that stretch along steep slopes (Fig. 25.6). This morphology has been mapped below sea level up to depths of 1,800 and 2,000 m for the southern and northern zones, respectively. The rift zones correspond to areas of preferential magma ingestion for the volcano summit towards the outside flanks. Between 1708 and 1998, seven eruptions (three on the northeast and four on the southeast rift zones) occurred in these areas. Their implication in terms of volcanologic risk is important. For instance, in 2012, 11,809 permanent residents lived in areas located within these rift zones.

25.4.3 The Main River Scars

On Piton de la Fournaise, 19 drainage basins have been identified, but only Rivière de l'Est, Rivière Langevin and Rivière des Remparts have perennial flow. All three rivers are fed by springs resulting from rapid infiltration due to the high permeability of the rocks and the high rainfall.

Four main hydrologic units are well identifiable on the Piton de la Fournaise edifice. Rivière de l'Est (24.5 km^2) is the only one cut into the north flank of the volcano. The Ravine Basse Vallée (9 km^2), the Rivière Langevin (23 km^2) and the Rivière des Remparts (56 km^2) drain the southern flank of the volcano. They are affected by different geomorphic processes (collapses and topographic inversion).

Erosion is clearly the main process involved in their formations, but a more complex history, with caldera and mass-wasting episodes, has been revealed by a detailed study of Rivière des Remparts and Rivière Langevin (Merle et al. 2010). Rockfalls and landslides are common in these valleys. They result from (1) differential resistance of materials, (2) regressive erosion mechanisms and (3) hydrogeological erosion mechanisms as described by Fèvre (2005). In addition, several lava flows may also locally cover the riverbeds, showing that volcanic activity may continue after the erosion of the river.

25.4.4 The Saint Pierre-Les Plaines-Saint Benoit corridor

The Saint Pierre-Les Plaines-Saint Benoit corridor is a smooth surface that has been covered by relatively recent lava flows. These areas compose the morphological transition between the Piton de la Fournaise and Piton des Neiges volcanoes. The Plaine des Palmistes is the most complex area. It is a depression open to the sea, and its northern, western and southern cliff slopes are steep. The current morphology of this area certainly reveals the preponderance of erosional mechanisms.

25.5 Main Geomorphological Features of Coastlines

La Réunion Island is composed of 250 km of coastline that can be classified into three main groups (Fig. 25.8):

- 140 km of volcanic coastline
- 85 km of alluvial coastline
- 25 km of coral sand coastline

The volcanic coastlines can be divided into two types: (1) low volcanic platforms and (2) volcanic cliff shorelines that can be also called wave-straightened cliffs in volcanic areas. The first type has its origin in lava flows entering into the sea and can also be residual platforms. The second type develops under wave action at the bottom of cliffs that generates instabilities, such as rockfalls and landslides. From Grand Bois to Sainte Rose, on Piton de la Fournaise, the coastline is composed of 65 km of uninterrupted, generally high volcanic cliffs. The possible cliff collapses are a threat for the population. On Piton des Neiges, along "La Corniche" which is the highest and longest coastal cliff (~13 km long and 100–200 m high) and along the Cap La Houssaye cliff, very high-traffic roads have been built at the foot of both cliffs. Several collapses have occurred, causing dramatic accidents, making these roads the most hazardous and at-risk areas.

Fig. 25.8 Synthesis of different coastal morphologies observable around La Réunion. Coastline colour code: *red* for volcanic cliff, *yellow* for coral sand beaches, *black* for black sand beaches and *grey* for rolling stone beaches (*Courtesy of Olivier David)

The alluvial coastlines include two types – (1) the black sand beaches and (2) the rolling stone beaches – directly linked to the transfer of fluvial erosion products and grain-size sorting. Along the eastern coast (from Sainte Rose to Saint Denis), the 51 km long beach of rolling stones formed after hydraulic processes have washed away fine sediments. The same process occurred in small river deltas around La Réunion. In contrast, the Saint Paul and Etang Salé black sand beaches are mostly composed of fine sediments transported by marine currents. Here, strong winds have also transported fine particles to develop black sand dunes. In the Grand Brûlé area, two black sand beaches developed under a specific process: when the lava flows reached the sea, the contrast of temperature produced explosions generating volcaniclastic materials. The accumulation of these products usually builds non-permanent beaches.

The coral sand beaches are localised (1) from the south of Cap La Houssaye to La Souris Chaude in the Saint-Gilles area, (2) from La Pointe des Chateaux to La Pointe au Sel in the Saint Leu area, (3) from Ravine Blanche to Terre Sainte in the Saint Pierre area and (4) in the Grand Bois area. These have been built recently, less than 8,000 years ago (Camoin et al. 1997), by accumulation of the biodetritic elements from the reefs. The Saint-Gilles units are the largest ones. Around 12 linear kilometres are composed by the sand beach (160 m wide at its maximum), and sand dune accumulation is limited by old volcanic cliffs. The accumulation of sand can reach a maximum of 200 m in thickness.

25.6 Conclusion, Cultural Value and Touristic Promotion

Lava flows, landslides, floods and rockfalls have built La Réunion Island with its incredibly diverse landscapes. The high cliffs, steep valleys and variety of microclimates have allowed 236 endemic plant species and 92 endemic animal species (23 of which are extinct and 4 of which are threatened) to grow locally. For these reasons, 40 % of La Réunion Island is included in the UNESCO Nature World Heritage list. This international label covers the core of the National Park of La Réunion. This is not only a source of pride for residents but a true added value for the touristic promotion and the environment protection.

The great paradox is that the magnificent morphologies and areas celebrated here are also those for which natural hazard and risk are the highest.

References

Bachèlery P, Robineau B, Courteaud M, Savin C (2003) Avalanches de débris sur le flanc occidental du volcan-bouclier Piton des Neiges (Réunion). Bull Soc Géol Fr 174:125–140

Camoin GF, Colonna M, Montaggioni LF, Casanova J, Faure G, Thomassin BA (1997) Holocene sea level changes and reef development in southwestern Indian Ocean. Coral Reef 16: 247–259

Courteaud M (1996) Etudes des structures géologiques et hydrogéologiques du massif de La Fournaise par la méthode audiomagnétollurique. PhD thesis, Université de La Réunion, Réunion, p 212

Deniel C, Kieffer G, Lecointre J (1992) New 230Th-238U and 14C age determinations from Piton des Neiges volcano, Reunion – a revised chronology for the differentiated series. J Volcanol Geotherm Res 51:253–267

Fèvre Y (2005) Mécanismes et vitesses d'érosion à l'échelle géologique sur une île volcanique jeune à relief élevé - La Réunion (Océan Indien). PhD thesis, Université de La Réunion, p 217

Garcin M, Poisson B, Pouget R (2005) High rates of geomorphological processes in a tropical area: the Remparts River case study (Réunion Island, Indian Ocean). Geomorphology 67(3):335–350. doi:10.1016/j.geomorph.2004.11.002

Le Friant A, Lebas E, Clément V, Boudon G, Deplus C, De Voogd B, Bachèlery P (2011) A new model for the evolution of La Réunion volcanic complex from complete marine geophysical surveys. Geophys Res Lett 38:L09312. doi:10.1029/2011GL047489

Lénat J-F, Bachèlery P, Desmulier F (2001) Genèse du champ de lave de l'Enclos Fouqué: une eruption d'envergure exceptionnelle au Piton de la Fournaise (Réunion) au 18ème siècle. Bull Soc Géol Fr 172:177–188

Lénat J-F, Bachèlery P, Merle O (2012) Anatomy of Piton de la Fournaise volcano (La Réunion, Indian Ocean). Bull Volcanol 74:1945–1961. doi:10.1007/s00445-012-0640-y

Merle O, Mairine P, Michon L, Bachèlery P, Smietana M (2010) Calderas, landslides and paleo-canyons on Piton de la Fournaise volcano (La Réunion Island, Indian Ocean). J Volcanol Geotherm Res 189:131–142

Michon L, Di Muro A, Villeneuve N, Saint-Marc C, Fadda P, Manta F (in Press) Explosive activity of the summit cone of Piton de la Fournaise volcano (La Réunion island): a historical and geological review. Submitted to J Volcanol Geotherm Res

Oehler J-F, Lénat J-F, Labazuy P (2008) Growth and collapse of the Reunion Island volcanoes. Bull Volcanol 70:717–742. doi:10.1007/s00445-007-0163-0

Ollier C (1988) Volcanoes. Blackwell, Oxford

Saint-Ange F, Savoye B, Michon L, Bachèlery P, Deplus C, De Voogd B, Dyment J, Le Drezen E, Voisset M, Le Friant A, Boudon G (2011) A volcaniclastic deep-sea fan off La Réunion Island (Indian Ocean). Grad Vs Catastr Geol 39:271–274. doi:10.1130/G31478.1

Salvany T, Lahitte P, Nativel P, Gillot P-Y (2012) Geomorphic evolution of the Piton des Neiges volcano (Réunion Island, Indian Ocean): competition between volcanic construction and erosion since 1.4 Ma. Geomorphology 136:132–147. doi:10.1016/j.geomorph.2011.06.009

Index

M. Fort and M.-F. André (eds.), *Landscapes and Landforms of France*, World Geomorphological Landscapes,
DOI 10.1007/978-94-007-7022-5, © Springer Science+Business Media Dordrecht 2014

MIX
Papier aus verantwortungsvollen Quellen
Paper from responsible sources
FSC® C105338

If you have any concerns about our products,
you can contact us on
ProductSafety@springernature.com

In case Publisher is established outside the EU,
the EU authorized representative is:
Springer Nature Customer Service Center GmbH
Europaplatz 3, 69115 Heidelberg, Germany

Printed by Libri Plureos GmbH
in Hamburg, Germany